Engineering Manager

By

Joseph Panicello

This book is a work of non-fiction.

ISBN: 1-4107-2379-8 (e-book)
ISBN: 1-4107-2378-X (Paperback)

Library of Congress Control Number: 2003091829

This book is printed on acid free paper.

Printed in the United States of America
Bloomington, IN

1st Books - rev. 5/13/03

Acknowledgement

I would like to thank my wife, Barbara, for her patience and encouragement to complete this book on *Engineering Mananager.*

PREFACE

Engineeering Manager is a reference book and guide to educate prospective engineers into becoming a good Engineering Manager. There is no other reference book similar to this that provides this information.

There are many fields in engineering that require key managers such as the auto industry where the key manager is a Mechanical Engineer. In the Chemical industry the key manager is the Chemical Engineer, in the aircraft industry he's the Aerodynamics Engineer, in the building industry he's the Civil Engineer, in the computer industry he's the Computer Science Engineer, in the radio and television industry he's the Electronic Engineer, and in the power industry he's the Electrical Engineer. Several of these industries require specialized electronics engineers to support them. In the aircraft industry they are called Avionic Engineers which encompass the main ingredients of this book. Although this literature emphasizes Electronic Management, the information provided is essential to all engineering managers.

Before I prepared this book I discovered there was no formal training or written material available to create a new Avionic Manager in the aerospace industry. Generally, when an engineer is promoted from within a company, he's given no prior instructions on how to manage his new organization. This had happened to me when I was promoted to supervisor of a very sophisticated Avionic Department with no prior training. I was told, "You're now the Manager of the Avionics Department responsible for designing electronic black boxes for our aircraft."

Designing aircraft avionics is one thing, but managing a large group of engineers who have as much experience as I had was not an easy task. It was no longer just technical ability and experience that allowed me to be a design leader, but now I had to deal with personalities. Not only did I have to monitor the designs, I also had to be concerned with budgets, schedules, deliveries, purchasing, meetings, etc.

This book provides a different approach on a subject that has not been fully documented or thoroughly explained before. The method used here covers all aspects of Engineering Management, mainly from an experienced point of view. Over the forty years in the electronic design business, I have learned many executive techniques, and by combining these experiences with my own ideas I believe I have created the ideal text that can be used to teach most engineers to become a manager.

The text may be used by companies to assist upper-management to monitor their programs, or to train potential supervisors in the basic art of managing a department. It can be used as a guide by the graduating student, or the entrepreneur who is interested in starting up a new company. As I mentioned, this comprehensive book can be used by all types of engineers and not exclusively in the electronics field. The principles are basically the same. The military would find the information in this book an ideal text to train their personnel on how to monitor military programs. It would help them in the process of selecting vendors and evaluating quotations.

Chapter I covers what I consider to be the proper structure of a design team. It consists of the Electronic Design Manager, Electronic Engineers, System Engineers, Mechanical Engineers, Software Engineers, Printed Circuit Engineers, and Technicians. I thoroughly explain the responsibilities of each of these positions. To illustrate the management structure, I walk the reader through the procedures required to design and produce a black box step by step. I discuss the complete electronic design approach and its mechanical enclosure. I introduce a unique budget tracking system showing man-hours spread charts that would assist the Electronic Design Manager (EDM) to monitor all of his programs.

Chapter II covers the different support organizations that are needed to make up the structure of a complete engineering company. It explains the relationship these organizations have with the EDM's design team and with the Engineering Project Manager (EPM). Examples of some of these support organizations are Reliability, Maintainability, Thermal, etc.

Chapter III covers the classical company structures of upper-management. It explains the different types of organizations such as Matrix and Projectize. It provides a complete Organizational Interface Chart and explains their relationship with upper-management. This chapter goes into explaining the duties of a Program Manager (PM) and the Engineering Project Manager and how they interface with the Design Manager. It covers the non-technical responsibilities of such organizations as Finance, Administrations, Contracts, Legal, and Purchasing. This book is essential for all managers to have in their possession.

Chapter IV provides examples of actual design practices. In my design approach, I do not use the classical approach because there are many text books available at Universities and Libraries that provide technical formulas in designs. The method used here is to provide several design examples using my own experiences I gathered over 40 years. This approach may help the prospective engineer when he comes across similar design problems.

"Doing The Design" in Chapter IV, is more of a biography of some of my engineering design experiences. I do, on occasions, become pretty technical mainly to prove a point. A lot of my experiences may be out-dated because of new technologies and new Integrated Circuits on the market today, but some of the thoughts still apply where "tricks of the trade" can still be utilized. I do not cover all of my designs but only those that I feel are informative.

Throughout the book, I use expressions such as "learn from experience" to understand what really goes on in the engineering world. By using my own experiences as a tool, I believe the book provides a better idea of how this industry works in the real world. It provides all the "Do's and Don'ts" in engineering. Also in the book, I stress the term "Check and Balance" as a key way of doing business, in budget tracking and in engineering design practices. A Manager can be a rewarding vocation as long as he follows a few basic rules.

I also want to point out that this book was not written like a normal text book. I decided to write the text in the first person so that I may relate to the reader on how I learned to become an Electronic Engineering Manager from my own experiences. Hopefully, this approach would be very helpful in understanding the enormous responsibilities associated with becoming an Engineering Manager.

An Engineering Management course outline is also provided which can be used to teach this subject to prospective managers or executives in a company. It would be ideal for military personnel. The course requires 33 hours of lecture using this book as the text. It may be taken over 11 weeks at 3 hours per week or 11 days at 3 hours a day which ever is convenient. The course outline covers mainly Chapters I & II, but does makes some references to Chapters III & IV which, by themselves, would require another course.

TABLE OF CONTENTS

ILLUSTRATIONS

TABLES

INTRODUCTION

This book provides a unique approach on the subject of Engineering Management. *Engineering Manager* covers the subject of managing an Electronic organization that has not been fully documented or thoroughly explained before. The method used here applies to all aspects of Engineering Management from the experienced point of view. The techniques that I utilized provides many viewpoints of engineering management and not just in the electronics field.

I've been involved in hardware design for over 40 years and 10 of those as manager of an Electronic Design Organization. Over those years, with different companies, I have learned many management techniques, and by combining these with my own methods, I can now provide the ideal approach to become a Manager from the supervisor to the president. This book can be used by companies as a text to assist executives, to train potential managers in the basic art of management, as a guide for the graduating student, or for the entrepreneur who is interested in starting up a new company. This comprehensive book can be used by all types of engineers and not exclusively the Electronics Engineer. The Military would find this information invaluable in training their personnel to contract military programs and monitoring their progress.

Chapter I is what I considered to be the proper structure for a Design Team. My team consists of a Design Manager, Electronic Engineers, System Engineers, Mechanical Engineers, Software Engineers, Printed Circuit Engineers, Technicians and the Development Shops. I explained the responsibilities of these engineers, and present a design approach of an electronic enclosure. I introduced a unique budget tracking system with manhours spread charts to assist the Design Manager.

In those 40 years that I was a designer and manager of electronic equipment for military programs, I had witnessed engineers who assume they were hardware designers just because they had a degree in engineering, or were a Systems Engineer or Project Engineer. This assumption is also true of Packaging Engineers, where every graduate ME thinks he can design an electronic enclosure. In reality, it takes many years of experience in military design before one develops the skills and knowledge of becoming a hardware designer. Many techniques that are accumulated over the years by an Electronic Engineer are not taught at the Universities. It also takes a special type of individual to be an engineer. The engineer has to have the mental attitude and desire to work in an atmosphere with his hands under a tedious laboratory environment.

Electronic Managers also require many years of design experience themselves. The old saying "Good engineers make poor managers" is not always true. Some good engineers can be taught to be good managers if they are given the proper training and tools. The problem is, most companies that promote from within give the supervisors job to their most talented engineer with no prior management training, and tell him to simply take over. This approach usually destroys a prospective manager and he would go back to being a designer and not try managing again. That's what this book is basically about "The Making of a New Manager."

Sometimes a new manager may even have a good business background but is not technically competent to make the proper technical decisions. Because of this new acclaimed power, he would eventually make a bad decision because he would invariably disregard the inputs from his peers. This occurrence has happen time and time again. Also, when a new supervisor is hired in, he hasn't the slightest idea of how the new company does business, but he would now impose his own practices on his new employees because that's the way it was done with his old company. This custom would disrupt a design team to no end.

This book is an attempt to help provide a shortcut to the many years of experience that it normally takes to be a manager. Being a good electronic designer, however, is another story. I can only provide some tools and suggestions in that respect to ensure good sound design practices are being used. But, nothing replaces on-the-job training or "learning it the hard way." It usually takes 5 to 10 years for an undergraduate to learn all the "do's and don'ts" in the electronic design business.

This text mainly addresses military electronics, but much of the information provided can be applied to the commercial world as well. It must be stressed that in military electronics, reliability is more important than cost, especially if the equipment is to be used under extreme environments. This does not preclude that commercial equipment is not reliable or suggest that the costs of military programs should be ignored. It simply means that one has to consider the priorities, and in military electronics "Mean Time Between Failure" has the highest priority.

The following literary endeavor reflects my actual experiences in the field of electronics. It is not expected that `all' engineers would agree with everything that I have written. I want to point out that I have not expanded on subjects that I have little experience with, because I did not want to rely on inputs from others and have to account for their accuracy. I do, however, discuss related areas of engineering that I'm familiar with, such as production. Detailed information would also be provided with regards to actual design practices and how the Electronic Design Manager interfaces with other organizations as well as with the customer.

If the female engineers would forgive me, I am using the male gender throughout the book to avoid using such terms as `his/her' or `she/he' with respect to an individual.

Every time I discuss an important issue I would try to use actual examples that had happened to me, or to someone I know. I use this technique throughout the book and sometimes I repeat them to stress a point that may be related but on different topics. Hopefully, when the student finishes this book or takes the related course, he would have a complete picture of what Engineering Design Management is all about; the good, the bad, and the ugly.

CHAPTER I

ENGINEERING MANAGERS

In the world of black box designs for military programs, there are three key staff members used by most companies to produce and deliver their products. These staff members are the Program Manager (PM), Engineering Project Manager (EPM), and the Electronic Design Manager (EDM). Many companies follow this approach. Upper-management are, of course, involved but usually they do not deal directly with the details of developing a product. They are more concerned with costs, schedules and overview.

The Program Manager is the Administrative force of a program. He is in charge of the contract and has control or interfaces with many departments. The PM deals directly with such departments as Finance, Administrations, Master Scheduling, and even Production. He is the only person that communicates directly with the customer to avoid legal disputes. On technical discussions with the customer, the PM may depend on the EPM or even the EDM. Not all programs require a PM such as in-house designs. The PM deals only with contracts associated with an outside customer (Military) or a supplier. He should not communicate directly with the EDM nor with his engineers.

The Engineering Project Manager is the Overall Technical force of a program and he deals directly with all of the technical support departments such as reliability, maintenance, testability, and the EDM. The EPM may only communicate with the customer if the subject matter is on a technical level, but he should report any technical changes or requirements to the PM. These two managers must communicate together on all subject matters. The EPM deals directly with the EDM but not with his Design Engineers.

The Electronic Design Manager and his organization Designs the Black Boxes and develops the equipment for production. The EDM doesn't work directly for the PM or the EPM but he has to interface with them, because they are in control of the budget and are basically in charge of a program.

The PM, especially, is the man that allocates the budget to all organizations and to the EPM. The EPM then distributes his budget to all of the support organizations including the EDM's department. So, the `indirect' chain of command is led by the PM, then the EPM and finally the EDM. However, neither the PM nor the EPM have any direct responsibility over the actual design. They must correspond through the EDM on any DESIGN matters.

ELECTRONIC DESIGN MANAGER

The EDM rarely communicates with the customer. He may be asked to present the design concept to a customer, but only when the PM and EPM are present. It must be understood that the EDM may have other contracts at the same time on several unrelated programs, and he could be dealing with different EPMs or PMs at the same time.

As an EDM, the author had dealt with four different PMs on four different aircraft. Some companies may use several EDMs on different program. The author prefers using one EDM for all programs with, possibly, a supervisor or leadman running the different programs.

An Electronic Design Manager is a special kind of a person in the electronic business. His position can be very complicated at times, because it is his organization that is responsible for the design of the electronic equipment. The personality of the EDM is also very important. He has to be the father, mother, brother, sister, and uncle to all departments including his own engineers. He has to be able to coordinate with the Engineering Project Manager, with the Program Manager, with Upper-Management, with all of the other support organizations, and sometimes with the customer.

When there is a problem in the field, the EDM seems to be the only one on the firing line, because his organization is the last one to come down the equipment pike. The whole design is on his shoulders, and he is the buffer between upper-management and his design engineers.

A good EDM is a person with a lot of design experience who has management talent. Before I became an EDM, I had always felt that I was a pretty good electronic designer, but I really found my niche when I became an EDM. I actually enjoyed this position because it had so many different challenges to it rather than just designing equipment.

A good EDM has to be involved in all aspects of a program. Many times I had to get involved in the packaging. As an EDM, I enjoyed this new challenge and, as they say, enjoyment is the key to any successful profession. As far as the firing line goes, I was so used to being reprimanded for a minor schedule slippage or being over budget, I learned to shrug it off because I felt "it came with the territory." The main thing is, I had control of the program and I knew that my department would produce a good design within a reasonable time frame.

My upper-managers always became excited when there's a schedule slippage because they really don't understand the electronic business. How could they? In my case, they weren't Electronic Engineers but were Aeronautical Engineers at Lockheed. They thought that once a design is completed the program is over. Unfortunately, there would always be design problems, and yet, I have never come across a design predicament that was un-solvable. This was because my engineers knew how to design and we always began with a solid fundamental foundation. Electronic design is rather simple if one knows what he is doing, and follows the basic rules and practices. On one update program that I was on, a female Master Scheduler became very frustrated because there were so many obstacles and problems with an ECP (Engineering Change Proposal), which was requested by the Navy to modify an old complicated black box. I assured her that there was no problem and everything would be corrected. She couldn't understand how I could be so calm about it. She didn't know that I had already figured out a solution in my mind, but it's nice to play the hero.

Most design engineers have this intuition of knowing or pre-determining the solutions without having all the facts, because they have been there so many times. It becomes second nature after awhile, just like an auto mechanic immediately knows the repair solution of a car as soon as a customer explains the problem.

There are many different types of electronic equipment that go on an aircraft or on a military vehicle. For example, there are the standard types such as Radar, Transmitters, Receivers, Navigation, Main Computer, Armament, etc. On most vehicles, especially on an aircraft, there is one other essential black box that is required which I will now referred to as the "Interface Box." An Interface box is usually the last item required by the System Engineers, because it is this box that provides the connection and communication between all of the different sub-systems.

The design techniques described in this book, could apply to all electronic equipment mentioned above. An example Interface Box will be used to illustrate one approach in the design.

Interface boxes are unique and different than other electronic boxes, because they have to interconnect Government Furnished Equipment (GFE) with other sub-contractor's equipment.

The Input/Output (I/O) information that is provided from these different manufacturers may not always be compatible or true, so a design engineer may be designing to an incorrect Equipment Specification (ES) that was prepared by a Systems Engineering.

For example, on a Teletype Interface box that I designed years ago, I had to interface with a new High Speed Printer (HSP) but it still had to work over the same wires that were used on another older aircraft which used a Low Speed Printer (LSP). The ES I/O requirements for the LSP were Bi-level signals of +12 volts and zero, but the I/O for the HSP was +5 volts using an RS-232-C bus line which is a special format to transmit data from one box to an other. The error was caused by two different System Engineers in my company, who did not confer with one another and each System Engineer created his own ES separately.

My I/O requirements in the design were correct because it was the exact requirements of the old LSP. When the HSP company was approached to modify their input they wanted an additional $100,000, because they would have to redesign their I/O package, so they claimed. That was in 1964, and in today's market that redesign would have cost a million dollars. I never believed that price in the first place but to save the program I was stuck with incorporating the modification in my Teletype Interface box where it didn't belong and it was never a perfect fix.

Interface is very important and, if I had my way in a management organization, I would have preferred to have both of those Equipment System Engineers matrixed to the EDM to avoid this type of error, and to insure that the same I/O requirements are in the Equipment Specification for both boxes.

Another problem with interface is, many companies know that they have the market cornered for their product, and would dictate what the I/O requirements shall be. On another program I had to interface with a complicated Transceiver that transmitted data through my Communications Interface box that was to be used on a modified airplane for Canada. This well known radio company changed the I/O interface requirements weekly. Their lack of decisions played havoc on my design and my schedule. I finally left both types of I/O interfaces on my input/output Printed Circuit Card and whenever that radio company's engineer made up his mind I would select the proper I/O by simply changing a jumper wire on the card. That was over 20 years ago and to my knowledge both of those I/O interface jumpers are still available on that card but only one is being used.

On a new aircraft, such as the Advanced Tactical Fighter now known as F-22, Interface Boxes would not be required, because all sub-systems would communicate over a dual Mil-Std 1553 hi-speed bus. The sub-systems are now modules instead of black boxes. Sounds good, but there would always be static control requirements, and years later the government would want to add another module that would not interface with the existing modules, so an Interface Module would eventually be required.

An inexperienced Electronic Design Manager could destroy a program. Unfortunately, several engineers were selected for the EDM's job after my boss retired because they had management experiences in other departments. I had been transferred at the time and was not available. They did not last long because an EDM's job is not an easy assignment in the electronics world. If an EDM had never designed and delivered an electronic unit, or has never quoted a job before, he would invariably ruin the program.

I was finally transferred back and promoted to the EDM's position when the last manager wasn't doing well, because he never had designed a thing in his life. He came up through the company ranks as a Test Engineer. Everyone thinks they can handle the Electronic Design Manager's job until they become one. This Test Engineer later became a good Test Manager when he was transferred back, but to put him in charge of a design organization was a poor decision. Conversely, I wouldn't make a good Test Manager.

The first problem that I encountered as the new manager was that this previous EDM made a poor quote on a program that was supposed to be over. The design phase on that program was completed, but the budget was already used up and the program was only half over. The previous EDM didn't know that he had to include in his quote the cost of drawing check, rework, drawing release and non-standard parts submittal. I ended up spending twice the hours that were originally quoted by the previous EDM.

I sure looked bad on that program because my upper-management didn't care whose fault it was, the last manager involved is always responsible for the job and he becomes the escape goat. My company had to eat those hours, because it was a fixed price contract. That was one of the reasons that I was asked to take over this job in the first place. The other EDM went back to being a Test Manager and, as I mentioned, he did a excellent job for the next 10 years before he retired. In this position he was now properly assigned and knew his business.

Another problem with many managers is they like to rule with an iron fist. This does not work so well as an EDM. Just because one has that position doesn't give him the authority to dominate his people, especially when working with experienced and mature engineers. It may work in the factory but not in engineering.

A good EDM does not dictate the design but simply makes suggestions. He has to work with his designers. He should listen to all of his people and then make the final decision, if necessary. I have overruled mechanical engineers on occasions on their design package, but when they were shown the reasons for my decision they usually agreed. Sometimes, a weak design had to be overlooked to encourage the engineers to take some initiative and to give them confidence. I would only do that when the design alternatives were insignificant. I was hurt once by this approach, but fortunately, I was able to recuperate and save the design during an ECP change requested by the customer. That was not a good idea on my part.

An experienced engineer, who is promoted from within to become an EDM, would probably have a better chance to make it, if he is trained properly. He would be respected by his peers because of his experiences, more so than if an EDM is hired-in from the outside.

One error that I made as an EDM was I didn't have a backup replacement when I retired. That was because my original backup left the company when he saw our department slowly being dissolved. It became a difficult position for me to fill. It takes a special type of personality, and a person with a lot of talent to be an EDM. It wasn't easy to find the right replacement. Regardless, every manager should always have a backup for the times he is not available, on vacation, if he changes jobs, or when he retires.

1. DESIGN TEAM

This chapter is based on a company that would be using the Service Structure arrangement in an Electronic Design Organization (other structures are discussed in Chapter III). The following information would apply to those structures as well, but the Service Structure will be used here for simplicity.

By definition: a Service Structure is an organization that provides all of the electronic designs and associated services for a company in one department that is controlled by the EDM. This Design Organization consists of:

Electronic Design Manager (EDM)
Electronic Engineers (EE)
System Engineers (SE) (Matrixed)
Mechanical Engineers (ME)
Software Engineers (SWE)
Printed Wire Engineers (PWE)
CADD (Computer Aided Drafting)
Technicians
Development Shops
Liaison Engineer (LE)

With this Design Team, an organization can design almost any type of electronic equipment. In this structure, all sub-groups would report to the Electronic Design Manager even if they are matrixed. Matrixed basically means a person is on loan to the EDM, but still reports to his original department.

The responsibilities of each of the EDM's positions mentioned above will be discussed separately. This structure represents only engineering design, and not production or any other support organizations. The Design Team, as used here, is like a company within a company, which provides special services to the company. It has to operate this way in order to supply the electronic services needed throughout the company.

Most of the advantages of this type of structure vs others are discussed in Chapter III. Because this structure serves all departments in a company, a well experienced team would eventually be established over the years that can handle almost any type of design. Because this team would be working together for many years, it would have accumulated many talents and each engineer would know the capabilities of his associates.

I worked in this type of structure for 36 years, 10 in management. My department had all kinds of capabilities. We had experts in Communications, Transmitters, Television, Computers, Antenna theory, Radar technology, Tape Recording, Digital circuits, Analog circuits, Teletype, ASW, EW, Power Supply design, Displays, Software, Armament design, Infra Red design, Micro Wave theory, Micro Processor design, Navigation theory, Interface electronics, Aircraft Wiring design, Packaging design, Printed Wire Board design, and experts in just about any piece of electronic gear used on an aircraft.

This capability didn't just happen. It evolved over many years as the result of designing sophisticated electronic equipment to support my company's main product line, airplanes. By having a nucleus of talent in one location the department was able to bid and produce almost any type of electronic equipment. Unfortunately, all good things have to come to an end, and since the policy of my company had changed of purchasing electronic equipment rather than to produce them in-house, my design organization was eventually dissolved. As a result of this bad decision, I decided it was time to retired. I was not bitter because I had a good run, as they say, and I retained a lot of good memories. The feeling of knowing that many of our aircraft are still flying in the sky with my department's designs on board is very gratifying.

This nucleus of talent can only survive if the work is interesting and the engineers are busy. The average life of an employee in my organization was 15 years. There was very little turn over because the work was always interesting. One important feature of a service type design organization, is that there are very few lay-offs, because the design team does not have to depend on any one specific program or one airplane to stay alive. Twenty design engineers can produce an awful lot of equipment.

My department supported many different Vehicles such as Aircraft, Missiles, Helicopters, Surface Ships, Tracking Stations, and Ground Support Equipment. The department had a lot of talent, but it did not have the support of a new company president, and from upper-management, in order for the department to survive. The same old problem; new manager new ideas.

"We are now a new division, and we will no longer design hardware," was his announcement. "We will design systems and purchase the required hardware."

The main problem with that philosophy was, who was going to support the electronics that were already in the field, especially if there was no Depot repair station? Also who would make the changes required by the customer on Engineering Change Proposals (ECPs)? Some of these equipment were designed 30 years ago and are still flying. With the waving of a wand by one man, a 40 year old department, with enormous capabilities, was destroyed without even consulting me. I must admit it was hard to swallow but these things do happen.

I know when I discuss this subject, it sounds like I'm crying in my beer, but I want the reader to understand that things like this do happen, and should be aware of it. It's a cruel world in this business, and if any of my experiences helps the reader in any way in this crazy business I would have accomplished my goal.

1a. Quotations

First of all, there are several ways that a department would be asked to quote a job. It could be in the form of a Request For a Proposal (RFP) directly from the government, it could be an Engineering change proposal (ECP), it could come from an out-side supplier, it could come from another aerospace company, or it could come from another department within the company.

When a Request For Proposal (RFP) arrives from the government, it would first go to the Program Manager's department through the companies Administration department. Copies then would go to different division engineers who would decide whether it pertains to their organization or not. If it is determined to be in the company's product line of business in a particular division, the RFP is passed on to the EPM department.

The EPM department would send a copy to me, as the EDM, and ask me to quote the RFP. I would first go over the RFP with my constituents to determine if the RFP is part of my charter.

The question now remains, should I quote or by-pass it? It is sometimes a difficult decision to make. Some RFPs were geared to be done by a particular company in the first place, but because the government is required by law to put the RFP out for bids, it did give my department a chance to compete. I did win a few jobs this way by under bidding the original company. Sometimes, however, I submitted a quote just to receive Bid and Proposal (B&P) money. I was not sure for certain that I would win the job, but the B&P funds did provide additional budget for my organization and kept a few people working. Who knows, I might have won the contract, so the quote still had to be good.

One must understand that the RFP has to go through upper-managers first before the EPM can see it, which would delay the final release to my department. Also, my design department may have to compete with outside suppliers. My company had a "make or buy" committee and they had to decide whether it should be made in-house or it should be purchased from a vendor. Once the in-house decision has been established for make, the quoting begins, and each department would submit their bids to the EPM.

Before quoting, my recommendation is for the EDM to first have an internal department meeting with all of the design team leaders and, together, produce the quote. One must realize that this group has to come up with an overall concept, usually in a very short period of time before it can be quoted. This is because the RFP may have sat on an upper-managers desk for weeks or months before the EDM ever sees it. Due to this delay the quote may have to be a Rough Order of Magnitude (ROM) or a WAG (Wild Ass Guess). If it's a Firm Fixed Price quote, one may be stuck with that ROM quote.

I have never been told by any customer after the fact, "We understand the quote was only a ROM or WAG, so we will renegotiate the cost later." Unfortunately, the ROM or WAG usually turns out to be the final quote. This is why an EDM has to be an experienced manager. He has to be able to decipher the RFP and know how much it would cost even without all of the details. He may never get another shot at it.

I have often had to quote millions of dollars for a program in one day. I have quoted jobs with no RFP and only from a verbal statement such as, "We need a box to interface with all the Communication Equipment and how much would it cost?" Because I, and my engineers know the aircraft business, we had a good idea what that Interface Box would consist of.

Generally, we had to produce a quote in a short time. My engineers would have to determine the package size, the amount of Printed Wire Boards (PWBs) required, the interface requirements, the software requirements, and the environmental conditions without a written Equipment Specifications. This takes a team of experienced people who know the military electronic business.

It sometimes comes down to a gut feeling or "guess-timation" on what it should cost, because one doesn't have enough time to research all the requirements. Once in awhile we did have plenty of time to analyze and produce a good quote.

In any event, the following quotation procedure (which I created) provides a important tool for the EDM to produce a good quote. It is not the only method available, but if it is used properly it would be very helpful.

1b. Quotation Procedure

The following procedure is what I consider to be the first steps the EDM has to begin with in preparing a quotation:

o Investigate the Request For Proposal (RFP)

o Examine the customers Statement of Work (SOW) which may not always be available.

o Develop a configuration approach (box size, etc.)

o Review all Mil-Specs' required

o Review the drawing requirements

o Review all the data submittals

o Review the Non-Standard Parts requirements

o Review the Production requirements

o Review the Contract Deliverable Requirements List (CDRL)

o Prepare the quotation using the Standard Quotation Outline (described later)

o Provide schedules, manloading, manpower and budget spread charts

o Create level I schedule for the Proposal and level II schedules for the Master Scheduler

o Prepare View-graphs for an oral presentation

o Submit the quotation to the Administrator

o The Administrator submits the Quote to the EPM

All the subjects outlined above including the RFP, the SOW, and the Standard Quotation Outline will be explained in detail later, and are just highlighted here to illustrate the quotation responsibilities of the EDM.

1c. Request For Proposal (RFP)

The RFP is exactly what it says it is, a request from the customer for a company to make a proposal on the design and development of an Electronic Black Box. Because it is required by law, the government would send out many of these RFPs to different companies who wish to bid on the hardware.

Sometimes, the RFP is geared to only one company, as was mentioned, who may have previously submitted a non-solicited proposal, but the Government is still required to solicit other bids. They would not always select the lowest bidder, but sometimes the best approach.

I did bid on programs like this and did win several contracts by drastically under bidding other companies, or by providing the customer with a unique design approach which became the driving force in their selection.

An RFP can also come from within the company or from another company. There was one step that I purposely omitted in the above discussions and that is the RFQ or a Request For a Quotation. The RFP is supposed to be a document that comes directly from the customer to solicit ideas on a design approach, and the RFQ is the follow-on request for the cost. However, the cost is almost always included in the response of the RFP.

On a major airframe proposal, these two steps may be followed in their proper sequence, but not so much on smaller programs such as a single Black Box. On occasions, I have received only the RFQ for the design and development of a black box. It depends on the government agency and the size of the program.

If the box were a non-solicited product improvement design, for example, the Government may only issue an RFQ to establish a cost, but then decide later not to use the box at all because the quote is too expensive. They may replace it instead with a GFE. There are so many things that could go wrong on a proposal. I have always said, "The government is like a woman, who may change her mind after she had ordered a dress." The government may also assert that option, and many times they do.

1d. Statement Of Work

The Statement Of Work (SOW) is a technical document that the government issues to suppliers who are bidding on a contract. It tells the bidders what the equipment is expected to accomplish. It may be included with the RFP or RFQ. It may even come later after the company has already been awarded the job. It is usually a gross document and, in many cases, it was generated from an existing SOW that was modified from another similar program. These copied documents are sometimes referred to as "Boiler Plate" information.

There is always a lot of, "Boiler Plate" information in an SOW, especially in the Mil-Spec segment. When a company bids a job they should renegotiate some of these requirements out of the SOW before the quote is finalized. It is sometimes not necessary because the customer is aware of these errors and would be more that happy to negotiate any possible changes after the job has been awarded and may even provide compensation for an increase in cost it may have caused.

If the company knows its business, however, these errors would not increase the cost because they would have been pointed out to the customer in advance. The government is always interested in saving money as long as a contractor can justify it by presenting several alterations to the SOW without reducing the unit's performance or reliability.

A presentation to the customer may or may not be required. View-graphs during an oral presentation is important and should be used whenever possible, because this presentation may make or break the proposal. One has to remember that a presentation to the government is not only given to technical military personnel but also to many non-technical civil service workers who <u>would</u> have a lot of influence on which company wins the job.

It is important that the presentation contain enough pictorial drawings and sketches to illustrate the box configuration and its internal structure. Remember, these non-technical types need to see something so that they have an idea of what it looks like and compare it with other boxes. Many managers have made verbal presentations without illustrations while trying to explain what their box looks like. It is sometimes pitiful to watch them "weasel" their way out.

Be prepared to answer all technical questions and don't be afraid to refer them to a company expert. Never, under any circumstance, try to answer a question by "weasel" wording. If the answer is not known, write down the question and tell them that they would be informed as soon as possible, and do it even if it's the next day. If they are ignored they would remember, and simply mark the company down in their evaluation report as non-responsive. Be truthful as best as one can. These military people are shewed, they know their business and can spot a "Con" artist right away. They have heard and witnessed many proposals and they know every trick in the trade.

Sometimes the questions they ask are foolish but one should never ridicule or ignore them. Answer their questions as best that one can no matter how ridiculous they seem.

For some of these presentations, the government may hire a professional consultant who knows his business, and he would tear a presentation apart because that's what he is being paid to do. Don't underestimate him or belittle him, especially in front of the customer even though he may be incorrect, because behind closed doors he is in command and who knows what he is saying to the government. If this consultant is incorrect, use diplomacy to point out his mistake or have a side discussion with him away from the customer. He would appreciate not being made a fool of.

On one program that I was on, the EPM decided to go it alone in Washington without the technical assistance of the EDM. He tried to weasel word around some of the technical points, and when word got back to my company the EPM was quickly removed from the program and assigned to a minor program. That EPM, by the way, was a graduate Business Administrator who, some how, became an EPM. He thought he could handle all the technical questions but, instead, he was digging a hole for the company. It happens sometimes. Most EPMs are qualified to discuss technical questions because they were once a Systems Engineer or a Designer, but please not a business major or, at least, have a qualified person available to answer any technical questions.

1e. Fact Finding

Fact Finding is actually the customer's investigation over the cost of a program. It could be a Military person investigating a supplier's proposal, or it could be the prime contractor investigating another subcontractor's cost and expenditures.

When the Military is Fact Finding a vendor, they may send out a team of people to a company's facilities or they may request that the company send a team to Washington. They may only send one person or several people.

Fact Finding personnel are generally accountants who have been specially trained as investigators. They generally consist of non-technical types who would reject using technical experience for justifying a cost.

Fact Finders take special classes at such places as the Air Force Institute Of Technology that teaches them how to evaluate a quotation, to determine if the actual work that was quoted on is required. The Institute publishes a class document LS-24 titled, "Understanding And Evaluating Technical Data Prices." This document shows the Fact Finders how to determine the manhours needed to do a design, to develop the software, and to produce drawings. Fact Finders are generally very capable women accountants who have been in the business many years.

They sometimes would enter the company with a chip on their shoulders, because their sole purpose is to find faults with the proposal and reduce the costs. They are expected, by their managers, to find some faults in a quote, and they must show a cost savings. It's like a policeman who has to give out X number of tickets a month. These Fact Finders are not considered doing their job unless they can reduce some of the costs.

I have to confess that, on one occasion, I purposely introduced a cost error in my quote so that Fact

Finders could find it. In that way they can justify to their superiors that they have found an overcharge in the quote and that they are doing their job. It was a little underhanded of me but, in that way, I retained the hours that I actually needed to do the job. I don't recommend doing this because it could backfire. It may have been a little unethical of me but, then again, I had to use the same tactic on my uppermanager.

When I submitted a quote to my Division Manager, I knew that he would not be satisfied unless he found some financial fault with it. So, I would insert a menial task, such as rework hours required in the quote. When my manager went over my quote and caught this task he throw it out. He reprimanded me for trying to slip something by him.

He said, "Quoting rework hours is a sign of weakness." My manager was now satisfied because he caught me in a goof-up and so everyone was happy. If it were a perfect quote my manager would have gone though it with a fine tooth comb and may have deleted some important functions. It's like trying to sell a house. No one would pay the asking price so you up the price a little. Actually, re-work in a quote may be justified in some cases but my manager wouldn't tolerate it.

Fact Finders may come to the company at any time. It can be at the start of a program, in the middle or even after a year into the program. As was said before, they would come in with a chip on their shoulders ready to tear into the budget. They take the attitude that the company is guilty before being proved. They would spend countless hours studying the quotation ahead of time in Washington, and develop a plan of attack. They can sometimes be mean and would treat the EDM as a criminal. Even though they seem to be the enemy, one must be careful and treat them respectfully or the company is a dead pigeon.

In the Fact Finding courses that they have taken, and under the heading of drawings, for example, the drawings are classified as simple, medium, or complex. When manhours are allocate to a drawing, the complexity of the drawing is the first thing that they look at. They won't take the EDM's word for it, either. They want to see the drawing or a reasonable example, to determine its complexity. If they think it's a simple drawing and the EDM thinks it's complex, guess who wins.

As I mentioned before, there were many times I didn't have more than a day to deliver a quote, so Fact Finding could be disastrous if one accidently over quotes. The way I got around this problem, when I did a ROM quote, was to continue working over the details of the original ROM (Rough Order Magnitude) after it had already been submitted by using my quote format to make sure I'm not too far off.

When one has quoted on as many programs as I have, it become second nature, so I had a pretty good idea (gut feeling) on what the cost of a program should be. However, when I go over the ROM in more detail, I have to make sure that the numbers agreed with those that were submitted or, at least, be close. If they are close, the detailed quote should be made available to the Fact Finders. If the numbers are not close, then the reasons should be documented and ready to be explained to the Fact Finders. In any event, the EDM should not drop the quotation effort on a ROM or a WAG as soon as it's released. Follow it up with a detailed quote.

The Fact Finder's would never accept individual experiences as justification for the hours quoted on any task. They would ask, "How were the numbers arrived at?" and if the EDM says through engineering experience, he's dead. They would ask for a complete breakdown of each task and the hours required. If the EDM says, "I took actual hours that were used on a similar task on another program," they would probably accept this without question as long as the numbers can be justify by similarity.

This is one reason I created the Quotation Format. It can be used to show the detail break down on how the manager arrives at the hours he quoted, and used to justify future quotations by correlating the actual hours that were used on an older program. An EDM has to be one step ahead of the Fact Finders.

In another situation, new EDM went to Washington with his ME to justify the cost that was already spent on a program. He was being torn apart. The ME, however, saved the day because he brought with him, unknown to his EDM, all of the man-hours that were accumulated on the program using my quote format which I had recommended to him a year before I was transferred. When the Fact Finders saw how the hours were recorded for each sub-task, they were elated, and accepted the hours without any further questions. As a matter of fact, they said the ME's record of man-hours was the best presentation that they have ever witnessed.

Material dollars are another sticky item that Fact Finders will cover. In this business, I have never been reprimanded or questioned by my upper-management on the cost of material dollars because those costs are usually insignificant in comparison to the cost of manhours. One thousand dollars worth of parts is nothing compared to one thousand manhours at a possible $100 an hour charge, including overhead. But not to the Fact Finder. Remember they are accountants and they only go by the numbers.

I had a female Fact Finding (nice looking) question me about a $20 component that was used in several places in our black box, and she wanted to know how I arrived at the price. I told her it was taken from the manufacturer's catalog. She insisted that I show her the catalog. So I had to go down the hall to our parts library where we keep our supplier catalogs for all of our programs. She tagged right along very close behind me to make sure I didn't pull a fast one. As an EDM, one has to be very patient with Fact Finders because they are only doing their job.

As I said, they just go by the numbers, so I had to show her either the manufacture's cost or the bill for that part. There was no way that I could have convince that lady that $20 was insignificant on a two million dollar program. The time that I spent with her ate up those 20 bucks up in no time at all, but that's the way it goes during a Fact Finding investigation. When she saw the price was accurate she was satisfied.

1f. Quotation Format Approach

The Quotation Format I am now proposing provides a more detailed analysis of the man-hours and monetary costs other formats do not provide. I used it on every program that I have ever bid on. Once it is accepted by the customer, it may also be used as reference for quoting future proposals. Most companies do not require a detailed tracking tabulation for every insignificant assignment as I'm now proposing, because it's another accounting task. Most companies do keep a running tab of manhours used on the program but not for each task. This new format is really not all that complicated, especially in today's computer oriented financial market.

The main advantage of this format is, the quote provides a list of manhours on every task performed, and after the contract is awarded the tracking hours may be used to compare it with the original quote. In other words, the quoted hours can now be used as a guideline to track the program hours as they are being used on each assignment. If the hours are way out of line the reasons may be made available to satisfy the Fact Finders or upper-management.

Each person that works on a specific task number adds the hours used on each task to his time card or into a computer. All of the numbers would be recorded by the Administrator with the man-hours used and a print-out would be provided every week. If this time card keeping or computer is not available, it wouldn't be too much of an effort for a secretary to record them.

The ideal approach is to have a central computer system with a terminal for all engineers. This is the way of the future anyway. Every engineer would enter the task number for each assignment he worked on and the hours used each day. It would automatically be recorded and accumulated in the main computer. This allows the EDM to be able to examine the status of any task on his own monitor each Monday morning or late that Friday. Manhours Spread Curves would be automatically updated each week. With this information, the EDM can present the status of his program to upper-management every Monday.

1f1. Work Breakdown Structure Quotation

The requirements for a quotation come in several different formats. The customer may require that all quotes be submitted in a Work Breakdown Structure (WBS). If this format is required, my new quote approach can easily be applied by using an additional numbering system for each major task. These additional tasks for the WBS are development tasks, testing tasks, assembly tasks, production tasks, etc. The WBS gives the customer a method of separating the costs of each sub-task which allows them the options of eliminating or delaying any task in the WBS such as production, for example.

1f2. Technical Level Quotation

Another type of quotation that may be required by the customer is the Technical Level Quotation (TLQ), where a list is needed for each classifications on various work levels. This is done by assigning grades to the work force such as, Engineering salary grades 1, 2, & 3, technicians grades 1, 2 3, etc. In this way the Fact Finders can break down the costs right down to the workers pay check rather than just manhours. Most of the time the final quotation consists of manhours, material dollars, overhead, and profit. Regardless of what quote method is required the quoting format that follows is still applicable.

1g. Quotation Format Procedure

There are basically two ways to produce a quote. One is by listing the number of people and the length of time they would be on the program (man-months), or by a detailed break-down (man-hours) of each task. The manhour method will be described in the following format procedure. Both of these methods should agree in the final analysis. I have used the man-month method many times, especially on a ROM, but backed it up with the detail manhour breakdown later.

The man-month quote approach is done by estimating how many men would be required for a program. For example, on a medium size program I might estimate using two design engineers, two mechanical engineers, two printed wire engineers, and two technicians over a six month period. This totals to 32 man-months.

In actuality, I may start with one designer and one ME over the first month then increase as necessary. At the peak of the program there may be 6 engineers working on the program which would gradually be reduced by two in the last month. Of cause, a detailed breakdown would be required later to convince upper-management and the Fact Finder by using a more precise budget report. This particular man-hours spread I just described is known as the Bell curve which will be explained later.

The following quote format is strictly an example to illustrate a method of quoting, to track an existing job and should only be used by the design team. Notice in the format that follows, I use task numbers of 0300 for EE tasks, 0400 for ME tasks, 0500 to 0700 for PC tasks, and 0800 for Software tasks. In this way each engineer knows his own task by the one hundred number for a particular program, and the EDM can quickly tell at a glance the budget status for each classification simply by comparing the 100 Task Numbers.

Increasing the numbers such as 0301 or 0302, for example, are used for EE sub-tasks. These sub-task numbers are assigned by the EDM or a lead man, whichever suits the program. One doesn't have to follow the numbering system, however, and can use worker's names instead, but remember different people may work on the same task, which can be confusing to the EDM as to who did the work. One man could work on a drawing first and later it could be finished by another.

On some programs, the detailed sub-task number breakdown for each classification may not be necessary. On most small jobs, one EE or one ME task number or a name may be sufficient. If there is only two PC boards required, for example, only one task number is adequate for both boards. It all depends on how much detail the EDM wants to use on each program.

Some engineers had resented this method at first, because to them it was like having a report card. In a way that was true, but later they found out they could use it for their own benefit when they were being evaluated for raises. They can point out, to their manager, all the work they had done on each task and how fast they did it. A task number is usually assigned to one person per program, but not always.

When the quote is sent to the customer, only the totals of each major task are submitted, not the sub-tasks. For the Fact Finder and for tracking purposes, all of the sub-tasks should be recorded and be available.

As I pointed out, in a previous discussion, when the ME went to Washington to justify the hours used, he presented this format showing the accumulated hours for each task down to the details of drawings. The Navy was pleased to be able to use this same breakdown structure to justify the cost of the program to their own superiors. They said they never saw any company provide such a complete run down, and they accepted this tracking data with no further questions.

Military managers should impose a similar type of format on all new contractors. In this way, there would be a standard format for all programs, which would provide a simple method of monitoring the suppliers costs. At the present time, no two companies follow the same standard format, even within the same company. With this format, if a man had spent 18 hours on a drawing, for example, it would be recorded. If another man spends only 6 hours on a similar drawing, some sort of explanation would be required. It could be because of rework, which can easily be justified, but if one man is too slow it could be reflected in his performance and raises. This procedure provides management with quite a tool.

1h. TABLE 1. QUOTATION/TRACKING FORMAT

Notice: The first two digits (01) are the Program's Code number. If there are more than 99 programs the letter "A" may be used ahead of the two digits (A01),

TASK NO. TASK DESCRIPTION	HOURS QUOTED	HOURS USED
PRELIMINARY TASKS		
01 0001 DESIGN MANAGEMENT (EDM) TASKS Totals	______	______

Review Quotation (It may be a year since it was last quoted)
Review the Statement of Work (SOW)
Review the Equipment Specification (ES)
Review all Mil-Specifications
Review the configuration with others
Review schedules and Manloading curves (bell)
Review Manhour and Budget spread
Create level III (Upper-Manage) & IV (EDM Dept.) schedules
Prepare view-graphs for first "kick off" design review
Prepare for the critical design review with the customer
Select lead design engineers in EE, ME, SWE, & PCE
Assign tasks to all engineers together with the lead men
Assign task numbers in detail. Use first 2 digits for program or WBS
Monitor the Program to the end of the design phase
Have weekly internal department meetings
Have weekly program design meetings with selected departments
Assist the EPM for his weekly design reviews with Upper-Management
Prepare for quarterly design reviews with customer
Provide status for upper-management
Work with the EPM and the PM
Interface with the other support organizations

Note: The EDM would use one task number per program unless he charges to the company Pool System. Pool charges are overhead charges set aside for upper-management and Administration.

TASK NO. TASK DESCRIPTION	HOURS QUOTED	HOURS USED
01 0050 SYSTEM ENGINEER Totals	______	______

Prepare systems block diagram
Prepare Equipment Specification (ES)
Attend all design reviews

Note: The System's Engineer's quote may be listed under the PM's quote.

TASK NO.	TASK DESCRIPTION	HOURS QUOTED	HOURS USED
01 0100	CONCEPTIONAL DESIGN (Lead-EE, ME, PCE, & SWE)	Totals______	______
	Review SOW (all)		
	Review ES (all)		
	Review configuration (EE, ME, PCE)		
	Develop a design block diagram (EE)		
	Develop Software Approach (EE, SWE)		
	Prepare for kick off design review (ALL)		

Note: All Lead men would use this pre-design task number 01 0100 for conceptual design by adding an additional number such as 01 0102 for the ME.

TASK NO.	TASK DESCRIPTION	HOURS QUOTED	HOURS USED
01 0200	PRIMARY DESIGN (EE, ME, & SWE)	Totals______	______
01 0230	Design only critical electronics (EE)	______	______
	Breadboard critical electronics (EE)		
	Record all test data for submittal (EE)		
01 0240	Partition mechanical design (ME)	______	______
	Create a paper mock-up (ME)		
01 0250	Partition PWB (PCE) preliminary	______	______
01 0280	Preliminary Software design (SWE)	______	______
01 0300	ELECTRONIC DESIGNERS (EEs)	Totals______	______
01 0301	Circuit A	______	______
01 0302	Circuit B, etc.	______	______
01 0320	Enclosure electronics design	______	______
01 0330	BITE design	______	______
01 0340	Power Supply design	______	______
01 0350	Schematics (in Auto CADD)	______	______
01 0370	Partition Circuits on PWB	______	______
01 0380	Simulations in Auto CADD	______	______
01 0390	PC Thermal and Short lead data	______	______
01 0400	MECHANICAL DESIGNERS (MEs)	Totals______	______
01 0401	Enclosure design	______	______
01 0403	Internal structure design	______	______
01 0405	Any Modules to design	______	______
01 0420	Power Supply Module design	______	______
01 0425	Thermal design	______	______
01 0430	I/O design	______	______
01 0435	Internal Harness design	______	______
01 0450	Interconnect Diagram or Pin List	______	______

TASK NO.	TASK DESCRIPTION		HOURS QUOTED	HOURS USED
01 0500	PRINTED WIRE DESIGN (PCEs)	Totals	______	______
01 0511	PWB A Layout		______	______
01 0512	Assembly Drawing		______	______
01 0513	Fabrication Drawing		______	______
01 0514	Master Pattern		______	______
01 0521	PWB B Layout		______	______
Note: Four drawings for each PWB				
01 0600	Motherboard Layout		______	______
01 0601	M/B Assembly Drawing		______	______
01 0602	M/B Fabrication Drawing		______	______
01 0603	M/B Master Pattern		______	______
01 0610	Motherboard Pin List		______	______
01 0620	Motherboard wire wrap design (if used)		______	______
01 0700	Interconnect Diagram or table (to I/O connectors)		______	______
01 0800	SOFTWARE DESIGN (SWEs)	Totals	______	______
01 0810	Micro-Processor SWE-A		______	______
01 0820	SWE-B		______	______
01 0830	Bite SWE-C		______	______
01 0840	Integration SWE-D		______	______
01 0900	PROCUREMENT DEVELOPMENT (LE)	Totals	______	______
01 0910	Ordering Prototype Parts		______	______
01 0920	Parts Listing		______	______
01 0930	Tracking Prototype Parts		______	______
01 0940	Non-Standard Parts List & Submittal		______	______
Note: LE is the Liaison Engineer				
01 1000	TEST FIXTURE (EE, ME, & SWE)	Totals	______	______
01 1010	Electronic Design (EE)		______	______
01 1020	Mechanical Design (ME)		______	______
01 1030	Software Design (SWE)		______	______
01 1040	Debug and Test Fixture		______	______
01 1100	DEBUG & TEST PROTOTYPE (EE & SWE)	Totals	______	______
01 1200	PRE-QUALIFICATION (EE & ME)	Totals	______	______
01 1210	Temperature Test (EE)		______	______
01 1220	Vibration Search (ME)		______	______

TASK NO.	TASK DESCRIPTION		HOURS QUOTED	HOURS USED
01 1300	QUALIFICATION (EE, ME, & LE)	Totals	______	______
01 1400	DRAWINGS (Not including PWB)	Totals	______	______
01 1410	Mechanical Drawings		______	______
01 1450	Electrical Drawing		______	______
01 1470	Engineering Drawing List (EDL)		______	______
01 1500	Drawing Check/Corrections		______	______
01 1550	Drawing Release/Rework		______	______
01 1560	ECO's, Project Slips, or MRDR's		______	______
01 1600	LIAISON (LE) Totals		______	______
01 1610	Engineering Interface		______	______
01 1620	Project Office Interface		______	______
01 1630	Production		______	______
01 1640	Customer		______	______
01 1650	Support Organizations		______	______
01 1700	FUNCTIONAL TEST (EE, SWE, & LE)	Totals	______	______
01 1800	SUPPORT INTERFACE (ALL)	Totals	______	______

EMI, Thermal, Components, Safety, Stress, Testability, Reliability, Maintainability, Material, Quality Assurance, Human Factors, Computer Services, Master Scheduling, Contracts, Weights, Functional Test, Production, Finance, Checkers, Shops, Integration, Flight Test, Purchasing, Produce-ability, & Electrical Wiring.

Note: This task is the budget for the design organization to interface with the support departments.

TASK NO.	TASK DESCRIPTION		HOURS QUOTED	HOURS USED
01 1900	INTEGRATION DEPT's (EE, ME, & LE)	Totals	______	______
01 1910	Integration		______	______
01 1920	Flight Test		______	______
01 1930	Customers Test Facilities		______	______
01 2000	PRODUCT SUPPORT (EE, ME, LE)	Totals	______	______
01 2010	Field Service		______	______
01 2020	Training		______	______
01 2030	Logistics		______	______

TASK NO.	TASK DESCRIPTION		HOURS QUOTED	HOURS USED
01 2100	MANUALS (ALL)	Totals	_______	_______
01 2110	Description		_______	_______
01 2120	Theory of Operation		_______	_______
01 2130	Flow Diagrams		_______	_______
01 2140	Wiring Diagram		_______	_______
01 2150	Mechanical Drawings		_______	_______
01 2160	Software		_______	_______
01 2170	Test Procedures		_______	_______
01 2200	Documentation (ALL)	Totals	_______	_______
01 2210	Final Report		_______	_______
01 2220	List Of Materials		_______	_______
01 2230	Drawing Submittal		_______	_______
01 2240	Software		_______	_______
01 2300	TRAVEL Quoted $______ Used $______			
01 2400	DESIGN REVIEWS ALL)	Totals	_______	_______
01 2410	Customer (number of reviews)		_______	_______
01 2420	Internal (number)		_______	_______
01 2500	CDRL's (ALL)	Totals	_______	_______

TASK NO.	TASK DESCRIPTION		HOURS QUOTED	HOURS USED
01 2600	MATERIAL Quoted $______ Used $______			
01 2610	Material Cost Breakdown List	Totals	_______	_______
01 2700	TECHNICIANS	Totals	_______	_______
01 2710	Mechanical Assemble		_______	_______
01 2720	PWB Assemble		_______	_______
01 2730	Main Harness Assemble		_______	_______
01 2740	Test Harness Assemble		_______	_______
01 2750	Mother Board Assemble		_______	_______
01 2760	Prototype Assemble (3 Units)		_______	_______
01 2765	Debug & Test Prototypes		_______	_______
01 2770	Test Fixture Assemble		_______	_______
01 2775	Debug & Test the Test Fixture		_______	_______
01 2780	Engineering Support		_______	_______
01 2800	MANUFACTURING SUPPORT (ALL)	Totals	_______	_______

TASK NO.	TASK DESCRIPTION		HOURS QUOTED	HOURS USED
01 300000	REWORK	Totals	______	______
01 310000	After Integration		______	______
01 320000	After Production		______	______
01 330000	After Flight Test		______	______

Note: Repeat 4 Digit Task No. above for rework as required. Also rework may not be required, or it may require new budget and not in the initial quote to the customer.

Total Non-Recurring ManHours for the Program Totals______ ______

01 4000 TOTALS NON-RECURRING Quoted $_______ Used $______

01 5000 RECURRING EFFORT $_______ $______

Non-recurring cost in dollars is for Design and Development only, by converting manhours to dollars not including overhead, production, support and profit. In 1962 it was $14 per manhour. In 1990 it was $80 per manhour. Who knows what it is today?

Recurring Effort is the manhours required for engineering to support the follow-on production units and Field Services. This support may be over a two year period after the Design phase is completed and at an average cost of one man per month depending on the size of the program. Production and other departments submit their costs separately from engineering.

Converting these two costs are usually the responsibility of the Administrator. The EDM may never see these converted dollar costs. I was shocked once when I discovered that the cost for engineering alone on one job was only $2 million but the quote that went out was $30 million. Production usually submits the highest price in a quote depending on how many boxes are to be produced.

1i. Prototypes

The normal and best approach in the development of an electronic box, is to use four (4) phases of fabrication. The first phase is the Breadboard stage, the second is to build a Brassboard, the third is to build Prototypes, and the fourth is build the First Production Unit.

1i1. Breadboard Phase

I recommend that the EE should breadboard only those circuits that appear to be complicated. Analog circuits, especially Power Supplies and RF circuits, should always be breadboarded. High speed digital circuits should also be breadboarded but must be laid out on a simulated PWB to cut down pick up, radiation and crosstalk. High speed or RF haywired breadboards would not play the same as they do on PWBs.

Low speed combinational logic circuits may not have to be breadboarded. Breadboarding is a very important phase in the program because electronic circuits don't always work the way the paper design says it should. There are so many unknowns that cannot be foreseen until the design is breadboarded and tried out. Crosstalk, pick-up, EMI, and race problems cannot be calculated on a paper design especially with high speed chips. An example of a race problem is, when the gate of one chip is enabled before its turn. Delays may have to be added.

If one has the capability of CADD electronic simulations, many of the low speed logic circuits can be checked out with this system and may not have to be breadboarded. This is the kind of a decision the designer has to establish by himself. If he is uncomfortable with a paper design, by all means breadboard it first.

It is also a good idea to place the breadboard in the Laboratory's oven to ensure it would meet the environmental temperature conditions, especially analog circuits. If the breadboard circuit is not to be tested in an oven over the temperature range, commercial chips may be used for breadboarding which would save the cost of expensive breadboard components.

Generally, in breadboarding, one is only interested in performance of the circuit and not its reliability. Do not use pre-screened chips for breadboarding or on prototype development units, because they are too costly and the design may invariably change later, which would increase the cost further. If pre-screened chips are used one may be stuck with these expensive chips if the design has changed. Pre-screening also takes a lot of time for chips to be delivered, so it could even cause delays in the program.

In our department, we had our own stock room which we used to obtain parts for our breadboards. This helped a lot, but sometimes even the stockroom was out of certain parts. To get around this shortage, I would accumulate components in my desk from previous programs. I sometimes had to buy parts at a surplus store, or directly from a parts house on "Will Call" with my own money, and try to recover those costs from a petty cash fund from our department. Many times, I had to eat the costs because of the politics involved trying to recoup using the petty cash fund. Whatever it took, I was going to get my breadboarding done to meet my schedule.

1i2. Brassboard Phase

Brassboard, by definition, is a Lab model of the unit. It only resembles the final unit, but it may be put together with screws and rivets. It provides a very fast way of having an experimental unit for preliminary testing. I have often skipped the brassboard phase because of budget, but in a properly run program this phase should not be excluded. A brassboard could be tested in the Integration Lab, on a hot bench or even on the aircraft which could help resolve some early unknown deficiencies. See Figure 35, page 328 as an example of an Integration Laboratory.

When the brassboard is hooked up in the integration Lab, many unforeseen problems somehow crop up. On many of my programs, as an EDM, I had to bypassed this phase because upper-management didn't think it was necessary. They claimed it added to the cost, and I was using it as a crutch because I was not sure of myself. I could not convince my boss that it would save time and costs of doing rework in the long run.

1i3. Prototype Phase

The Prototype is to be built as close to the final unit as possible. It is fabricated with engineering drawings by Engineering Development Laboratory Technicians, if possible. For all practical purposes it is the final unit. The only difference between the prototype and the production unit is, who fabricates it.

Production may have the capability to produce a prototype but they generally require released drawings and complete inspections. Also, in the process of testing, many problems may crop up, which sometimes changes the final production package, and it may end up slightly different, so all that release drawing effort and inspections were wasted.

I have always recommended that at least three prototypes be built by engineering. At Lockheed Aircraft we called them XN1, XN2, and XN3 for experimental models. XN1, is the first prototype (not the brassboard), which should be used by engineering to debug and modify as necessary in the engineering Lab.

It then could go to the Integration Facility to be tested against other equipment that it interfaces with. This type of testing would be as close to the real world as possible, until the unit flies. XN1 should remain there until it is replaced by a permanent production unit. XN1 may later be refurbished as a production unit, if possible, to save some costs. This means replacing all the PWBs using pre-screened components and going through inspection with production testing.

XN2 is upgraded with the faults found in XN1 and with new PWBs containing military components, but not pre-screened parts. XN2 would then go to the Environmental Facility for a preliminary vibration search to see if there are any structural problem. It would, eventually, be used for final Qualification. After Qual, it goes back to the engineering laboratory permanently to be used for upgrading future ECP changes. XN2 would not be completely refurbished for delivery, because it has already exceeded its life cycle during the qualification period of shake, bake, shock, salt spray and humidity testing.

There would always be changes as a result of integration testing. Believe me, there's no getting around it. No one can foresee all the variables that can happen when a new unit is integrated with the rest of the ships equipment, especially if they are also new designs. The changes discovered in XN1 would be incorporated in XN3 while XN2 is being environmentally tested.

Engineering should always have a unit available in the Lab to continue their own testing, to incorporate changes and to try them out. XN3 would remain with engineering until XN2 has returned from the Environmental Facility. XN3 can then be used to develop the Functional Test Procedure for Production and, later, be made available for Flight Testing. When all the testing is completed, XN3 may be refurbished in production to save more costs.

EMI testing may be done with any box so long as it is close to the final configuration. EMI testing is done independent of the physical qualification. An EMI search may be done on the first prototype or any XN. This is a gross test to determine if there is any interference emitting from the box. By doing this test early, changes can be incorporated in XN2 & XN3 ahead of time to save a major cost of refurbishing.

The prototype method just described, has worked in my department quite well for many years. The cost of the program becomes a little higher up front, but in the end there would be a substantial cost savings in modifications, and the program would run much smoother.

By the time all the testing is done, the experimental units would be close to being one hundred percent correct. However, sometimes production problems or flight test problems crop up, which might alter the design once again. It's almost impossible to cover every situations.

My upper-management one day decided that they were going to save a lot of money by fabricating only one prototype XN unit, and it would be refurbished for production. This was a foolish decision for several reasons. It left engineering very often without a unit to work with. There was a lot of lost time in engineering waiting for the box to come back from the Integration Facility or Environmental Facility. EMI testing alone tied up the unit for more than a month. In addition there could be no refurbishing done on this one unit because it was a basket case after qualification. This is the kind of decisions that are made by upper-managers who have no experience in the electronic design field.

In another situation, I had an upper-manager who would not even allow any breadboarding. He felt that a properly educated and experienced engineer should be able to go from a paper design to the final product with no changes. I knew it was stupid decision, so we had to hide the breadboarding activity from him by doing it in the production laboratory in another building. The games one has to play just to get the job done correctly.

The Production Supervisor allowed us to use his facilities because he knew we would have problems. In a meeting, someone casually mentioned breadboard and my boss heard it. I had to weasel word around that remark and convince him that it was for testing production equipment only.

As I said before, most upper-managers don't know what goes on in the trenches unless they were once an EDM, but they sure no how to make decisions. The reader may find himself in a similar situation, where upper-managers wants help him to death. What can one do? They are the bosses and one must obey.

Another bad upper-managers decision was made when my Division Engineer decided to break up my design organization into different sub-groups. My new responsibility was to control only the electronic design portion of a black box. It happened when the lead ME squawked to our Division Engineer about my interfering with his mechanical design. So, I was ordered not to even talk to any of the MEs, PWEs, or SWE. My hands were tied.

On that program, with this arrangement, the department turned out one of the worst packaging jobs that had ever came out of our division. The Production manager and Integration manager, while trying to assemble and test this package, squawked so much that our ME had to completely redesign the package.

These Sub-departments cannot be separated this way. They have to work together as a team, with the lead manager in charge to produce the highest quality and most useful hardware. In this particular package, the ME tried to get cute with a far out design, but it could not be serviced. It cost the department an additional 1,000 hours to correct the design, not to mention schedule slippage. My company had to eat that cost. The key statement here is, "check and balance," and I will refer to this statement often in this book because it is so important.

A word about changes. Changes are the name of the game. If one thinks that he can design a perfect unit the first time, he is in for a major surprise. Accept this fact, and not be frustrated if changes have to be made. A German immigrant engineer once told me, changes are completely frowned upon in Germany and could cause one to be dismissed. What nonsense. No wonder they lost the war.

1i4. First Production Unit

The First Production Unit is exactly what it says it is. It's produced with production equipment using engineering drawings. The main reason for this is to allow production the opportunity to make their production test fixtures, Soft Ware designs and set-ups to meet early schedules.

It's surprising how many different problems arise in production. When the shop makes a prototype, for example, they may overlook some drawing errors and make the box as it should be made with no feedback to engineering to correct them. The prototype in the Engineering Lab was done by a lot of hand waving to get it done.

This approach won't work in production. The next production box and all follow-on boxes, would have to be done with released drawings, not with engineering drawings. The topic of Drawing Structure will be discussed later.

1j. Manhours Spread and Tracking Charts

The Quote Format Procedure described above provides a unique format for quoting and tracking electronic programs. Another part of the submittal in a Quote is a Manhours Spread chart, which shows Manhours versus months, as illustrated in Figure 1 page 31. This example chart shows the allocated hours for a program and the hours that are being used during the program on a monthly basis, which could also be done on a weekly basis. Obviously, the quote that is submitted to the customer would only include the allocated hours, with the word allocated left out.

As one can see from the chart, the accumulated used hours during the program weaves around the allocated hours, and eventually ends up together at the end of the program, if one is lucky. It is very difficult to follow a straight line as illustrated by the allocated hours. This chart serves as a quick overview of how the expenditures for a particular program is being spent. When the EDM compares this chart with the schedule, he can tell the status of the program at a glance, and can respond immediately if there are any excessive expenditures over or under the quote.

Once I had complained to my upper-manager over the many charts and data submittals that I was required to produce when I was an EPM. It became apparent to me that a Manhour Spread Chart and Schedule Chart are probably the only two graphs necessary to monitor a program. The rest are just for politics. How many different ways does an EPM have to tell his boss he's in trouble, over budget or slipping schedule. He may be doing a good job but it would take him days to prepare the many different charts and submittals required by his company.

The recommended tracking data and the charts of Figures 1, 2 and 3 should be sufficient to monitor any program. Anything else is just more work for the EDM.

The Tracking Chart of Figure 2 on page 32, is a print out summery of the manhours used, compared to the Allocated manhours. This chart shows the Task No., the Task Name, and the Accumulated Hours Used by the month and by assignment. The total accumulated hours are accounted for at the right side of the chart as they are used, and a total for each month is shown across the bottom of the chart. A good business administrator could very easily produce these charts from the Quotation/Tracking Format that I have recommended by using a central computer system or a desk top PC.

With this data, all the EDM has to do is look at his terminal's display and he could tell at a glance what the status of a program is for each major task, and for any other program he is responsible for. He would know when a task is completed and when the next task has to start. If there is a delay in one task, he has the option of reassigning the individual to another task. He can modify the internal schedules and tracking charts to reflect these knew assignments. The important thing is that the final delivery date does not change much.

The EDM has to have some latitude to make internal schedules changes because, after all, these schedules are just a guide line for his programs. Many things can happen during a design. Parts may be late, or a prototype assembly may be late. If the design is changed new parts have to be ordered. There are so many variables in the development of a black box one cannot predict the outcome.

The EDM has to be aware, however, that any changes in his internal schedule may also affect other department's schedules. This is why there should be weekly design review meetings, to make certain that all parties are aware of any changes in schedule, and to discuss and resolve any impact on production, Qual testing, drawing up-date, or shop schedules.

1k. Manloading Charts

Manloading charts of Figure 3 on page 33, are used to show the overall manpower needed on a program. The EDM would use these charts on each of his programs to show how the distribution of manpower is allocated for each program under his jurisdiction. He would upgrade these charts monthly, and would be part of his monthly report.

In some cases, individual names would be entered onto the charts to insure that the same name is not being duplicated on other jobs, or that two people are not assigned to the same task. This happens quite often when an engineer is assigned to several tasks. He could be working on a specific task and, for some reason, his name appears on the chart for another project which he's not working on. It happens.

These charts may be required to be submitted as part of the quotation. The customer may want to know how many men are assigned to a program and what kind of manloading curve is being used. They may want to know the shape the manpower curve takes, and may require classifications of each worker to be specified on the curve to help them judge the cost.

Figure 3, has examples of four different manpower curves that may be used for program manpower distribution. Figure 3A, has a 4.5 man level at the start and continues at that level almost to the end of the program, and then drops off rapidly. This type of curve is typical of small programs such as an ECP, where all of the participating engineers are required to start together and stay to the finish. This type of curve is also used by the support engineers such as Reliability.

Figure 3B, is probably the most widely used curve. It is generally referred to as the Bell curve. This chart starts out with a small crew, then gradually builds up to a peak and levels off for a period in the middle of the program. Depending on the contract, the peak may continue for a few months or over years. The slopes on the sides of this curve, are not as smooth as shown and are more like steps. Some of the steps may even extend over several months. In a quote, the customer wants to see the approximate curve shape and prefers a smooth curve.

Figure 3C is just the opposite of Figure 3A, where the manloading builds up slowly and remains at a high constant level until the program ends. This type of curve is usually used on research programs, where a fixed amount of budget is allocated and all the work stops at the end of the year, but may be renewed again the following year with new budget.

Figure 3D is sort of an upside down Bell curve. This typical curve is used when a portion of the program is completed first, then delayed for a period before it can start up again. The delay may be needed to allow for shop fabrication of a prototype, for example. The shop would have its own curve to follow. When the delay is over, the program resumes again but, sometimes, if the delay becomes too long, different engineers may be assigned to the program. This could cause overruns.

No matter what curve is selected it is important that they are upgraded periodically in order to monitor the budget. Some companies would insist that the EPM follow the predicted curve closely and must answer to upper-management if the program wonders from this curve, extensively.

On one program that I was on as an EPM, I had to meet a predicted Bell curve. Because of delays in procurement, I was not using enough men as the curve indicated, so I added more men to the program to meet company budget requirements, but then the job was completed too soon.

This early completion caused me to have another bad mark written against me. I was reprimanded for it by the Vice-President of Finance. A project completed too soon, in his words, was, "Just as bad as a job that is slipping schedule. The company loses money if it ends too soon." I don't believe in that baloney but that's the way it was.

These curves should only be used to assist the EPM and EDM to do their respective jobs. They cannot be forced to follow them exactly by management regardless of what problems arise during the design.

The upper-managers in the company I just mentioned, had their priorities reversed. The EPM and EDM cannot be expected to work a job by strictly adhering to a particular curve. The curve is simply a guide line, which may have to be altered periodically. The main priorities of any design are to complete the program on schedule, within budget and be reliable. Remember my previous statement, "Mean Time Between Failure" is still the primary goal of any military program.

One time I was told not to complete a program under budget because they may have to give the underruns back to the customer. What's wrong with that? The customer would be pleased if a job were done for less, and it would help establish a sound company integrity.

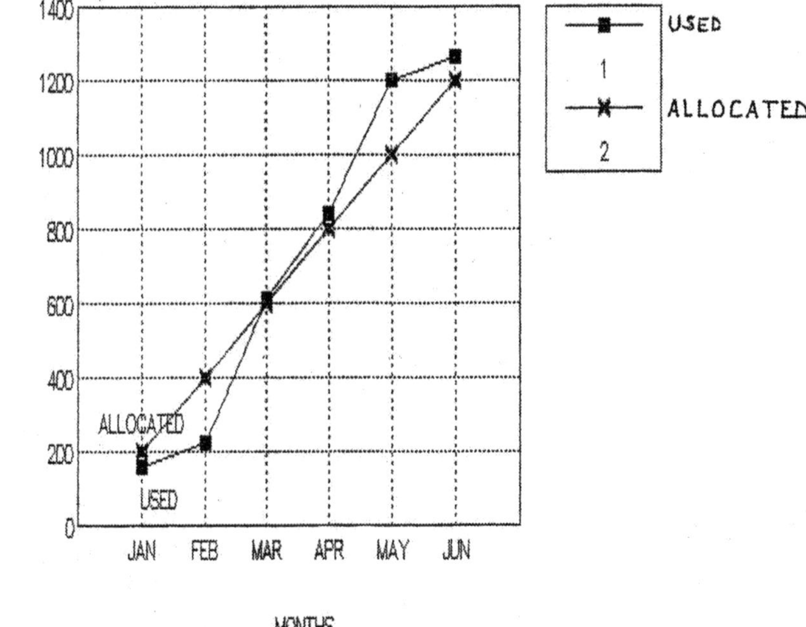

MANHOURS SPREAD

MONTHS	USED HOURS	ALLOCATED HOURS
JAN	160	200
FEB	220	400
MAR	610	600
APR	840	800
MAY	1200	1000
JUN	1270	1200

Figure 1. Manhours Spread Chart

MANHOURS USED NUMBER	TASK	JAN	FEB	MAR	APR	MAY	ACCUM TOTAL
01 0001	EDM	148	134	126	160	100	668
01 0100	CONCEPT	350	---	---	---	---	350
01 0200	PRELIMINARY						
01 0230	EE	--- 310	160	---	---		470
01 0300	EE DESIGN						
01 0301	CK-A	---	120	140	110	---	370
01 0400	ME DESIGN						
01 0401	BOX	---	180	178	158	163	679
01 0500	PWB DESIGN						
01 0501	PC-A	---	---	80	120	---	200
01 0502	PC-B	---	---	---	160	---	160
01 0503	PC-C	---	---	---	---	120	120
01 0800	SWE	---	148	305	364	310	1127
TOTALS		498	892	989	1072	693	4144

ALLOCATED MANHOURS

NUMBER	TASK	JAN	FEB	MAR	APR	MAY	ACCUM TOTAL
01 0001	EDM	160	160	160	160	160	800
01 0100	CONCEPT	320	---	---	---	---	320
01 0200	PRELIMINARY						
01 0230	EE	--- 320	320	---	---		640
01 0300	EE DESIGN						
01 0301	CK-A	---	80	160	160	160	560
01 0400	ME DESIGN						
01 0401	BOX	---	160	160	160	160	640
01 0500	PWB DESIGN						
01 0501	PC-A	---	---	---	160	---	160
01 0502	PC-B	---	---	---	160	---	160
01 0503	PC-C	---	---	---	---	160	160
01 0800	SWE	---	160	320	320	320	1120
TOTALS		480	880	1120	1120	960	4560

FIGURE 2. TRACKING CHARTS

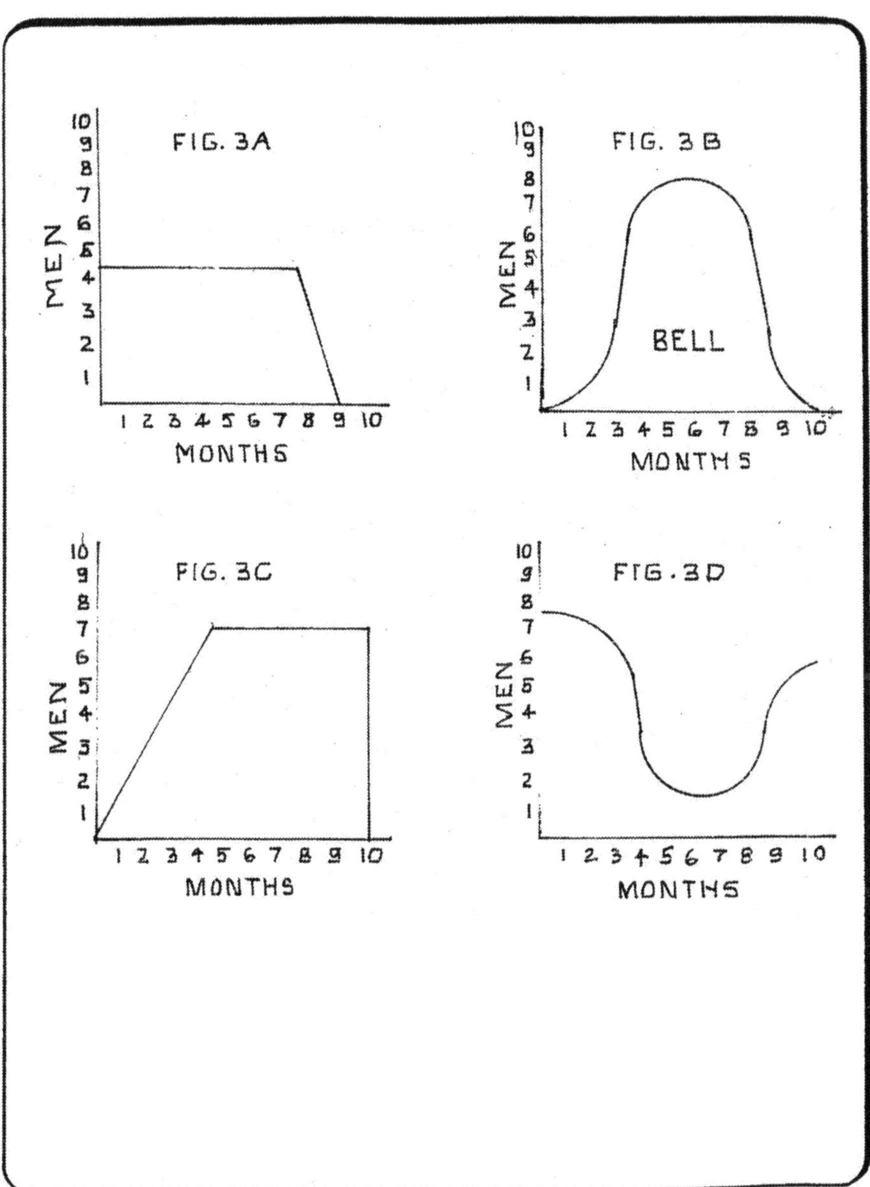

Figure 3. Man Loading Chart

2. ELECTRONIC ENGINEER

2a. Who Should be an Electronic Design Engineer?

Electronic design for military programs is a unique field. As was mentioned before, many engineers assume they are designers as soon as they graduate with a BS-EE. To illustrate this, I hired a new engineer with very high grades and as I was escorting him through the laboratory his first comment was, "How many technicians would be assigned to me?" I didn't want to discourage him, of course, so it took a little explaining on how things worked around the Lab. He, eventually, turned out to be a good designer.

A formal education provides the new graduate with the preliminary tools to be a designer, but it would take another five to ten years in design, if he is lucky, to understand the full significance in this field. Some can never become designers. Becoming a designer is very similar to becoming an apprentice machinist or an intern doctor. It takes years working with experienced engineers to learn all the "do's and don'ts."

Unfortunately, engineers are not required to serve this internship, and yet, many lives depend on the quality and capability of an experienced electronic designer, especially in military and commercial aircraft.

An Engineering State License is only required for State programs and not in the design field, so it becomes imperative that all designers meet or exceed the requirements that are dictated in the Equipment Specification, to provide the highest quality unit with the longest MTBF.

Not every engineer is capable or even wants to be a designer. Many EEs prefer to be a System Engineer. Many prefer to be an EE/Software Engineer. Some even prefer the business end of engineering. Still, many graduate EEs want to become designers but never get the opportunity to become one.

Being a designer in Military programs, is not all that glorified as people suspect. As a matter of fact, it is sometimes down right boring. In a design program I would estimate that the work required is 20% design, 20% testing, 40% paper work and 20% meetings. Only 20% of the design is fun, the rest is sheer dog work. Quoting a job alone could be tedious and boring. So the glory and fun isn't always there.

Being a designer can sometimes be nerve racking. A designer is the next person on the firing line next to the EDM. When a design is late or cannot pass a test, it is the designer who gets all the flack from the EDM or upper-management. If the equipment fails in the field or is not performing as it's supposed to, the designer would be reprimanded and may not be considered in the next raise period. The designer is always the last man in the loop. It is he that everyone comes down on when there is a problem.

If the reader is the type of engineer that cannot take reprimanding, he had better look for another profession. As was mentioned, my favorite expression for this situation is, "It comes with the territory." The designer would always be used as the escape goat. He could be doing a good job for the company, but if there are problems, guess who they blame? Remember one goof-up replaces ten atta-boys, and the only reward for doing a good job is probably a pay check and possibly a raise.

One unforeseen problem in military programs is, an error may not be discovered until after the equipment is delivered and in the field for a while. Everyone in this business, must understand and recognize this situation, and accept the fact that the design is not completed until it has proven itself in the field, and with many production versions over many years. Also, the first production model may work fine but the follow-on units may not.

I have never found a problem in the electronic field that was designed by my engineers that could not be resolved. I've seen, however, other companies go "belly up" because they didn't quote properly or they did not have the engineering capability. "Bid low and pray for ECP's," was their motto. Or bid any job and hire job shoppers to do the design. This is a not always good practice. I will get into the discussion of job shoppers later.

2b. Design Awards

In the design world there is usually no financial reward for doing a good job in engineering. As was said, a pay check every week is probably the only financial reward one would ever receive, but there is a gratifying reward. When I walk through an aircraft and observer our equipment performing their missions well, I get a little emotional, and I know deep down that I, or my engineers, have created a dependable electronic design for the aircraft. It is a feeling that stays with an engineer forever, and each time he sees a photo or discusses that particular vehicle he gets that nice contented feeling. The longer that vehicle remains operational with his equipment on board, he knows that he has done a respectable job and has produced a reliable design. What greater compensation can there be than knowing one has fulfilled his obligation and produced a dependable design.

Sometimes there are repercussions, especially if the equipment doesn't work so well in the field. When a design didn't perform to it's requirements on one program I was on, I had to defend my engineer from upper-management with the comment, "If the unit fails in the field consistently it's the our fault. If it doesn't fail and its MTBF is high but does not perform properly in the vehicle, then it's the System Engineers fault for producing a poor ES."

In this particular case, the equipment was very reliable with a high MTBF, but it had an operational problem on the aircraft. The System Engineer on that program was not under my jurisdiction and all we could do is meet his Equipment Specification (ES). There were many arguments between our departments over this unit.

This attitude from both sides, is wrong! Finger pointing has never solved any problem. This is one reason, I believe a Systems Engineer should work or be matrixed to the EDM and be available. In some companies, if a System Engineer turns out a poor ES, the designer may not communicate with him until the end of the program. This does happen often, especially when a vendor is contracted to do the design.

A good designer should work with the System Engineer, to point out any weakness in the Equipment Specification up front, and to reduce any risks. Unfortunately, too many designers accept the Specification as is, and would not challenge it. They take the attitude, "If that's what the System Engineer wants, that's what he's going to get, wrong or right."

That attitude happens often, especially when the System Department and the Design Department are in separate buildings, and they rarely communicate. This problem is precisely what this book is all about; team work, to produce the highest quality unit by using the "Check and balance approach." The designer should always interface with the System Engineer and with the many other support engineers that are involve in the program.

2c. The Hardware Design Engineer

There are too many engineers who have that false notion that just because they are System Engineers they think they are qualified to be hardware designers. These types are the worst offenders because they could compromise the design since they had also written the Specification. If they can't meet the Equipment Specification they may simply change the ES to meet the design. Also, the tedious tasks of doing drawings, exhaustive testing, and working with their hands in the LAB, is not what most System Engineers care to do.

System Engineers consider themselves the ultimate Engineer. Actually, that statement is not far from being wrong. I have always considered System Engineering as the most difficult profession in this business, but when they get into hardware design they show their weakness.

I can always tell who is a true hardware designer, just by having a conversation with him. It's like software people know software people, Doctors know Doctors, Secretaries know Secretaries, etc. System Engineers do not speak the same language as hardware designers do. Their conversations are a dead give away because they don't talk like a hardware designers.

Now, this does not mean that all System Engineers cannot do design work, and conversely not all hardware designers cannot do systems work. There are certain talented engineers that are capable of any task, and do it well. These individuals are, of course, the exceptions rather than the rule.

I have always consider myself a pretty good hardware engineer who can do a fairly good job on a systems design. I was also a Project Engineer, but I did not consider myself an expert in either of those two professions. Hardware design was my forte. I still say, however, it does take a special talent to be a System Engineer.

A true EE design engineer is the type of individual who likes to work with his hands. The hardware designer likes to get into the laboratory, work with meters and oscilloscopes, and doesn't mind the laborious testing, or even doing his own soldering and wiring. As a matter of fact, when I recruit young graduates, this is the type of person I look for. The highest grade student doesn't necessarily make the best designer. In reality, the high grade student would actually be board being a designer, because of all the "dog" work involved. He would be better suited as a System Engineer or in research.

The main difference is that the Design Engineer has to be creative. He has to have the capability to invent because many hardware designs have to be developed from scratch. He can't copy someone else's work, or look it up in a text. He also has to have the knowledge of packaging, wiring, printed wire board design, and drafting practices. He has to know systems engineering and integration. He has to understand how to read Military Specifications to design for environmental conditions. He has to be able to conduct verbal design reviews and submit documents to the customer.

He has to be able to quote a job from a Request For Proposal (RFP), sometimes in one day. He has to be able to interface with other organizations, especially with production. He has to provide tracking charts to the EDM with schedules, budget, and manpower. He has to know how to release drawings and how to deliver equipment.

He has to be able to research for parts, purchase them and submit non-standard parts to the customer. He has to know his companies policies on how they do business; dealing with Finance, Budgeting, and Master Scheduling. How does one learn all about this? Only from experience. Most of these tasks are the EDM's responsibility, but then again, what is an EDM but a design engineer with a lot of experience.

2d. Engineering Change Proposals

The EE design engineer has enormous responsibilities. He not only has to invent new designs, but would continuously be responsible for his old designs, that is if he stays with the same company long enough.

Before I retired, my department received an Engineering Change Proposal (ECP) from the Navy to modify a box that I designed 25 years before. There were several problems that had to be overcome to accomplish this task. The drawings were outdated, the parts were obsolete, and who can remember a design that far back. In this modification a new Printed Wire Board (PWB) had to be added using the latest parts on the market.

Now, a different experienced engineer had to be assigned to investigate the old design, modify it and see what effect it had on the existing electronics. In this particular case there was a domino failure effect, because the new electronics loaded the power supply down, it caused the existing circuitry not to work properly. In addition, the `Built In Test' functions did not work.

The point I'm making here is, the life of a designer does not stop at the end of development and delivery stages. An engineer is responsible for a design for life or as long as he stays with the same company. Some incompetent engineers have left the company for that very reason, to get rid of old responsibilities.

ECPs are a major source of income for most companies and cannot be ignored. So having the capability of reworking an old design is not only financially advantageous but provides creditability to the customer. They know they can depend on the company to support any ECP for whatever reason it's required.

2e. Designer Frauds

A favorite trick among many unqualified engineers is to hire in as an experienced designer an take on any difficult assignment given to them. After maybe one year into the design, he suddenly finds a new job.

I was the victim of this ploy several times. On one occasion, I was given the task to take over another man's design. When I investigated the details of the design I saw, right away, why the other engineer bailed out of a bad situation. There were so many weakness in his design, it was easy to conclude that this person was a fake, yet he got away with this sting for 15 years by moving from company to company.

A good way to tell if a man is a fraud, is to look at his resume. If he's been bouncing around from company to company every one to two years, his work habits would be very suspicious. There are some engineers who change jobs regularly for financial reasons, and are not fakes, but then a manager has to consider how long would that man stay with his company if he were hired and would he be available for ECPs.

Then there are designers who use handy-dandy design hand books to design complicated military electronics. I knew of a HAM operator who claimed he was an expert designer. He even fooled me at first with his verbal experiences on how he saved the whole missile crises. But when I saw him using a HAM helpful handy-dandy-book to design the electronics, I knew he was a fraud.

One problem with this handbook approach is that the engineer would be calling out parts that come out of this book rather than the Program Preferred Parts List (PPSL), which is part of the contract that is issued by the military. This means, many Non-Standard Parts would have to be submitted to the military, which could be rejected. A part rejection could cause a major redesign later on at the company's expense.

The second problem with the hand-book is, the engineer is trying to make the schematic from the handbook fit the design, instead of producing the design to fit the program. "Now let's see, can I use the schematic on page 23?" It's like going through a library of schematics to see which one fits the program. Amateurs!

2f. Consultants

Another problem in the design field is Job Shoppers or Consultants who are hired to do the design. There are advantages and disadvantages of using consultants. An advantage is that they are not as expensive as captive employees, because they don't receive benefits and are not part of the overhead costs. Many companies hire them for that purpose.

Another advantage is, when the program ends the manager can release them immediately and not have to go through a tedious employment layoff procedures. He can hire and fire them at will. They do serve a purpose, however, if they are used properly.

The main disadvantages are; the EDM is not sure how good they are, they don't know the companies policies, and they can leave at any time. Another disadvantage is, after they have completed a design and leave, they may not be available to correct any deficiencies that are later detected.

One vendor that my company dealt with was mainly a production house. They did not have a full time design engineering staff, yet they would be one of the lowest bidders. After they won a contract, they went out and hired many design Job Shoppers. The problem is, years later after the consultants left, there was no design engineer who could follow up or correct a flaw in the design.

As was mentioned, problems do arise in the field much later. In this one particular case, this company did originally under bid me, and became a sub-contractor for my company. But after they completed the design, and were in production, the Navy began having problems in the field. They kept on making modifications or "Band-Aiding" the design for over 13 years, and never really fixed the problem. The Job Shoppers for that company were not available any longer, and no one knew how to fix it. This system was for an armament program to drop torpedoes from our aircraft.

My organization was finally given the task to fix the design under a research program. What this other company couldn't do in 13 years my engineers fixed it in six months. Six PWBs had to be completely redesigned to fix the problem. The difference was, the people that worked for me had many years of experience in military design, especially with interface electronics for an aircraft and could spot a weakness very quickly.

I'm not trying to discourage the use of consultants, because if used properly they do provide a very important function in this field. If they are used mainly in a supporting roll, to assist rather than being the key designer, or if there is a good backup full-time employee who works with him so that his design can be followed-up after he leaves, then Job Shoppers might be a good investment.

The reason I'm discussing all of these side issues in design, is that the EDM must understand all of the weaknesses in this industry to be a good manager himself. This is all part of his maturing in the electronic world, and to understand all of the various responsibilities he would inherit.

2g. Design Learning

As a potential engineer progresses in this design field, he would develop certain skills through the normal process of learning as he makes a few mistakes or sees a better way of doing the design after the fact. He would become familiar with all the "Do's and Don'ts" associated with the electronic industry. He would learn there are certain laws that cannot be violated, and he would pick up a mountain of information from his own experiences that is not written in any text. When I examine a design by a novice engineer, I can easily spot a weakness in his design, because this person had violated the basic rules.

There was a situation, where my upper-manager assigned a young graduate to a design task without my approval. After this young engineer's design went into production, it continuously failed Functional Test. This individual almost had a nerves breakdown trying to correct the problems. I finally had to assign a seasoned engineer to correct some of the problems to avoid doing a complete redesign.

The unit became a "hanger queen" by the Navy (always on a hanger shelf waiting for repairs) for many years thereafter, even after it was modified. This young engineer finally left the company and was not heard from again. He had the potential of being a good engineer, but was mentally broken because he was given a difficult task too soon. He wasn't brought along slowly and trained properly. He did not develop his skills and "Do's and Don'ts" yet. This is another bad example of what happens when upper-managers by-passes the EDM.

2h. Maintaining Capability

Designing is very similar to many professions. It can even be made analogous to sports. One must practice by playing the game of his sport continuously to be sharp. If a baseball player, for example, sits on the bench too long he would become rusty and his batting average would go down. When a design engineer works every day at his profession, and is in a good frame of mind he goes about his business and applies the "Do's and Don'ts" automatically. His mind is kept fresh and he's trained to do things by reflex action. He automatically thinks reliability, component stressing, testability, thermal, environment, and overloading, just to mention a few basic design practices of the `Do's and Don'ts.' I had been away from doing design work for four years while doing EW Flight Analysis on the U-2 bird for the Lockheed Skunk Works, and when I returned it took me quite a while to reoriented myself. I would say to myself, "Now how did I use these Chips before?" It was a little scary.

The thing that frightened me most was, I was not aware of all the new Integrated Circuits chips that were now on the market, which comes down to the familiar problem of how does one upgrade himself? What provisions are there in the company's policy to improve engineering capability? Are there enough new designs to keep the designers sharp? Are there enough research programs for engineers to apply themselves to the latest technology? Are there any special classes within the company's education system for engineers to upgrade their capabilities? This is where proper upper-management and a progressive company makes the difference. Education doesn't stop after graduation. A company should have an in-house sponsored educational program and encourage advanced classes under a paid tuition program.

2i. Lead Engineers

The Electronic Designer is the key player in the electronic business, and usually takes the responsibility as the lead engineer. MEs have complained to me that they are treated as second class citizens in the electronic business, because the EE has so much authority. Why can't an ME be the lead engineer?

The answer is very simple. The contract is for an electronic box not a mechanical box, so the electronics is the primary function of that box. It comes with the territory. If it were strictly a panel with non-active components, such as switches and relays, then it would be consider a mechanical box and the ME would be the key player. Sometimes, the Software Engineer is the key player, especially when test equipment is purchased and programming is the essential requirement. It all depends on the contract and the unit's function as to who would be the overall lead engineer.

The EDM, however, could assign a lead engineer for each sub-group. In a large electronic program, he would assign an EE as the overall lead engineer, then assign sub-lead engineers for Mechanical, Software, and Printed Circuit designers. The EDM would probably be responsible for several jobs going on at the same time, so he has to depend on a lead engineer for each program.

If the job is small only an EE and ME may be assigned as the lead men. However, just because the EE is the overall lead man, he should not be allowed to dictate the enclosure. It should be a three-way decision between the EDM, the EE, and the ME. In any event, the ME must have the freedom to initiate and produce the package. He must be given the respect he deserves as an ME. Too much interference would only cause disruption, and the ME would just go through the motions rather than be involved in the design and possibly compromise the design.

I want to stress once again, that the design is the responsibility of the design department. The design engineers should not allow the EPM or the PM to dictate the design directly. It has happened too often that an EDM was by-passed by an EPM, who gave orders to the EDM's engineers without his knowledge. Under these circumstances, I told my engineers to do what they asked them to do but to report any changes to me immediately so that I can settle it with the EPM or PM later to make sure the changes don't degrade the design.

Changes without my knowledge, has happened to me when I was out of town on business. The EPM took charge of the design. Needless to say, when I returned I had to straighten out the situation and clean up the EPM's changes. This took an additional week of design and drawing changes. I had made the mistake of not assigning a back-up engineer when I was not available, but I didn't expect the EPM to make a major modification.

2j. Support Organizations

During the design phase, the design organization would constantly be monitored by different support organizations. The Thermal Department would analyze the PWB for hot spots and make recommendations. The Reliability Department would examine the schematics and the parts selections for design weaknesses. The EMI Department would examine the overall box for radiation, and the power lines for susceptibility. The Testability Department would examine the design for Built-In-Test (BIT) and test points. The Maintainability Department would examine the design for access and repairability.

All of these departments and more, would participate in the design reviews and present their findings and make recommendations. This sometimes turns into a shouting match because the support departments may not always agree with each other. It's up to the EPM to keep the peace, but some of these confrontations are necessary.

Engineers sometimes have a large ego and do not like their designs being torn apart by outsiders. This is why these support organizations have to get involved in the program early to minimize any modifications later. The support organizations and their functions are reviewed now briefly, and will be discussed in more detail in Chapter II.

An engineer should not always take the support engineers recommendation as gospel either, because each support department sees only his own responsibilities. He doesn't see the affect his recommendation may have on the other support departments.

My ME was once put into a Ping Pong situation over a black box design, where the Stress Department insisted that the bottom cover be beef up with rib stiffeners. At the same time, the Weights Department insisted that the weight be reduced in the bottom cover by eliminating the ribs and to use a lighter gauge metal.

The problem was finally resolve by putting the unit on a shake table. It was found that without the rib stiffeners, the bottom cover would "oil-can" and short out the wire wrap connector pins of the mother board. So this time the Stress department won the round.

Another consideration is that the support departments may increase the cost of the design. The EDM has to use diplomacy over this problem and sometimes work out a compromise. These support engineers are necessary for "check and balance" of a unit and should definitely be involved in the program.

2k. Thermal

Thermal is always a major problem in design and if the EDM incorporates everything the thermal people want, the box may grow beyond the design requirements. As a design manager, the EDM has to weigh all of their inputs but still has to cooperate with the thermal people. They do have a responsibility and shouldn't be ignored.

I was told by the thermal engineer in my younger days as a designer, that my box was too hot and that I would need a fan in it. Being young and stubborn, I fought this addition as hard as I could because my calculations showed that the box would be relatively cool and the components would not exceed their maximum temperature derating limits at their highest environmental levels.

The Thermal department finally wrote a memo to the President of our company about the design weakness of my box. Upper-management came down on me hard, and my boss forced me to install a fan in the box. I was told by my Division Engineer that I didn't know what I was doing. He even tried to prove how wrong I was using his own calculations which, of course, were amateurish. How does one tell his upper-manager is full of bull? My supervisor wouldn't back me up, either. I've been designing electronics for him for twenty years. He knew my capabilities, but he had to protect himself.

After the box was assembled I had to debug the electronics in the Lab. The Thermal Engineer, who had raised so much fuss, came into the Laboratory to see how hot the box was. Earlier, I had a technician disconnect the fan because I couldn't stand all the noise from the high pitched 400 Hz fan while I was debugging the design. The thermal engineer was ready to give me the "I told you so" statement, but when he examined the power supply temperature (with the back of his hand and no fan running) he walked away and never returned.

There is a lesson to be learned here. I should have had a better working relationship with the thermal organization and with all the other support departments. These departments should not play god either and make it a contest between departments. It should not be a situation, "I caught you in a goof-up," contest either but, instead, it should be a team effort working together.

Years later in another situation, I agreed with the thermal department to add fans to an Interface Device (ID). I was not the EDM but I had learned my lesson the hard way not to fight. Product Support did squawk the use of fans and went so far as to get the Navy to back them up.

This battle went on for months because Product Support's argument was that the IDs would be used only in an air condition environment and, in addition, the fans themselves reduced the MTBF of the box. I insisted that Engineering had to meet the Mil-Spec requirement that were imposed on the unit, and because the temperature requirements were not negotiated out of the contract in the first place by the EPM, I had to qualify the unit under extreme environmental condition.

The issue was finally resolved when Thermal Imaging was performed on the box under its laboratory environment, with and without fans. The resultant data was an eye opener. It was discovered that even in an air conditioned room, some of the components on the PC boards, with no fan, exceeded their maximum allowable IC junction temperatures. This proved beyond a doubt that a fan was required. This time Thermal was correct in their analysis. I'm glad I didn't fight them on this one.

The proper way this should have been handled was to first correct the SOW with the customer on the Mil-Spec requirements before the program began. This was not done. The disagreements came after the fact. If it was handled properly and there was still a conflict, Thermal Imaging on the equipment would solve many of these problems.

The true test is made when the unit is placed in a heat chamber with specific hot components being monitored. Thermal data should be recorded and always be available to the customer to prove the reliability of the design. Originally, the EPM on the above case did not want to pay for the Thermal Imaging test, but it was finally done on my insistence as the Staff Engineer to resolve the issue. It certainly did.

2l. Research Designers

It is obvious, that an Electronic Engineer has a lot of responsibilities besides doing the design. Unfortunately, many companies would groom young engineers just out of college as Research Engineers. This is fine if they remain there. I came across several individuals in my company who were excellent designers for research and development, but were never exposed to the frustration of delivering a box to the military. Research is fun because one does not have to worry about production, reliability, drawing release, testability, qualification, etc.

A Research Engineer may build one or two items just to prove a concept. He doesn't have to bid the job or interface with other organizations. The package can be of commercial quality. These engineers would go on designing like this for many years and never learn the technicalities for meeting military requirements.

Then one day, we had a reorganization in my company and a research engineer was transferred to my hardware group. He was a very capable man who had a lot of ideas. However, in this new position he was required to produce a fully qualified military box. He didn't know anything about non-standard parts or PPSL parts or our own company preferred parts list.

He finally came around, but at first he was dumbfounded with all the data submittals required. I was patient with him because I knew he was a good engineer. He said to me, "There should be a special course in designing for the military." I agreed with him and that's one reason why I wrote this book. Some requirements in designing for a commercial aircraft are more stringent than the military, believe it or not.

Later on, more research types were transferred into my organization during a major reduction of research programs. They could not believe all of the paper work and non-design efforts that were required just to design an electronic box. The drawing effort alone just baffled them. They didn't know that my company would purchase production components by the truck loads to save on the cost of parts, and these research engineers didn't like the idea of being limited to those components.

A designer should always take advantage of the company's parts storage facilities and use them first in his design if he can. The price savings of parts could be as much as 1/10 the catalog cost and they are available overnight. Selecting components that are not in the company's inventory could mean late deliveries and higher costs, especially with connectors which could take six months or more for delivery. This is the kind of limitations that research engineers were not exposed to before.

2m. Steps in Electronic Design

The Electronic Engineer can save himself a lot of grief by using the following steps in a new program:

2m1. Quotation Start-Up.

Examine the RFP. Develop a concept. Work with the ME & PC Engineers to develop a package, and quote the electronic portion.

2m2. After Contract is Awarded

After the contract is awarded (could be years later) begin the conceptual design phase. Examine the old quote, examine the SOW, and re-quote the job again to see if the old quote is still valid.

Re-quoting is absolutely necessary, because it may have been done by someone else and other factors have come into play. If it was under-quoted upper-management should be informed immediately. They may have to go back to the customer for more funds or reshuffle the budget.

Generally, what happens is upper-management would consider a budget distribution shift to help out. Most upper-managers hold back %10 of the budget, anyway, for unknown problems that may crop up which they can draw from to assist the program, if necessary. Also, they may even exclude a task or two from the proposal and transfer the hours to the EDM. They may transfer hours from other departments to support the EDM.

Prepare Level III & IV schedules, manloading spread curves, assign task numbers to subordinates, and prepare for a kick off meeting.

2m3. Breadboarding

Begin the preliminary phase of the program. Bread-boarding is done to only those circuits that are considered critical. If computer simulation is available, then many of the breadboards can be simulated on CADD, especially digital circuits. Analog circuits are difficult to simulate and it may be necessary to breadboard them.

High speed digital and clock frequencies greater than 10 mHz, may require more than a simple breadboard. For high speeds circuits, it is recommended that the engineer layout the circuit on a PC board. The PC board now becomes the breadboard. This is the only way one can "quite down" high frequency pick-up and crosstalk due to the fast rise times of the chips and the high frequency clock.

2m4. Electronic Design

Begin the detail electronic design. The EE engineers should draw all of their schematics in the computer aided CADD system. No more "chicken scratching" sketches. When the EE does his own schematics he has more control which reduces human errors. The procedure that follows, assumes that the company is equipped with an Auto Routing system that does the following:

Schematic, Simulations, Auto Placement, Auto Routing, PC Photo Plotting, provides Master Patterns, Assembly drawings, and Fabrication drawings. Each EE would do a simulation analysis of his design to insure that the design works and minimize breadboarding.

2m5. Printed Wire Board (PWB) Design

Note: The terms PWB Engineer and PC Engineer are use interchangeably throughout this book but they are the same person.

The EE should work with the PC Engineer on the placement of components on the PC board. In general, 70% of the circuitry can be auto-placed, but first the EE has to decide which chips are hot and placed at the edge of the board or placed on a cold plate for better heat transfer to the chassis. This also depends on the Thermal design requirements of the box.

The next input from the EE, is the trace length and width. For HF circuits, the traces may have to be short to reduce pick up and insertion loss. A long trace at high frequencies acts like a choke (coil).

If the lines have high currents, the power traces may have to be made wider. Also, in a multi-layer PWB, special Guards (shielding) may be required.

The EE has to work with PC designer on the Ground Plane, and the Voltage Plane on the mother board for the same reasons as the PWBs.

The PWB design will be covered in more detail later, when the requirements of the Printed Wire Board Engineer are discussed.

The EE cannot be just only an electronic designer, he has to be involved in the packaging. This is why an Independent Structure is not recommended, where the designer does only the electronic design and another department handles the PWB design independently, while still another department is responsible for the packaging.

2m6. Packaging

The following tasks are some of the key functions that the EE has to coordinate with the ME and other organizations, during the design of an enclosure.

Backplane Design, Partitioning, Wire Harness, Power Supplies, Crosstalk, EMI, TEMPEST, Thermal, Front panel, Repair Access, Testability, Plug-in Modules, Chassis mounted components, Coax cables, Test points, I/O connector selection, Pin List, non-standard parts, drawing release, and Human Factors. These tasks will now be covered as follows:

2m6a. Backplane selection (mother board)

If the clock frequency is less than 5 mHz, one may get by with a wire wrap backplane. Above 5 mhz, the engineer would be better off with a multi-layer PC mother board. For RF signals, an enclosed shielded module should be used for the PWB. Hard wire may be used on small units at low frequencies.

Microwave technology is a whole new ball game. Placing microwave circuits in a box is not that difficult. Most of the components, cables, and fittings for microwave circuits are purchased items. The EE simply has to select these parts from a catalog, but he must understand the properties of insertion loss and coupling.

If wire wrap is used, the EE has to make sure the wires take the shortest path. The EE may have to create the wire list when automatic wire wrapping is used.

If a PC mother-board is used the EE would have to provide the interconnect diagram or a pin list between modules. He also has to create the mother-board I/O pin list to the external connectors no matter which type of backplane is used. So, the EE has to be involved in many of the tasks required in packaging a unit.

2m6b. Partitioning within the box

In partitioning, the EE has to work with both the PC and ME Engineers. For example, the EE has to decide where on his schematic it is best to break a circuit and connect it to another board. He has to tell the ME the location of each board in the box; because of frequency, thermal, and crosstalk problems.

Usually, it is best to put the driver and receiver PC modules close to the I/O connectors at the back end of the box, and the power supplies at the front end. Micro-processor (Computer) electronics may be next in the box after the driver and receiver cards, and any special electronics in the middle.

2m6c. Wire Harness

For some reason, many Packaging Engineers forget about the wire harness. It has happened too often where the ME would design an enclosure, have it fabricated and not worry about the I/O wires. An ME once explained to me that he never worries about the harness. The technician in the Lab would fit them in, somehow.

Yes, but what about production? The wire harness is part of the packaging design. The ME should create drawings to illustrate how he wants the harness routed. The EE should work with the ME with regards to wire size, coax cable selection, wire length and he should approve the routing of the harness.

2m6d. Power Supply

The power supply may be an in-house design or a purchased item. In the old days power supplies were always designed in-house because they were not available on the market for military applications. In today's market, it is foolish to design a power supply when they can be purchased and fully Mil-Qualified.

The technology used today, in many cases, is using high voltage supplies at 270 volts dc. This dc is then reduced to any number of low voltages required. The 270 volts is the rectified voltage number that happens to come out of a three phase 220 volt ac power line. Because no step down transformers are required, it reduces the weight of the PS considerably. This type of a supply is about 80% efficient and is referred to as a converter.

Some System Engineers like to use one converter to feed 270 volts to several boxes. This means that each box would require a chopper type high frequency supply to reduce the 270 volts to as many lower voltages as required. The main advantage with this technique is thin wires can be used at 270 volts because the current ratio is 1/50 of that required of a 5 volt supply.

The Navy was promoting the idea of standardizing the low voltage supplies. If one needs +5 volts, he would draw current from a central standard supply. The two problems with this approach is that large wires would be required because there is no current ratio improvement, and there would be a large amount of decoupling required in each box. Without decoupling, there would be a lot of noise and pick up between boxes through the common dc bus. Decoupling means that large chokes and filters would be needed, which would cause insertion losses and may require complicated voltage regulators.

There is no advantage with the 5 volt approach because each box would need it's own input filters and regulator. The major squawk the Navy had, was that they were tired of spending so much money on fixing all the different individual power supplies, but now they would have to fix individual regulators. Where's the advantage?

2m6e. Crosstalk, TEMPEST, and EMI

These three subjects are closely related because they have to do with pick up, noise and interference. The EE has to work with the EMI department and with the ME to establish the packaging principles to control these interferences. ESD (Static) may also be included in this category. These subjects will be discussed in more detail when EMI is covered in Chapter II.

2m6f. Thermal Design

The most controversial subject in the design is usually thermal. Over the last 25 years the military has made enormous studies on the effects heat has on the life of Integrated Circuit chips and other components. They have found that the life expectancy of a chip can be predicted by the heat that the device is exposed to.

The Junction temperature of a chip is the parameter that governs the life of the IC. The EE and ME would be checked on this subject from the Thermal department, the Reliability department and the Military.

When the EE is designing a circuit and selecting components, thermal is one of the first considerations that must go into his design. The SOW and the ES would establish the ambient temperature limits that the unit would be involved in. From this, the ambient temperature inside the box has to be established. Reliability, Mil-Specs, Manufactures Catalogs, or company handbooks can be used to establish derating curves for each component.

A good practice, to insure high reliability, is reduce the derating level even further than required. This sometimes increases the cost but, as mentioned, in military programs reliability is more important than component costs.

To illustrate derating further, if a 1/10 watt resistor is sufficient in a circuit, then use a 1/8 watt. The cost is the same in this case but it may require more space on the PC board.

No one seems to worry about the low temperature affects, but the EE definitely has to be concerned with it because the transistor Beta effect (Gain) goes down to practically nothing at -55 degrees C. This effect is more of an operational analog problem rather than a reliability or failure problem.

Another consideration that the EE has to be aware of is the installation of the components on a PC board and on the chassis. It depends on the cooling method. Would the PC boards be cooled by radiation or convection? Would each PC board have a heat plate mounted on it to draw the heat away from each chip to the chassis? Would the box be air cooled? Would cold plate cooling be used? Would refrigeration cooling be used? All of these questions must be considered by the EE.

As was mentioned before, I designed an electronic box that I thought did not require a fan, but the Thermal department insisted there should be one. The reason I knew I didn't need a fan, was because the unit was being used in a low frequency situation, and I chose to use C-MOS chips for the logic circuits and transistors for the I/O circuits. C-MOS chips draw only micro-amps, so the I/O circuits were the only heat produces and they were spread throughout a large chassis. The power supply load was less than one ampere. A fan was not necessary.

The important point I'm trying to establish is that the thermal design should come after the initial electronic design. Too many programs have established the thermal philosophy on an aircraft before the electronic design has even started. This is not always the smartest way to go but sometimes there is no other choice.

For example, I worked on an aircraft program where Cold Plate cooling was the thermal design philosophy for all vendors. This is basically a good approach, but the problem was, the subcontractors used this to compromise their designs. In other words, they would reduce the cost of their designs so that their box could not function without the airplane engines running to operate the blowers to cool their boxes. When the plane was on the ground and engines off, the blowers had to be hooked up to ground power in order to pass air through the Cold Plates of each electronic box. These vendors were using the Cold Plate to meet reliability. They deliberately took advantage of it to reduce the cost by using low temperature parts and increase their profits.

This same crutch was allowed for shock mounts for vendor boxes. If one gives a supplier an option up front, he would always take advantage of it. I designed two boxes that were used on this same aircraft (Fig. 5) and when they asked me if I needed Cold Plate airflow and shock mounts my answer was a definite "No!"

Now it became a challenge for me to do the best design possible without using these crutches. I knew, in the back of my mind, if I needed them they would be available. It turned out that I didn't need them. Several hundred of these planes have been flying with my boxes for the last 25 years with no Cold Plate cooling and no shock mounts to support the box. It gives me a certain amount of satisfaction that I produced a reliable box with no crutches, and I didn't have to use ground power for the blowers to cool my boxes.

If the thermal philosophy of an aircraft is established first, sub-contractors would take advantage of it because they had originally under bid the job in the first place. This is one area where they can recoup on a low bid and may even make a profit. The worst part of using all these crutches was, heavier boxes. By adding heavy Cold Plates and Shock mounts to all of these boxes the weight of the aircraft had increased and the payload decreased, not to mention the addition of large blowers with air ducting weight, and the electronics that cannot operate without airflow.

Unfortunately, when I was in the field, I personally witnessed several vendor supplied electronic boxes burn out and had to be replaced because a white hat (sailor) forgot to turn on ground power to run the blowers during his testing. If it can happen it will, (Murphy's law).

2m6g. Testability, Test Points, and Repair Access.

These three items are all associated with maintainability. The EE has to work with the support group on these subjects as well as with the PC and ME engineers.

Built In Test (BIT) is almost always required. In the old days BIT was never used but many test points were required as well as performing a manual self test using front panel switches. With the advent of micro-processors BIT is required in today's market. The military has to have the capability of repairing their equipment in the field, especially when they are in a combat situation and the planes have to fly. They cannot wait for repairs by shipping the equipment back to Depot.

There are three levels of maintenance in the military; the Organizational level "O", the Intermediate level "I", and the Depot. The "O" level is an on board replacement repair. At the "I" level, repairs are made in a hanger or below decks of an aircraft carrier. The Depot is generally at some military center located at different parts of the world or at the original manufacturer's facility.

There are two test philosophies on "O" level maintenance. First, there is a diagnostic test performed on the equipment using the aircraft's on board computer. Second, an internal automatic BIT test is made on the Electronic box that has failed the test. The object is to make a readiness test before the flight and be able to make repairs by modular replacement on board. If a PC board replacement doesn't solve the problem, for example, the box is then replaced with a known good one and the failed unit is sent to the "I" level.

At the "I" level the unit is tested in more detail with Standard Test Equipment or Automatic Test Equipment (ATE). If the problem cannot be solved at the "I" level the unit would be sent to the Depot or back to the manufacturer.

Testability and Maintenance, are cover further in Chapter II. The EE has to understand all of these different test philosophies in order to provide the proper BIT. He has to design the BIT in such a way that it can detect a fault to a module at the "O" level. He has to design the electronics and PC cards so that the unit and PC cards can be tested on an ATE and be able to detect a fault to a small group of components on the PC board at the "I" level.

The maintenance philosophy has to be designed for technicians to be able to replace the modules and components in a short period of time. The maintenance procedure and the time it takes to test and repair would have to be demonstrated in front of the customer.

The military's latest philosophy is to bypass the "I" level repair. New aircraft would have enough on board test capability to allow most of the repairs be made at the "O" level. All failed modules would bypass the "I" level facility and are sent back to the Depot. Sounds great, but how many spare modules would the military store at each base or on each aircraft carrier. PC cards would not be repaired in the field.

The problem with this philosophy is, there would be a constant stream of modules running back and forth from the Depot to the different bases or aircraft carriers all over the world. Half the planes may be grounded because they ran out of modules. Any seasoned maintenance Chief would say, "I have to be able to repair all modules at each base to keep the planes flying." In combat one has to be able to repair a box in the field and not depend on the Depot.

The second problem is that the aircraft is now being used as the test bed. Imagine trying to solve a failure using an aircraft to perform the testing when it should be flying. Maybe the military should leave one aircraft on the ground as the test bed to replace the "I" level maintenance.

I have always preached that a piece of equipment is only as good as it's repairability. This includes automobiles or home appliances. If one cannot repair it easily, it is a poor design and a poor product.

I'm, obviously, not in favor of eliminating the "I" level repair station. I cannot see using the aircraft as the test bench, either. If a module blows out on the aircraft, does one plug another module in its place and then proceed to blow the new good one out also? This is known as trial and error maintenance or the substitution method of repair. With this new military philosophy there would be no way to check a module on site except using the aircraft as the test bench. I can just visualize a pilot being aggravated because the technicians have to delay his flight to test a module using his aircraft.

3. SYSTEM ENGINEER

In a large Aerospace Company, there are several distinctive types of System Engineers. These are categorized as Overall Systems, Computer Systems, Avionic Systems, Navigation Systems, Armament Systems, Equipment Systems, Communication Systems, Instrumentation Systems, Electronic Warfare Systems, Display Systems, and Electrical Systems. In a small Electronic company only an Electronics System Engineers may be required. It all depends on the company's line of business.

3a. Overall System Engineer

The Overall Systems Engineer furnishes the design requirements for the entire aircraft or the vehicle being built by the company. He would produce block diagrams consisting of all the sub-assemblies required to operate that vehicle to perform its missions, and provide a memo explaining the functions of each sub-system or WRA and how they should interface. These block diagrams would then be used by other organizations for quoting the entire vehicle and for final configuration purposes. All of the other Systems Engineers would use these block diagrams and the memo to develop their own Black Box Equipment Specification. The overall System Engineer would have to determine where these boxes will be located.

3b. Computer System Engineer

The Computer System Engineer is responsible for the Equipment Specification for the main computer and the I/O software for all the sub-systems that interface with the main computer. This is a major undertaken in today's world of sophisticated equipment. The main computer is the key apparatus in the operation of a vehicle to perform its mission. The Computer System Engineer has to interface with all of the other System Engineers, the Software System Engineer and with the vehicle's installation Hardware Engineer. The interface bus lines could be RS-232, 1553, Manchester code, or by discrete lines.

3c. Avionics System Engineer

The Avionic System Engineer is responsible for all the Avionic Sub-Systems on board and provide the requirements and block diagram on how these Sub-Systems are connected.

There are several specialized Sub-System Engineers that have to interface with the Avionic System Engineer. They are responsible to produce the ES for many of the standard sub-systems such as Navigation System, Communication System, Armament Systems, RADAR Systems, EW Systems, Displays Systems, etc. They would also provide the requirements for each sub-system on such items as Antenna's, controls, etc. They would also have to interface with the Equipment System Engineers.

3d. Equipment System Engineer

The Equipment System Engineer would create the Equipment Specification for each black box that is <u>not</u> standard and called out by the Overall System Engineer, especially Interface Boxes. The Equipment System Engineer should have a fair knowledge of hardware design and may have been a designer himself. He has to be familiar with Mil-Specs and environmental conditions. He may even be responsible for more than one black box on the vehicle.

The Equipment System Engineer may be matrixed to the Electronic Design Manager, but should not be controlled by the EDM so that he can maintain the freedom he needs to work with the hardware designer or with a sub-contractor. Even though he is matrixed to the EDM he still works independently, and has the authority to interface with all the support organizations and with the EPM directly.

He should attend all in-house design reviews and the customer reviews. He still has to report his progress to his own Systems Department Manager. In my company we referred to these individuals as Project Engineers who sometimes were doubled as EPMs. An EPM should not wear two hats. It would become obvious later that the EPM would be too busy to have both responsibilities.

Too many companies would ignore the System Engineer's title and use the EPM to be the System Engineer. Some EPMs are qualified to do systems work but if he is qualified, then maybe he should be one. The PM would sometimes by-passes the System Engineer and would make systems decisions directly with the customer. This could be disastrous, but it happens.

The design team should work very closely with the Equipment Systems Engineer because the designer may spend too much time trying to meet a poor ES when a simple conversation with the Equipment System Engineer over a minor modification to the ES could easily resolve the problem.

In some cases the Equipment System Engineer and the Avionics System Engineer may be the same person, depending on the program.

3e. Instrumentation System Engineers

The Instrumentations Systems engineer is basically just that. He has to provide the requirements and specifications for the cockpit on an aircraft or other vehicle panels. These Systems types are unique and are very specialized in their field. Most of these individuals are private pilots themselves, or ex-military pilots. They know what is needed to fly an aircraft and the equipment to navigate it. They also should work very closely with the Avionics System Engineer.

3f. Electrical System Engineers

The Electrical System Engineer has one of the most important function on any vehicle, because it is up to him to provide the wiring and the electrical interface requirements between equipment. He is also responsible for the power management of the vehicle to provide AC and DC power to all sub-system, and power to operate the vehicle.

The Electrical System Engineer has to determine wire sizes, wire type, micro-wave plumbing, antenna wiring, coax wiring, and fiber optical wiring. He is responsible for providing the requirements for junction boxes, relay panels and switching panels. He has to work very closely with the Avionic System Engineer, the Equipment Systems Engineer and the EPM on each black box.

3g. Experience

As mentioned before, the technical ability of any System Engineer, in my opinion, is the highest level of engineering and should be composed of the most qualified personnel. System Engineering is mostly theoretical in nature and has to be a person with tremendous experience and technical capability. He has to be highly educated, preferably with a Master's degree or the equivalent experience.

As was mentioned, there are all types of specialties in System Engineering. There is Communications, RADAR, Computers, EW, ASW, IR, Antennas, Armament, Aircraft Instrumentations, SONAR, TV, Displays, System Controls, Navigation, Teletype, just to name a few. Some of these specialties may overlap and be covered by one System Engineer. The key function of all of the above System Engineers is theoretical in nature. The System engineer has to have the theoretical capability to be able to calculate and design a complete system, including Black Box requirements.

A RADAR System Engineer, for example, has to know RADAR theory, Pulse Transmission, Antenna theory, Communications, Transceivers, Displays, Computers, and know all the mathematics for the above. He has to put the complete system together and create block diagrams to illustrate how that equipment works and how it interfaces with other equipment. He usually works with a company that specializes and produces RADAR equipment.

3h. Responsibilities

As it often happens, the Overall System Engineer has to sell a new design approach first to his managers and then to the customer. He would be traveling to Washington or to other companies making presentations with view graphs or slides. The original requirements could have been generated by the customer but, in many cases, the Overall System Engineer would probably have to modify the requirements to produce a practical system, and then he has to sell his ideas.

When he is through with his systems block diagram and write-up, the customer may want to use GFE to save cost of any new equipment that the System Engineer has allocated in his design. The Government has many similar old units in their inventory and in the field which are still being produced for other vehicles. In such cases the Overall System Engineer would have to go back and modify his block diagram to integrate the GFE in his design.

Now the Equipment System Engineer has to upgrade or modify his Equipment Specification for each of the new boxes that are left in the overall block diagram design to interface with the GFE.

This effort alone may require a six months to a year turn-around. After the Equipment System Engineer examines the interface requirements for the GFE and the Sub-Contracted boxes assigned to the aircraft, he may discover that they won't play together. So, now he needs an Interface Box or a Module to make them comparable. The problem is where does he go to get the interface box made.

The Equipment System Engineer has the option of going outside or have it designed in-house, depending on how his company does its business. This is how my department was created in the first place. The problem I had, however, was that the Project Engineers in my company preferred to go outside because it was easier for him. When a design is done in-house the System or Project Engineer has to become involved with the details of the design. He would be more responsible for the outcome of the design and he would have to monitor the program more closely. I will discuss this situation in detail later under Make or Buy Committee.

If the design does not perform as expected whose fault is it? As was mentioned, if the box is not reliable it's the Design Engineers fault but if it does not perform its function in the vehicle, it's the System Engineers fault. This is the kind of finger pointing that should be avoided.

Some Systems Engineers do not like to be put in that position and by going outside they can always blame the supplier. I have often seen that ploy being used before. It's unbelievable the pettiness that goes on when a box doesn't do its job. This trivia does go on, however, and it's called "saving one's butt."

I feel that if the customer would penalize the Prime Contractor for a bad supplier's design the Prime Contractor would be more careful in selecting a vendor, and they would monitor their progress more closely to insure they are receiving a proper designed unit. It's too easy to shift the blame onto the supplier when, in reality, it is the Prime Contractor's System Engineer's fault for not doing his homework and providing a proper ES, or monitoring the design of a vendor.

One Equipment System Engineer told me, "I only get involved with the vendor's design when it fails to perform on the aircraft. I cannot spend countless hours checking their schematics and test reports." My opinion is, that attitude is wrong because by then it's too late!

My department had to modify many supplier boxes, after the fact, because the vendors didn't know the aircraft business. Many vendors try to cut corners, and were not monitored properly, or were designing to a poor ES. Selecting the lowest bidder invariably results in a poor design. Also, using a company that is suspect of going bankrupt could be disastrous.

Bid low and pray for ECPs to recoup their loses is the name of the game for many suppliers, but not all. ECP corrections could be very costly.

There are, of course, many consciences suppliers that are very capable of designing black boxes and should not be overlooked. This does not mean that they do not have to be monitored. All designs should be monitored. It's the old "check and balance" routine that I consider most important.

The Equipment System Engineer should be involved in selecting the Supplier. One thing that must be made clear again at this point is, the System Engineer and the EPM, in some companies, may be the same person. This was the case in my former company, so the capability of this person may not be of the highest quality. This overlapping of responsibilities has resulted in poor selections of suppliers or creating poor Equipment Specifications.

Many of the jobs that I had to take over from suppliers were the direct result of a poor ES. I have mentioned these problems before and would repeat them again throughout this book because I believe that these bad selections and bad ESs are the fundamental weaknesses that most Prime Contractors are faced with when they are dealing with suppliers.

3i. In-House Design

A good System Engineer is not interested in this game of blame shifting because he wants to produce a dependable box and doesn't care who designs it. As a matter of fact, some System Engineers prefer the in-house challenge and favor getting involved with the design. They want to be part of the internal box design and to watch the package grow into a working WRA.

The main disadvantage with going outside is time. The program may have already lost a year in delays due to preparing the Equipment Specification of the equipment, competitive bidding, selecting the supplier, and now, the System's Engineer has to go through the same process all over again for an Interface Box which all of a sudden has popped up after the fact.

One advantage of an Interface box designed in-house is that the design organization can immediately start on the design even though the ES has not been completed. On the other hand, a supplier must have a finished ES because he has to bid the job against competition by using this document to make it legal and binding.

It may take another year before the supplier can begin the design of a new interface box. Also, if there is any change to the ES after the start of program, the supplier may ask for a new contract and more budget because it was not in the original contract. Because there is no legal contract for in-house design, this sort of a problem doesn't exist.

The way I handled in-house designs in the past, was to allow any systems changes to be made at no extra cost for approximately the first three months in the design phase, or at a time when all departments have agreed that the system design is set. After this point, each change should be carefully examined to try other alternatives so as not increase the budget drastically.

Sometimes, I was lucky enough that I didn't need more budget to incorporate the new changes, but, then again, sometimes I did. I would always keep track of the changes to protect myself because years later upper-management forgets why the increase and then the finger pointing begins because I was over budget.

With a supplier, every major change requires a updated contract and more budget. The updated contract has to be re-negotiated which, again, requires more time. This can run into an uncontrolled expense in delays, especially in the design of an Interface Box where the requirements in the ES are constantly being altered by other suppliers.

As was mentioned before, sometimes a supplier would create the ES themselves first, because they are the experts in the field of that equipment. They could conceivably make changes six months into the program.

I had designed a Communication Interface Box that was connected to seven different supplier boxes, and they were all making changes throughout the program. It sometimes gets out of control. This is why a good System Engineer is required on the program to monitor and control these changes. It should be understood that if any electronic equipment that is the main product line of a particular supplier, it is probably better to deal with that supplier as long as their bids are reasonable and they are monitored. It would be foolish to "reinvent the wheel," as they say.

My conclusion on this subject is, every aerospace company should have the capability of producing in-house designs. This department should be of the highest quality and be directly controlled by a Division Engineer or higher. What was wrong in my company was we were treated as access baggage with no home. My boss was an Aeronautical Engineer and all he said to me was "keep me out of trouble." The Systems Engineers only came to me when they were having trouble with an outside supplier or there was no one else to turn to.

3j. Equipment Specification

The Equipment Specification is the most important document to an in-house designer or to any sub-contractor. Even though there was an RFP and an SOW originally written by the customer for a particular program, the ES is the final negotiating document that specifies the configuration of the equipment which was agreed upon by the company, the supplier and the customer.

The normal procedure of creating an ES is by extracting the requirements from the SOW or RFP and other documents such as the quotation. The System Engineer should require approval of his ES from the EDM (if in-house) and the EPM. This Equipment Specification now becomes the legal document that is contractually binding, and any changes made may have to be re-negotiated with the supplier. Depending on the contract, in some cases, the customer (Military) may insist on having the option of approving the ES before it is released to a sub-contractor.

If the ES is a document for an in-house design the approval is not required by the Military because the Prime Contractor now becomes the customer for that equipment. Some companies would produce only a "Requirements" document instead of an ES on in-house designs. The problem with a Requirements document is, many of the items called out may not have been negotiated with the military and later, in the design phase, it may be necessary to ask the customer for waivers to meet some poor requirements.

A requirements document provides the general operation of a unit. It doesn't go into the details as an ES does.

I still prefer a well written ES rather than a Requirements document because it protects the designer, the System Engineer, and the company from embarrassment over any financial losses. As long as the unit meets the ES that was approved by all, then engineering has done its job. Then again, if the unit does not perform properly on an aircraft, an ECP from the customer may be justified with more budget.

There would always be something unforeseen or unaccounted for in a major aircraft program, especially when GFE are used. Some GFE were made twenty or thirty years ago and not all is known about its interface or operation. Documents and drawings get lost. Sometimes, designers go bananas trying to figure out how to hook up to an old GFE box.

The Equipment System Engineer, as apposed to other systems types, has to be hardware oriented and has to have a great deal of experience in the design world. Ex-designers who prefer theoretical work rather than laboratory work make good Equipment System Engineers.

I have always recommended that an Equipment System Engineer should get his feet wet in the electronic design field first before he attempts to be an Equipment System Engineer. In this way, when he prepares the Equipment Specification, he would have a better understanding of what kinds of problems the designer would be faced with and he would be more realistic in preparing his ES.

Too many bad Equipment Specifications have caused small companies to go "belly up" or to produce poor equipment. I personally had to bail out several companies or other organizations within my company and redesign their programs over a 30 year period. The cause could have been that the Equipment Specification was unrealistic, the unit was designed by an unqualified in-house department, or because a poor supplier was selected.

The other reason for having an ES under all circumstances is, other departments within the company depend on it and would use it as reference. It becomes the bible. Maintainability, for example, would refer to specific pages in the ES for their final submittal to the customer. Testability would use the ES to create their test program. Production would use this document as reference to sell off the unit by using an Acceptance Test Procedure (ATP) which was written from the ES.

If there is a controversial in production the Military Inspector would always refer to the ES for proof of requirements before he buys off the unit.

The Drawing Checker would refer to the ES on certain drawing requirements before he would sign off and release a drawing. I cannot stress enough how important the ES is. It is the governing equipment document in this business. It is definitely the Bible.

This is why the ES has to be properly prepared by an experienced Equipment System Engineer and not by an EPM or another proposal engineer. Too many companies treat this document loosely and allow this task to be done by inexperienced engineers.

When I worked for an Electronic company as an EPM, the systems engineer was an ex-pilot navigator in the Navy. He was an intelligent person and did the best he could to prepare the ES for my communications box but I had to help him constantly. He knew absolutely nothing about electronic hardware because he was a graduate accountant. When I complained to his manager, he told me that he had no one else available.

Helping him create an ES took a lot of my own time and I was almost tempted to write the ES myself but that would be a conflict of interest. By me preparing my own ES leaves the door wide open for cutting corners on my design requirements and it could cause other political problems. That ES was finally completed and approved by the Navy, but it was a chore to get it done.

3k. Equipment Specification Outline:

The following outline is only an example of the requirements covered in an Equipment Specification. These items would vary from job to job and from company to company, but the principles apply. The outline gives examples in certain categories to illustrate the methodology used in preparing this document. Every item is essential and contractually bound.

EQUIPMENT SPECIFICATION

1.0 SCOPE
1.1 Intended Use

2.0 APPLICABLE DOCUMENTS
2.1 Referenced Documents
2.2 Issues of Documents
2.2.1 Government Documents

SPECIFICATIONS
MILITARY
MIL-E-5400 - General Design Requirements, Electronic Equipment, Aerospace General Specification
DoD-D-1000 - Drawings, Engineering and Associated Lists
Note: Many of the MIL-Specs would be covered later in Chapter II.

STANDARDS
FEDERAL, MILITARY
OTHER PUBLICATIONS

2.2.2 Non-Government Documents

3.0 REQUIREMENTS

3.1 Item Definition
3.1.1 Associated Equipment
3.1.2 Functional Block Diagram
3.1.3 Interface Definition

3.2 Characteristics
3.2.1 Performance
3.2.1.1 Performance Tests
3.2.1.2 Diagnostic Fault Isolation (DFI) Tests

3.2.2 Physical Characteristics
3.2.2.1 Weight
The lifting weight shall not be greater than a two person lift (105 lb female).
3.2.2.2 Size
3.2.2.3 Electrical Incompatibilities
3.2.2.4 Complexity
3.2.2.5 Optimization
3.2.2.6 Expansion
There shall be a 20% expansion capability
3.2.2.7 Signature Resistors

Signature Resistors shall be used for ATE

3.2.2.8 <u>Accessibility</u>

3.2.2.9 <u>Enclosure</u>

The enclosure shall meet the requirements of MIL-STD-108. The enclosure shall be an aluminum weldment.

3.2.2.10 <u>Handles</u>

3.2.2.11 <u>External Connectors</u>

3.2.2.12 <u>Circuit Cards</u>

3.2.2.13 <u>Test Fixture</u>

3.2.3 <u>Reliability</u>

3.2.3.1 <u>Mean Time Between Failure (MTBF)</u>

Each unit shall have a minimum MTBF of 2500 hours calculated in accordance with MIL-HDBK-217.

3.2.3.2 <u>Useful Life</u>

3.2.3.2.1 <u>Service Life</u>

The unit shall have an operating life of ten thousand (10,000) hours under the operating environmental conditions specified in 3.2.5.

3.2.3.2.2 <u>Storage Life</u>

3.2.4 <u>Maintainability</u>

The unit shall be designed to the requirements of MIL-STD-2080, MIL-STD-2076, MIL-T-28800, and for Maintainability of MIL-STD-1472.

3.2.4.1 <u>Testability</u>

The unit shall have the capability of Self Test and meet the following fault isolation requirements.

3.2.4.1.1 <u>WRA Fault Isolation</u>

3.2.4.1.2 <u>SRA Fault Isolation</u>

3.2.4.2 <u>Mean Time To Repair (MTTR)</u>

3.2.4.3 <u>Access</u>

3.2.4.4 <u>Test Points</u>

3.2.4.5 <u>Environmental Conditions</u>

3.2.4.5.1 <u>Temperature</u>

3.2.4.5.2 <u>Altitude</u>

3.2.4.5.3 <u>Humidity</u>

3.2.4.5.4 <u>Vibration</u>

3.2.4.5.5 <u>Shock</u>

3.2.4.5.6 <u>Electrical Magnetic Interference (EMI)</u>

3.2.4.5.7 <u>Salt Spray</u>

3.2.4.5.8 <u>Drop Test</u>

3.2.4.5.9 <u>Transportability</u>

3.2.4.5.10 <u>Bench Handling</u>

3.3 <u>Design and Construction</u>

3.3.1 Parts and Materials Approval
3.3.1.1 Non-Standard Parts
3.3.1.1.1 Definition of Standard and Non-Standard Parts
3.3.1.2 Cables
3.3.1.2.1 Strain Relief
3.3.1.2.2 Wire Coding
3.3.1.2.3 Cable Connectors
3.3.1.3 Electrical Connectors
3.3.1.3.1 Contacts
3.3.1.3.2 Spacing and Quality
3.3.1.3.3 Protective Caps
3.3.1.3.4 Internal Connectors
3.3.1.3.5 Connector Keying
3.3.1.4 Standard Buffering Devices
3.3.1.5 Finish
3.3.1.5.1 Protective Finish
3.3.1.5.2 Color
3.3.1.5.3 Lacquer
3.3.1.6 Terminal Blocks
3.3.1.7 Component Re-screening
3.3.1.8 Soldering
Soldering shall be in accordance with WS6536.
3.3.1.9 Printed Wire Boards
Printed Wire Boards shall be in accord with MIL-P 55110 and MIL-STD-275.

3.3.2 Electromagnetic Interference

The unit shall operate satisfactorily in a government environment, and not degrade nor be degraded by any other adjacent equipment installed in that environment. The unit shall comply with MIL-STD-461.

3.3.2.1 Grounding

3.3.3 Nameplates and Product Markings

3.3.4 Workmanship

3.3.5 Inter-changeability

3.3.6 Safety

3.3.7 Human Engineering

3.4 Documentation

3.5 Logistics

3.6 Personnel and Training

3.7 Major Component Characteristics

3.8 Precedence

4.0 QUALITY ASSURANCE PROVISIONS
4.1 General
4.1.1 Responsibility for Inspection
4.1.2 Standard Conditions
4.1.3 Special Tests
4.1.3.1 Visual Inspection
4.1.3.2 Inspection by Analysis
4.1.3.3 Tests
4.1.3.4 Environmental Tests
4.1.3.4.1 Low Temperature Test
4.1.3.4.2 High Temperature Test
4.1.3.4.3 Altitude
4.1.3.4.4 Vibration
4.1.3.4.5 Humidity
4.1.3.4.6 Shock
4.1.3.4.7 Salt Spray

4.2 Quality Conformance Tests
4.2.1 Acceptance Tests
4.2.2 Manufacturing Screening
4.2.2.1 Random Vibration
4.2.2.2 Burn-In
4.2.2.2.1 Temperature Stabilization
4.2.2.2.2 Temperature Rate of Change
4.2.2.2.3 Functional Tests

5.0 PREPARATION FOR DELIVERY
5.1 Preservation and Packaging
5.2 Shipping
5.3 Marking

6.0 DEFINITIONS

6.2 ALTERNATE REQUIREMENTS

6.3 ACRONYMS
6.4 ILLUSTRATIONS AND FIGURES

4.0 MECHANICAL ENGINEER

The Mechanical Engineer plays a key role in the design of a military black box. He is, actually, the Packaging Engineer even though he prefers not to be called that. To him packaging sounds like someone in the shipping department packing crates. When a design is in the proposal stages, it is the ME that comes up with the initial configuration. The lead ME is the person that an EDM relies on because most of the work in the program would be done under the ME's jurisdiction.

The reason the lead ME is so important is that he would be responsible for most of the program's activities, especially the "dog" work. He would have control of all the drawings, deal with the parts submittal to the customer, handle the non-standard parts specifications, deal with the environmental testing, and deal with the prototypes and test equipment being made and assembled in the shop.

In addition, he is the expert in interpreting the Military Specifications. He would purchase most of the parts for the prototypes, quote the mechanical effort, be in-charge of the technicians that build the prototypes, have the wire harness run properly, follow up on drawing release and rework, provide the internal harness design, prepare the I/O requirements list with the EE, control the EDL, and design the test fixture working with the EE. Besides all of this responsibility he still has to design the black box.

This is a lot of effort required by the lead ME and, as was mentioned, drawing effort alone would be 1/4 of all of the work required. The three hardest things to accomplish in a black box design for military programs are releasing the drawings, Non-Standard Parts submittal, and producing Name Plates. They all have two things in common, rework and re-submit, over and over again. It may be an overstatement, but it's true.

4a. Configuration

When an RFP comes into the company the EDM, the EE and the ME are the key bidders for engineering. They have to come up with a design approach and a configuration before any other department can submit their quotations. This is why the EDM, the lead EE, and the lead ME require many years of experience to be able to analyze an RFP of a new program and create the configuration.

There are many considerations that have to be taken into account before deciding the configuration. There is no set rule. The most important consideration is where would the unit be installed on the vehicle. If it is to be installed on an aircraft, for example, one of the first things one must determine, is it for a bomber or a fighter? How would it be cooled? Would it require shock mounting? Is it a high frequency box, digital, or a low frequency analog unit? The EDM and EE have to establish all the electronic functions first to set the stage for the ME.

As a EDM, I had to come up with the package configuration many times by myself because the quote became a WAG. However, I would always consult with my lead ME first before I released the quote. He would, invariably, make some suggestions. I was probably more unique than most EDMs in this respect because I had 40 years experience in the business of quoting electronic boxes. Package configuration became second nature to me, but I would recommend that the EDM always work with the lead ME in this endeavor.

One problem with selecting a configuration of an Interface Box is, it is usually the last item to be accounted for by the System engineer. The Interface box is generally an after thought or an "oops I forgot" box and sometimes the aircraft Electrical Systems Engineer and Structures Engineer have to figure out where to put it on the vehicle.

This is where the experience of my department came into play. The first thing my ME would do is go to the main files and withdraw all of the pertinent aircraft drawings to possibly help the Electrical System Engineer select a location. The drawings are examined to be sure there is enough room for the box, and if there would be any obstructions from other units or from the aircraft structure. Would heat be a problem? Would the other units close by be heated from our box or visa-versa?

On a Fighter, locating a new box is always a major problem because there is never enough room. Because of this lack of space the box may not be square and may have to take an "L" shape. As was mentioned, in one instance I had only a couple of days to get a quote out because the RFP was sitting on some upper-manager's desk waiting for his signature. Many upper-managers are "gun" shy to release a quotation. They are afraid that if they sign the document they would be held liable if anything goes wrong. So they delay and delay hoping it would go away.

In another situation, I had to release a quote ahead of time without knowing whether the unit would fit on the aircraft. One has to take some chances in this business, but the decision would have to be made from experience or by "guess-timation."

Some of my boxes are located in the weirdest places which resulted in the oddest shape enclosures. I've been reprimanded for these odd shapes after the fact (like ten years later) because more room became available down stream. It's so easy to reprimand someone after the fact, especially when a manager doesn't know the score or the history of an enclosure.

4b. ARINC 404, ATR Enclosure

In some Military and Commercial aircraft, Standard ATR case sizes are used as shown in Figure 4 on page 67. The reason for standardization is to allow aircraft planners a way to set sizes and provide some sort of advanced allocations on the aircraft for each electronic box. Also, this standardization does help reduce the number of different sheet metal sizes that manufacturing has to produce for the various enclosures and aircraft structures.

The way the ATR box is formed, is that the enclosure shell is more of frame with a dust cover. The electronics is mounted on a chassis attached to the front panel. A dust cover slides over the chassis from the rear. The front panel is fastened down with quick release screws at the sides of the front panel. This package is very similar to a standard oscilloscope bench test equipment. The rear provides space for connectors and mounting screws on the vehicle.

As one can see on Fig.4, the maximum height is almost fixed for all ATR sizes, but narrower widths and different lengths are acceptable.

I am not in favor of using this package configuration as shown in a military vehicle because it is difficult to meet the vibration and shock tests, but if it is called out in the RFP, there is no choice.

Holding down the PC Cards in the chassis from vibrating loose using this ATR configuration is also a problem. The ATR or Arinc box, as described in Figure 4, would be better suited in a military test station or laboratory. The Military planner or System Engineer prefer the ATR sizes to determine the number of boxes required in the vehicle, so they would include this requirement in their RFP. Naturally, I have to go along with this approach or I would be considered non-responsive in my quote, and I would automatically be disqualified from the bidding.

To work around this requirement, however, my engineers would use the ATR sizes that were called out, but instead of sliding the dust cover over the chassis they would use screwed down covers on the top and on the bottom. In this way all of the electronics are accessible from the top or bottom, and the box would be much more rigid because the side panels are permanent. In this way we conform to the ATR required dimensions and still meet vibration.

Many times the RFP would state that the enclosure shall not be greater than a full ATR size. When I examine the electronics required on a particular box, I knew that it would be very easy to put it all into a 3/4 ATR instead. Does one tell the System Engineer this? The rule of thumb, of course, is "Take all that they give you because you may need it in the end."

On one occasions, however, I did tell the System Engineer that I could reduce the box to a 3/4 ATR which might help me win the job. In that particular case I knew that I could actually fit the electronic into a 1/2 ATR. So, one has to play the game but be sure he doesn't get hurt.

The size of the enclosure determines the size of the PC boards. In a full size ATR box, one may get by with the standard 9 x 6 inch PC Card. This is a large size card and suppliers have used it for years. After awhile, it became the standard card size for military programs. On the smaller ATR sizes the width of the board would be smaller but the height would always remain 6 inches.

In the last few years the military have re-standardized the board size to approximately 6 x 5 inches as the Standard Electronic Module (SEM-E) size. This would work in the 3/4 ATR box alright but for smaller ATR boxes it would have to be side mounted so that the 5 inch dimension is along the length of the box. This design would require several additional sections of PC cards in the middle if more cards are needed.

It is preferable to mount the PC boards on the width and height dimensions of the enclosure so that the sides of the box can be used for card guides and heat sink to draw the heat away from the PC boards to the metal sides of the chassis.

There are practical limits on the width of a module along the length of the ATR box, but I have seen power supply modules 4 inches wide.

Drawing the heat away from a power supply package is less of a problem when it is mounted along the width and height of an ATR box.

Height dimensions less than 7.625 maximum are acceptable under certain circumstances.

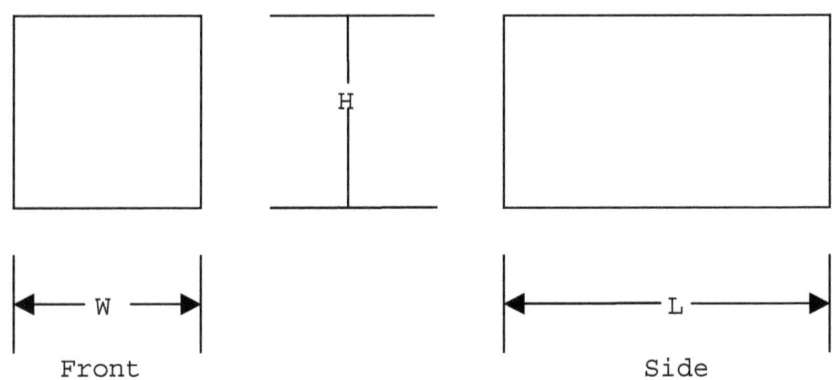

Military Spec. MIL-C-172

	Width inches	Length inches	Height Inches
Short Quarter ATR	2.250	12.5625	7.625
Long Quarter ATR	2.250	19.5625	7.625
Short 3/8 ATR	3.5625	12.5625	7.625
Long 3/8 ATR	3.5625	19.5625	7.625
Short 1/2 ATR	4.875	12.5625	7.625
Long 1/2 ATR	4.875	19.5625	7.625
Short 3/4 ATR	7.50	12.5625	7.625
Long 3/4 ATR	7.50	19.5625	7.625
One ATR	10.125	19.5625	7.625
One and 1/2 ATR	15.375	19.5625	7.625

FIGURE 4

STANDARD ATR/ARINC CASE SIZES

4c. Modular Design

I have always stressed modular packaging for military applications versus one large single PC board that contains all the electronics. MEs prefer not to use large boards because it requires more design time and it's harder to meet structural requirements. Also, changes made on a large board can be very expensive. MEs prefer to build the package like a Sherman Tank so that they can meet any environmental requirements. Obvious, a compromise has to be reached, but not with a large PWB.

In the 1960s when MEs didn't know too much about electronic packaging, they would use over 50 screws on the cover to hold down the cards, or every plate would be riveted with little or no access. My philosophy is that whenever I design something I assume that I would be the person who would have to repair it. I don't want to spend hours unscrewing 50 or more screws just to remove a cover to get at the boards. I prefer to be able to remove any item in the box for repair quickly. In addition, I want the box to be assembled easily in production. If automobile manufactures used this philosophy of designing for repairability, for example, it would not cost the consumer so much in auto repairs.

A large PC board is used exclusively in commercial home computers. The advantage there, of course, is that it's cheaper to produce. But because a military unit would be in a hostile environment, the large board could easily fail due to vibration, shock, rain or by just dropping the box.

The advantage of modular design, is that all the cards and back-plane sections are easily removable from the chassis. These individual section would also be much easier to wire and assembled as separate parts in production. The only items that would be bolted to the chassis are large components such as transformers, connectors, and parts that would have to dissipate a lot of heat such as power transistors.

The top cover should have quick disconnecting fasteners. The bottom cover may be assembled with screws. If a motherboard is used, all the PC cards should be plugged into it. To make connections from the outside world to the PC boards, chassis I/O connectors should be used, which are hand wired to the motherboard's I/O connectors.

If the box has a front panel, then the control wires should be bundled to a separate motherboard I/O connector. The power supply wiring may be the only wires that would be soldered to the motherboard. All other harnesses should be removable. To remove the motherboard it may be necessary to un-solder those power wires, however, the motherboard rarely needs to be removed.

Military requirements are that all PC boards shall use pin type connectors (versus knife edge) and shall not be hard mounted to the chassis. When the ME is designing a package he must think repairability. As part of the final acceptance, the military requires that the supplier shall demonstrate the speed and ease in which modules can be replaced in the chassis.

4d. Enclosure Construction

There are several ways to design an enclosure; sheet metal riveted, with screws, welded, Dip brazed, hogged out aluminum block, and cast aluminum. Of these methods I favor the Dip Braze approach. Cast aluminum would be fine for large quantities of production orders. For smaller quantities (less than 200 pieces) Dip brazing is more economical. On very small quantities, sheet metal with either screws or rivets are acceptable.

The reason I prefer Dip Braze aluminum is that it is strong, light, holds more accurate tolerances, and much easier to design. Dip Braze is a welding technique where all the parts of the sheet metal are pinned together with tabs, and the unit is dipped into a solution for brazing. It seems like a lot of effort but, like many jobs, it is actually easy once one knows the procedure.

I was introduced to this technique in 1962 and began using it exclusively on almost all of my enclosures ever since. A comparison study was made on the different methods mentioned above, and they were all tested on a shake table. Except for the hogged out approach, the Dip Braze box and the Cast aluminum box held up the best. The hogged out approach was not tested because it is the most expensive method of packaging and it was assumed that this box would hold up the best, anyway.

The main advantage using Dip Brazed aluminum is the simplicity in design. All of the sheet metal parts are laid out flat separately and the ME doesn't have to worry about bend radius. He doesn't have to show rivet or screw holes. There is a lot less assembly time in production. It takes a lot of manhours in production to rivet a box containing 200 or more rivets.

The Dip Braze unit is put together first with tabs, as mentioned, and then assembled in one operation. The tabs are broken off, sanded and welded. The box is then painted in accordance to the ES. All production has to do is mount the parts. The sheet metal covers, for example, can be made of very light aluminum, and can be strengthened with dip-brazed ribs, if necessary. If one wants to add or remove a strip of aluminum, simply add it to the drawing. It's that simple. No bend radius, screw holes, or assembly instructions to contend with. Most MEs prefer the Dip Braze approach for the enclosure because it is so easy to modify, and it can be produced in the shop in two or three weeks after the mechanical design has been completed.

My department has used the hogged out aluminum method on several occasions. On one particular case a strong platform was needed that would hold several heavy pieces of electronic equipment from different suppliers. The electronics would be mounted on the top and bottom of the platform. This platform was to be used in a tight area in a Drone and had to withstand a lot of weight and heat. The interconnect wiring between the different equipment was routed inside the platform. The hogging out was done in the shop using programmed NC Machinery. It was quite a task, especially when it came to the provisions that required cooling.

My design department has also designed enclosures using sheet metal with screws. The point I'm making is, before one makes a decision on the package approach, he must consider the application, the cost, the number of units, and the schedule. In most cases we found that a Cast Aluminum enclosure was cheaper when there are high production runs. The only problem with cast aluminum is, it is hard to make changes in the early development stages.

4e. Standard Enclosure

Over the many years in the design area, and with the advent of Micro-processors, I tried to convince my Division Manager that we should adopt standard enclosures in our department, which could be used for all future interface boxes. He thought it was a good idea but it interfered with his own plans of using only one large PC board in a box, which was against all military specifications. In order for my engineers to design my standard box, I would need research money and my Division Manager was not about to part with any of it, because he wanted it for his pet programs. So I never got the opportunity to design such a boxes.

The reason I still think it's a sound idea, is that by having a series of standard box sizes we could lower our bids for interface boxes, and we wouldn't have to "reinvent the enclosure" every time we designed a new box. The enclosures would then be fully Mil approved, eliminating the need to qualify a box every time there is a new design.

My approach was to design an enclosure that would accept the standard SEM-E size boards. The front of the box would be pre-designed leaving room for switches and potentiometers if required. The power supply would be a standard +5 volt supply that was generated from 110 volt 400 Hz or a 270 volt type with a 5 v converter depending on the need. This supply would be mounted close to the panel.

The back panel could have regular round connectors or DPX connectors whichever is called out for the aircraft. After the power supply on the chassis, there would be an area for several PC boards. The first set of PWBs, after the PS, would be a Micro-processor card and several other associated cards, including programmable and fixed memory cards. Several card spaces that follow, after the memory cards, are reserved for special electronic boards as needed. The next set of cards are the driver and receiver I/O cards, which would be mounted close to the connectors at the back of the chassis.

With this design, the box would already have 70% of its electronics and sheet metal designed. These would include the power supply, the processor cards, the memory cards, and the I/O cards. The only cards left to design would be the mother board and any special purpose cards. Of course, new software would have to be programmed, and there could be a minor modification of the front panel. With this approach I could have out-bid any outside vendor on interface boxes. But, alas, it didn't happen and we continued to bid from scratch as we always did and produced a completely new design packages every time we received an RFP. What a waste.

4f. Power Control Assembly

Figure 5 on page 77, shows an example of a package that my organization designed over 30 years ago. I know it seems a long time ago and may be out-dated, but I will use this particular unit to illustrate different engineering design techniques and approaches. The main function of these boxes (two were required in the aircraft) are to provide all of the power control functions of the Lockheed built S3-A aircraft, such as operating the flaps, landing gear, bomb bay doors, etc. This was the first aircraft that was ever built in production that used electronics to control all the power functions of a airplane.

By using combinational logic in the box, the pilot could now select any power function he desires, and doesn't have to worry about the proper sequences of switches in the cockpit as he normally does because the electronics ensures him that many of the initial functions had to be completed first before an operation can be executed.

This technique was accomplished by using sensors located all over the aircraft, and timing devices to allow an operation to be completed before another operation can take place, such as the bomb bay doors have to be completely opened before a bomb can be releases, or completely unfolding the wings on a carrier before the plane can be flown. Because the box is a life supporting unit, a redundancy system was used throughout the box and between boxes, so that any one failure would not cause the aircraft to fail.

When the go-ahead for this design was received my lead ME was concerned with the amount of room that was allowed on the aircraft for these two boxes. To illustrate how the in-fighting sometimes goes within a company, originally, my organization was given only 1/2 the dimensions of the box as shown in Figure 5. I had to convince the Project Engineer that it was impossible to put the required electronics in that 1/2 size box.

Then they asked me if cold plate cooling and shock mounts would be required. I said no. This box was installed in an aircraft that would have to land on an aircraft carrier and had to be rugged. Landing on the deck of a carrier is very close to a crash landing.

All the other suppliers of electronic boxes for this aircraft, took advantage of the cold plate cooling and shock mounts. As was mentioned, the other suppliers used this advantage as a crutch to reduce the cost of their design. When I had rejected the use of cold plate cooling or shock mounts, the Project Manager allowed the box to become the size as shown in Figure 5. The extra room was now available by not needing cold plates and shock mounts.

After the initial design review, my ME discovered that an additional inch in length was needed to install a pair of purchased full Mil power supplies. The extra inch was denied. Their philosophy was not to give-in to any supplier which they considered as unnecessary changes without just cause. "If one gives an inch the supplier would want another inch later."

I was told there was absolutely no room. I didn't know it at the time, but I found out later that they could have easily allowed us 5 more inches. There was plenty of room. Because the Project Engineer wanted to play "God" he refused to allow us this extra inch that we needed. So instead of purchasing a standard low cost military approved power supply, my engineers had to design a special, expensive, compact power supply to fit in the only space available.

A year later, when a new EPM was assigned to the program, the power supply cost was one of the first things he complained about. He couldn't understand why we didn't purchase a cheaper, full Mil power supply, especially with all the room available in the rack where the box was mounted in. His complaint went right up to the Vice-President and back down to me. One can never win in this business, and no explanation could ever be accepted since the original EPM left the company.

The height of the box was another restriction imposed on us. The original EPM would not allow the box to be any higher. This limited the height of the PC cards which forced us to use smaller cards rather than the standard ARINC SEM-E sizes. This limitation caused us to use many more PC cards to contain the electronics. I ended up as the bad engineer on both of these accounts, after the fact.

If a proper organizational structure had been used with a good Program Manager controlling the program and a good System Engineer on the job, these kind's of decisions could have been avoided. The Design team and the Project team have to work together and give and take a little to produce the best designed package for the aircraft. In this particular case ARINC size boards of 6 by 5 inches could have been used because there was ample room in the aircraft to increase the height. With less PWBs required, purchased power supplies would have easily fit in the box without that extra inch in length I asked for.

As one can see from the Figure 5, the enclosure is Dip Brazed light weight aluminum that is beefed up with ribs on both sides. Because the electronics is in a low frequency environment a wire wrap back plane was used instead of a PWB motherboard. See Figure 31 on page 324.

The back plane is a separate dip brazes assembly which also contains the mating PC connectors for the cards. In this way the back plane can be sent out separately to a professional wire wrap house and be automatically wire wrapped using a Gardener Denver machine.

A wire wrap back plane was selected because of the expected changes. At the time of the design there was no Equipment Specification available and these units had to interface with many unknown vendor boxes. If a PW Mother Board were used the changes would have run the cost out of sight.

The four I/O connectors shown, are hand assembled separately and are connected to the wire wrap chassis mating connectors. The two power supplies are mounted at the front end of the box and can easily be removed. Not shown are both top and bottom covers which have ribs on them for stiffening.

The partitioning of the electronics is such that the interface driver and receiver cards are close to the I/O connectors to reduce lead length. The top cover uses 9 quick disconnect Zeus fasteners and rubber strips to hold down the PC Cards. The bottom cover is similar except it uses screws since it doesn't have to be removed often. When the bottom cover is removed all of the wire wrap wires and power lines are exposed for test and repair. This box could be completely disassembled in less than 30 minutes.

Figure 6 on page 78, is the Ground Support Test Equipment which automatically tests the above Power Control Assembly. It uses the same basic packaging approach as the unit in Figure 5.

4g. Drawings

All mechanical and electrical drawings should be done on some type of Automatic CADD system. In this way the drawing history can be recorded and controlled better. During the design phase, the first set of complete drawings that are produced should be locked in as Read Only. If there are any changes after that, the drawings may be copied and new revision numbers are assigned to them. All the changes would be described in the upper right hand corner of each drawing with a revision number and dated. When the design is completed and ready for production, the final released drawings are wiped clean of all the revision numbers and all of the notes, but all the previous drawings are left in Read Only and cannot be destroyed.

If there are any more changes after the drawings are released, a new revision table is produced in the same location except, now, revision letters instead of numbers are used, and dated. With this procedure the history of the design changes are easily traced, because drawing updates are locked into the CADD Read Only memory system and cannot be altered.

If there are quick production changes required, the Engineering Change Orders (ECO) or the Project Slip methods can still be used. These loose 8 x 11 change notices shall be incorporated into the final drawing with new revision letters no later than 90 days. Otherwise, the drawing changes may get lost and never be incorporated. If an Inspector accepts an ECO to buy off a particular production box, that ECO cannot be used again to buy off the next box on the production line. Unless there is a compelling reason, the next box shall be sold off with all of the outstanding ECOs incorporated into the main drawing package.

By using this procedure the EPM and EDM would have control of the drawings and the history can be maintained in CADD on how the design has evolved. If there is a design deficiency, the customer would want to know how engineering arrived at the latest design. The historical data of a box is usually a requirement in the contract, anyway.

Drawing control is always a major problem. When I worked later on for another electronic company, they were still using hand drawings. ECOs were accumulated over many years and there were ECOs that canceled ECOs. Some of these 8 x 11 sheets got lost in the drawing files. What a mess. There were five versions of the electronics and, as an EPM, I didn't know which revision went with what version of the box.

I was new to this program and I couldn't convince my upper-management that this was not the way to do business. For this and other reasons, which I will discuss later, I finally left that company because I couldn't do my job under these outmoded drawing practices, and for other stressful reasons.

4h. Engineering Drawing List

The Engineering Drawing List (EDL) is actually one of the first items that the ME lead engineer shall create. All drawings would require drawing numbers assigned to them. These would be issued, by request, from a Document Control Center (DCC). The ME has to estimate how many drawings would be required, so he would request a block of drawing numbers from the DCC for his program. It is not uncommon to request 200 numbers for a black box.

It would be ideal to have all the drawings under one set of sequential numbers, but sometimes the numbers run out. If the ME needs more numbers later, they would not be in the same sequence as before because other programs would also be requesting numbers. The document control person would supply numbers on a first come first serve basis for all programs associated with the same aircraft.

If the revision letter and number procedure that I have outlined above is not used one would lose control of the history of the drawings with the latest revisions. On one program that I was the Staff Engineer, the ME would use a brand new drawing number every time there was a revision in the box instead of using revision numbers. The design engineers were working with one set of revision drawings not knowing if they were the latest. This is why an EDL document has to be created as the controlling document for all drawings.

The EDL is a list of drawings by their functions. It is usually on 8 x 11 sheets showing the main function first. For example, a PC card requires 4 drawings; Assembly, Schematic, Fabrication, and Master Pattern. The EDL would have these drawings listed in sequence by their numbers with the assembly number first, and the rest of the drawings would be indented below in columns. These four drawings would have the same basic drawing numbers except the last digit would be increased by one.

There is no reason to change the drawing numbers if the revision number procedure is used. Revision letters or numbers would also be updated and listed on the EDL and dated. The EDL is the controlling document and should be monitored by the lead ME or a designated person. In this way any drawing changes have to be cleared through the lead ME and the latest revisions would be recorded and updated by him on the EDL.

Red marked drawings are acceptable on prototype units only when quick response is required in the shop and for technicians, but these drawing shall be stamped Preliminary. This would provide some latitude for the design engineers to try different design approaches during development. The EE or ME engineer would decide when these red lines shall be incorporated, and update the drawing with revision numbers.

Red-marking is the only weakness in the system that I have just described because the EE or an ME may inadvertently forget to update the drawings. The way to avoid this, is for the lead ME to have the EE and another ME check their CADD drawings periodically or when there is a revision change. The ME should make a copy of the latest schematic and hand it to the EE and have him check it, sign it and date it.

Even this procedure could still be a problem, especially with EEs who tend to forget what they did in the Lab and didn't update their schematics in CADD. I have seen EEs make their breadboards work perfectly but when the first prototype is fired up they can't understand why it doesn't work. They forgot to update their schematics, so the PC Card designer never incorporated his latest changes in the PC board.

I had to constantly remind and ask the EEs in the design reviews, "Are all the schematic drawings updated?" Invariable, an EE would say he forgot. I sometimes felt like I was dealing with children instead of adult engineers. But, that's the life of a Engineering Design Manager.

An example of an Engineering Drawing List (EDL) is shown on Figure 32 on page 325.

4i. Drawing Release and Checkers

Drawing Release is another major effort in this crazy world of documentation. First of all, the drawings have to meet all of the required military specifications in the SOW and the company's standard drafting practices, if they have one. If the company doesn't have a standard drafting practice procedure, each department would probably do it their own way and one would end up with a conglomeration of drawings that no one can read. The Chief Draftsman of a company would usually dictate what these standard practices should be, and provide a published book on it to each department. In order to assure that a proper drawing format is being used, special checkers are utilized.

A checker's job is to make sure that all of the drafting practices are incorporated in the drawings and that the box would go together properly with the allowed tolerances. The checker has to be an experienced person who was probably a draftsman at one time. He has to know materials, screws, nuts, metals, paints, electronic components, schematics, and everything the ME has to know about drawing requirements. Drawings cannot be released without the checker's signature.

I once worked on a job as Staff Engineer and the EPM tried to by-pass the check organization using only ECOs to save some rework costs. When it got to the Military Inspector in production it created havoc. The delivery of the first box was then delayed more than a month because the drawings where not properly checked, corrected, back checked, and re-released.

The Lead ME is usually responsible for this task, to make sure all of the CADD operators and other MEs understand the proper procedures. This responsibility was taken away from the Lead ME, on one program, by an EPM. Needless to say the cost of the program went way up. That EPM should have been fired but he was just transferred. Oh well.

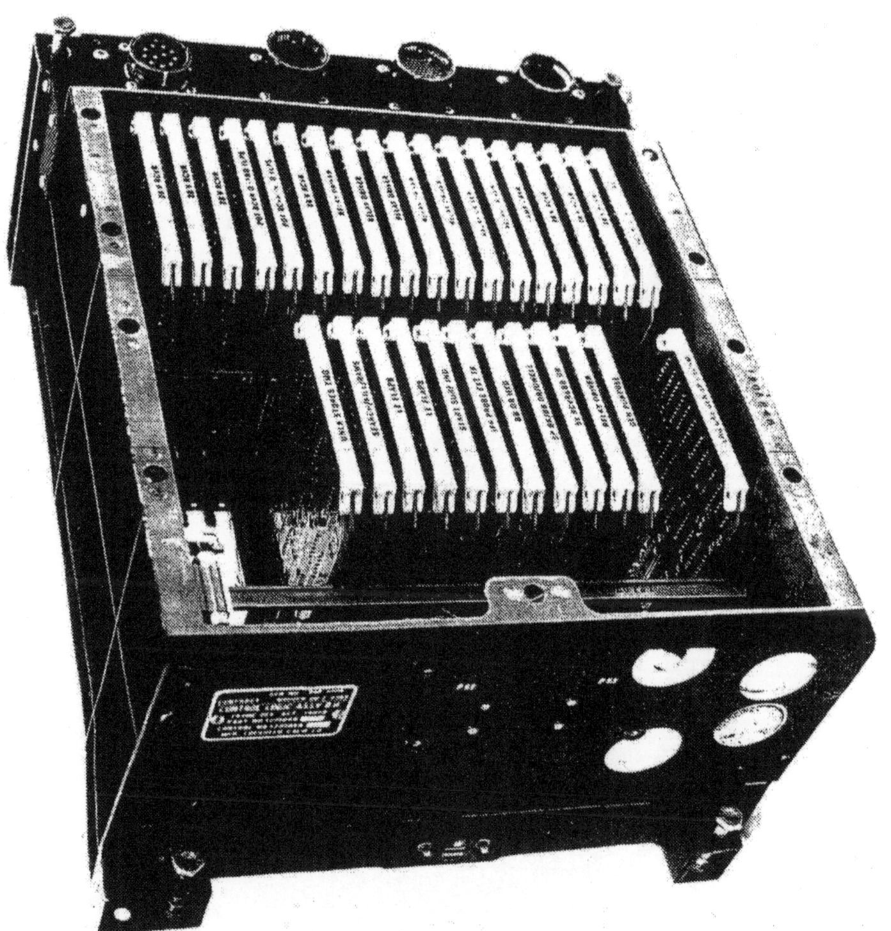

Figure 5. Power Control Assembly

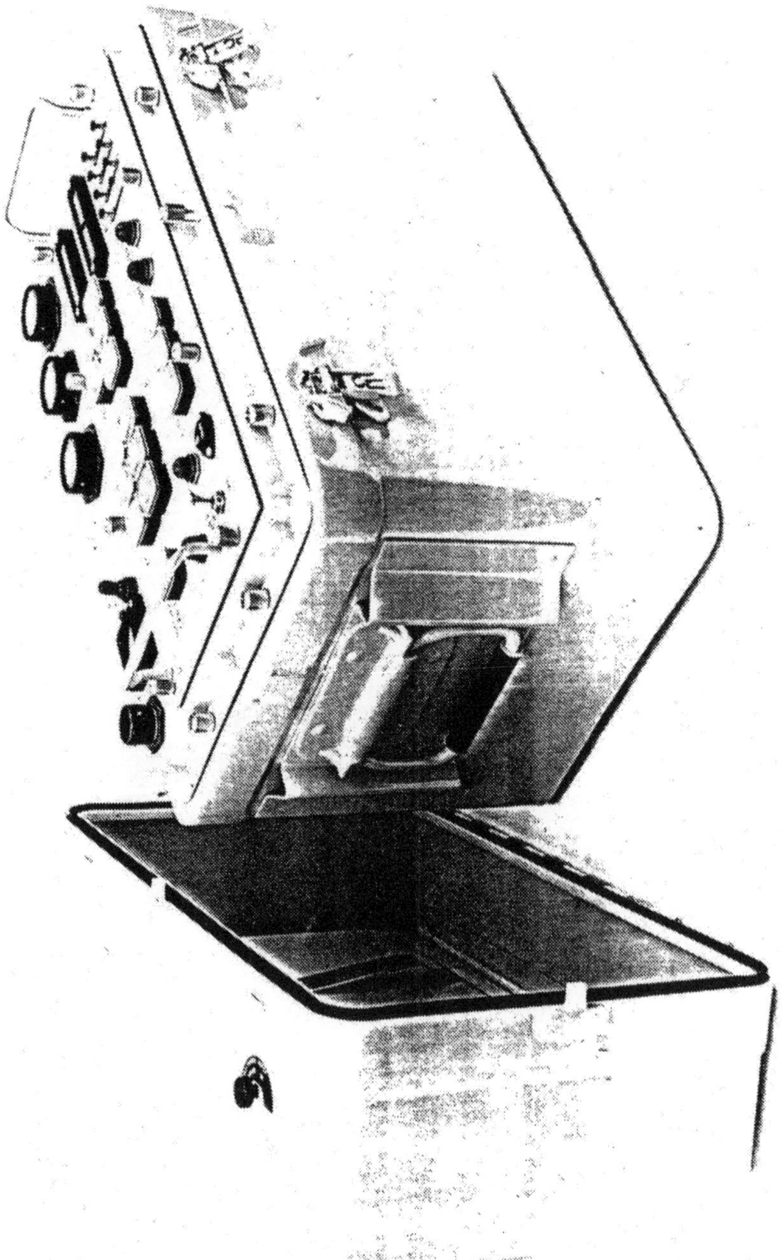

Figure 6. Control Logic Test Set

4j. Non-Standard Parts

The lead ME is also responsible for non-standard parts that are to be submittal to the Parts Department and to the customer. When an EE or ME is selecting parts for his design, he should first try to use approved standard military parts that are called out in the companies parts standards manual, if there is one, or from the Programs Parts Section List (PPSL) provided by the military and called out in the SOW.

EEs are not very good about following these requirements, because they like to use the latest chips on the market. They don't like to be tide down to using old approved chips or to a PPSL document. To control these engineers "check and balance" a parts list should be prepared early in the program (preferably on the PC card assembly drawing) and submitted to the Parts Department as soon as possible for their evaluation.

Non-Standard Parts could destroy a program. The military doesn't mind engineers using new chips as long as they can be justified, especially if there is a dual sauce. All one has to do is submit it to the government as a non-standard part and prove the technical advantages or possibly a cost savings for using that part versus an existing part. The government doesn't want to be stuck with a component that may go out of production later on, and then would have to pay for a redesign.

On one program that I was a Staff Engineer, the EPM did not want to include the Parts Department to save budget. When I asked the Parts Manager what was going on and why was he being by-passed, he said, "I'm not going to fight it. I'll just sit back and let the damage happen."

Two years later, it did happen. Because the Navy did not receive a non-standard parts list they threatened to cancel the production contract unless it was straighten out. This bad decision really cost my company a bundle because it was on a fixed price contract. All together, on this program, 30% of the non-standard parts were later rejected and the rework and redesign cost my company a fortune. I was the Staff Liaison Engineer on this program, and I tried to convince the EPM to submit the Non-standard parts list to the Parts Department but he ignored me, and so, I also let it happen. This EPM should have been fired but they allowed him to finish the job.

An ME also has to interface with Printed Wire Board designers, CADD operators, technicians, the shops, production, and the Environmental Lab. He certainly has no easy task.

5. SOFTWARE ENGINEERS

The Aircraft Software Engineer has to understand enough about electronics to be able to write the required software programs for military operations. The Aircraft Software Engineer is responsible for the overall aircraft using the ships Main Computer.

There are several software engineers on the aircraft that specialize in different fields. These are Electronic Warfare (EW), Anti-Submarine Warfare (ASW), Vehicle Controls, Communications, Navigation, Displays, Armament, RADAR, Infra Red Detection and Infra-Red Armament, Flight Controls, Automatic Test Equipment, Mission Recorders, etc.

In all of these fields the Aircraft Software Engineer would have to interface with several support computers and micro-processors (firmware) located within the so called black boxes. Not only would he have to interface with the other software engineers but he would have to provide the communications to other computers over a bus systems such as Manchester, RS-232, MIL-STD 1553, Discrete lines, Fiber Optics, high speed receiver/transmitter IC technology or combinations of the above. All of these fields require sophisticated computer calculations and specialized knowledge of the military environment and its missions. See Figure 33, page 326 for the front cover publication of RS-232-C, by Electronic Industries Association.

5a. Electronic Warfare

Electronic Warfare (EW) is a military action involving the use of electromagnetic energy to determine, exploit, reduce, or prevent hostile action against our country. Electronic Countermeasures (ECM) is a division of EW, involving actions taken to prevent an enemy's effective use of such systems as RADAR. ECM including Jamming and Deception.

Other sub-divisions of EW are Communications Intelligence (COMINT), which is intelligence information of voice from foreign communications, and Electronic Intelligence (ELINT), which is intelligence information from foreign RADAR, for example.

EW is a specialized field and requires sophisticated software to be able to collect data, analyze it, record it, and provide the proper counter measure to protect the pilot and his aircraft. Most of the information involved in EW is classified. What I'm trying to get across right now, is that the hardware designer has to work with these software engineers to provide the necessary tools for an EW program. It is a major effort and requires the association of many companies using different black boxes and interface equipment.

As was mentioned, I worked on the U-2 bird analyzing purchased EW equipment from several different suppliers. That was some experience, because we had to used the aircraft as our test bench.

5b. Anti-Submarine Warfare (ASW)

The Submarine is one of the greatest threats to the USA, especially if they carry ballistic missiles on board. They can hide in deep depths of the ocean and move about undetected. It's no wonder it is the most active field for many companies.

There are several systems that are used to counter the Submarine threat, and because of security I wont discuss them. This field is similar to EW in the sense that our country is trying to detect the enemy to provide some sort of countermeasure, except it is through water rather than transmission through the atmosphere. ASW is a very large field and requires a lot of computations from software personnel and with many interface boxes utilizing micro-processors.

5c. Communications

Communications is no longer a system whereby an operator manually selects a channel to receive or transmit voice or data. This selection is now done by a computerized hardware which not only selects different channels but provides cryptic information for voice and teletype equpment. In the ASW systems, the computer is used to select communication channels for Sono buoys and for torpedo pre-launch settings to be dropped from an aircraft.

Voice is being digitized more and more in Intercom systems and communication over the airwaves. So now, software plays an important role in the communications field. My organization has designed several different communications switching matrixes for different aircraft. In the old days switching was done with manual controls and electronic switching.

Today it is done with computer controls and with synthesized voice using micro-processors.

5d. Automatic Test Equipment (ATE)

ATE is another field that the software engineer plays an important roll. Maintaining electronic equipment in the field using ATE is a major task. An ATE station must have the capability of providing signals to be able to service many Weapon Replaceable Assemblies (WRA) or Black Boxes from different Aircraft or Vehicles. These Software engineer, however, have to have special talents over and above those SWE mentioned above.

These software types are generally EEs with a lot of software experience, because they have to be able to read electronic schematics and understand its functions. They have to provide the test program to stimulate the Unit Under Test (UUT) and to detect a failure to a Sub-Replaceable Assemble (SRA), to a Printed Wire Board which is an SRA, or to a group of chips on a SRA. I call them Electronic Software Engineers. More about ATE is discussed later.

5e. Navigation

Navigation is another major interface system that requires extremely skilled Software Engineers. The Navigation unit is normally produced by a supplier who specializes in this field. What we are discussing now is not the supplier software engineer, but the software required to interface the navigation unit with other electronic boxes via a bus lines such as 1553.

My department once designed a Interface Box for a special RADAR system (Search Water Radar) made in England that had the capability of locating submarines right down to the periscope. This Interface box had a micro-processor in it that was programmed to determine the location of the submarine by working with the on board Navigation system to provide the proper displays for the tactical crew. See the experimental model on Figure 27 on page 320.

When an aircraft is flying in circles, for example, the location of the aircraft relative to the Sub is difficult to calculate, because the flight's axis varies with respect to the ships roll, yaw, and tilt. So, the software engineer has his hands full.

These ships variations, by the way, also have to be accounted for and displayed by the Navigation unit for the EW and ASW systems mentioned above. Navigation also plays an important part in almost all fields, and many different black boxes depend on its accuracy.

No matter what system is being used; whether it be EW, ASW, IR, or Communications, they all have to know the location of the aircraft and where the enemy is. Navigation information has to be transferred to all of these sub-systems whether it be from Inertial, Doppler, or a Global Satellite type of navigation equipment.

An EE could also double as the software engineer in this case. It depends on the arrangements the company has and what type of equipment they use to program the micro-processor memory.

The best approach is to first program a main computer with Fortran, C, or other programming tools and load the micro-processor as required. Don't blow (program) any PROMs (Programmable Read Only Memory chips) until the software is completely operational in the Lab. Test data accumulated during this period should be recorded and documented because it may be required by the customer.

5f. Other Software Requirements

My division designed a Display Control Panel (DCP), and the software was produced on a SUN work-station using the C software program. The software was tested in the Lab using the end item equipment to insure it worked properly. Modifications were then made to the electronics as required to make the final display work. See Figure 28 page 321.

It was quite an education for me in this design, because I had never experienced software engineers recommending hardware design changes before. The test engineers were actually Software Computer Science majors that learned electronic through experience. One software person said to me that she never thought electronics could be so much fun. When she came out of school as a computer science major, electronics was a mystery to her, and now she was working with it. She was impressed.

In today's world of computers, an engineer can no longer be just an EE designer, a software man or even an ME designer. He has to be able to overlap with other engineering specialties. In the Design Team that I had mentioned, a Software Engineer is matrixed to the Design Manager. As one can see from the requirements from the above programs, a black box designer needs the assistance of experienced software engineers. I prefer them being matrixed to an EDM so that the EEs are not tide down doing software and can concentrate on the hardware portion, which is a major chore by itself.

One other important item that requires the services of an experienced software engineer, is Built In Test (BIT) and BITE (Built In Test Equipment). All new equipment that are designed for the Military today would require BIT.

There are several BIT requirements that would be imposed on the designer of a black box. The unit would be diagnostically tested by the main computer on board the vehicle to determine operational readiness, and provide WRA fault isolation (which black box is at fault). The unit may also have self test capabilities or a BIT to locate a fault to an SRA (PWB) in the box.

The software electronics required for BIT would be within the equipment, but must be physically isolated from the rest of the electronics so that it does not cause an operational failure. The main computer may require that the BIT data be made available for analysis and recording.

6. PRINTED WIRE ENGINEER

The Printed Wire Engineer (PWE) or PC Engineer is a special type of person. In the last company I worked for, they had an engineering union and the layoff procedure was determined by ability as well as seniority. As a Department Manager I had to stack the engineers in each classification according to there ability in case there was a layoff. But my PWEs were being compared with MEs having the same classification.

When the arguing started over the stacking, the first thing my opponents would say was that my PWEs do not have a college degree. They tried to compare a graduate ME with 2 years of experience with my PWE who had 20 years of experience. What a joke.

The point I'm trying to make, is that there is no such thing as a college degree for a PWE. This person probably went through a trade school and came up through the ranks as a design draftsman who learned Printed Wire Board layout the hard way by hand, taping PC boards on a drafting table over many years.

Most of these individuals had bounced around from company to company as Job Shoppers gaining a lot of experience. As a result, many of them turned out to be extremely good Packaging engineers as well, because they also had to learn how to design electronic enclosures.

6a. Experience of a PWE

In the 40 years that I was in this business I worked with over 100 PWB designers. They all have several characteristics in common. They know the drafting procedures required for military release, they know mechanical design requirements, and they know electronic designations. An engineer could call out a chip on his schematic and they would know exactly how to lay it out on a PC board. They know how to read component values like resistors and capacitors. They know about thermal problems, high frequency problems, environmental problems, and ESD. They especially know how to read MIL-Specifications.

The greatest attribute about these PWEs is that they have the perseverance to tackle any assignment no matter how dull or laborious it is, without questions. I had always felt guilty about asking them to do menial things like "go-for" tasks such as picking up drawings in another building or picking up parts from the stock room. These PWE are really special people. Naturally, there were a few "dogs" who claimed to be PW board designers, but they were eventually discovered, released or laid off.

In the early days, everything was done by hand, but when the Computer Aided Design Drafting (CADD) machines became available it helped tremendously. Most of these early military programs required inking the drawings by hand because they would have to be micro-filmed. When using a pencil, drawing the line thickness was difficult to maintain and would smudge so ink became essential. These problems were all solved with CADD. The line thickness on a drawing can easily be maintained very accurately with a CADD computer.

Many of the older PWEs were reluctant to use the CADD system for drafting. It was so hard to upgrade a person's ability when he has done PWBs by hand for over 20 years. We had bitter arguments between these individuals over the new CADD systems. Most of the young engineers took to CADD right away but the older guys fought it. One of my best MEs refused to even take a CADD class. The lead ME was not worried. He told me to leave him alone because he knew this ME would retire in a year or so.

When my group finally was given Auto-Routing capability the problem became even worse (the ability to lay out PC boards in CADD). One PWE was so egotistical over his capabilities as a hand layout man, he challenged an auto-route person. He claimed he could lay-out a board and tape it faster than the CADD operator could using auto-routing.

The contest was on. They selected a simple double-sided board that neither man was familiar with, and they both came in on a Saturday on their own time with no extra pay. They went at it as fast as they could and would you believe it, the hand layout PWE was ahead at the end of the day.

Now how could I convince the other PWEs that the Auto-routing system is the way of the future and they all must learn it. I asked for another contest, but this time they were given a 10 multi-layered PC board to layout. The next Saturday, after examining their efforts at the end of the day, the auto-router was so far ahead of the hand layout person, he quickly conceded. He was finally convinced that auto-routing was much faster.

When a PWE had to quote the hours to do a PC card by hand, the manhours would be on the order of 80 to 100 hours for a one sided standard board (6 x 5), 120 to 160 hours for a two sided board, and 200 to 300 hours for a multi-layered board - depending on the number of layers and the complication of the board. This included all drawings associated with the PC card.

With Auto-Routing it took about 2/3 the hours for a double sided board and sometimes just as long for the simple boards. But, for multi-layer board, it took 1/10 the time to lay it out. One has to remember, that in the early stages of the electronic design phase, the EE would probably be making changes constantly. Because of this, the PC cards may be modified often, which could cause over-runs and delays. Making changes using Auto-CADD is more cost effective, because they could be made much faster and easier, but a program could be a bogged down considerably if they were done by hand.

6b. Preliminary Set-up

There are a few things that have to be established first before a PWB can be designed. The card sizes would have to be decided together between the EE, ME, and the PWE. If the standard board SEM-E is called out in the SOW, then that size is set at approximately 6 x 5 inches. If the SOW calls out for a full size ATR box then the board size could be as large as 9 x 6 inches.

The box size usually determines the board size. Every once in awhile two sizes are needed. The PWB in the box of Figure 5, had to be odd in shape and made small because the EPM refused to raise the height of the box. We did not conform to the SEM-E requirements in the SOW and we had to ask the Navy for a waiver.

6b1. PC connector selection

There are several military PC connectors on the market to choose from. The selection depends on the number of I/O pins required, insertion force, board spacing, and board width. In the early 1950s knife edge or edge board connectors were used. This type has traces on both sides at the edge of the board for the male connector, and a spring loaded connector for the female end.

The problem with this method was that in a military environment the traces would corrode or become dirty. When there was a failure on board, the technician would first remove and re-insert the boards to clean its edges. Sometimes this cleared up a problem. It became a practice for the military to go through this routine before a flight of pulling boards and re-insert them. It was a joke. The military finally outlawed this type of connector even though it is still being used in the commercial world. This re-insertion method was one way I repaired my home television set.

In the 60's the only PC connector approved by the military was the tuning fork type. This type had two prongs on the board which engaged into a similar tuning fork mate receptacle but at a 90 degree angle. The insertion force and removal force was very high.

If only 10 pins or less were used the force wasn't too bad, but when more than 10 pins were used it became a chore and special extracting tools had to be provided. In one of our programs a technicians had to use a rubber mallet to insert the boards. These boards, unfortunately, are still flying in our black boxes since 1964. They also require a special crow bar as an extracting tool. Thank God, for new connectors.

In the 70's, when I designed the box of Figure 5, the PC engineer would have ran out of board space if the above conventional connectors had been used. I decided to use 1/2 inch spacing between boards.

I had to sub-contract for a new connector design by a company that specializes in manufacturing connectors. This new connector straddled the edge of the board. It provided connector spacing of 0.4 inches or more on both sides. The electronics in this box was low speed digital so I didn't have to worry about pick-up or cross-talk. I selected the spacing between boards to be 0.5 inches and it worked out quite well.

In today's world many designs are requiring 300 pins or more and they are now using modern low insertion PC connectors. These are pretty wide connectors which are intended to be used in high density modules with board spacings of 1.0 inch or more. These low insertion connectors are better adapted when used with a PW-Mother board. Wire wrap as a mother board back plane may also influence the type of connector that is selected.

6b2. Card extractor selection

I prefer the boards to have built in card extractors. Loose extractors tend to get lost. The extractor handle used in Figure 5 is a complete strap across the top of the board and the board identification is stamped on it as shown. These handles are no longer available because the easier extraction type placed at each end had become more in demand. The board identification can be stamped on these easy lifting type extractors. The lifting is done by placing a thumb on the board side of the extractor at both ends of the board and by lifting it up the board is easily removed. Insertion is no longer a problem.

6b3. Test points

The Testability department usually require test points at the top of the boards. This requirement depends on the contract. Testability usually gets their way, so don't fight it and allow for it in the PWB design. In some incidents they would allow test points to be placed in the middle of the board to be tested with an extender board.

There are several ways of handling test points. One may purchase the test points individually or in blocks. One may use the stand-offs types or just run the traces up to the top of the board. A PC connector placed on the top of the board is sometimes used for test points. I prefer the PC connector approach because the card now lends itself to be automatically tested in production and in the field with ATE by using a special harness that mates with the PC test connector. The EE has to decide which lines should be brought up to the test connector for ATE and production testing.

6b4. Decoupling filters

It is standard practice to use many 0.01uf ceramic disk capacitors throughout the board and should be placed on the Integrated Circuit (IC) dc power supply pins to ground. I am now suggesting to use 0.1 uf capacitors, instead, on all of these pins because they don't require much more room and it provides better decoupling. These 0.1 uf disks come in a CK06 package instead of the CK05 for the 0.01uf. It is my position that one can never use enough capacitors.

I also recommend that at least a +120 uf tantalum capacitor be used close to the connector dc input of the card in parallel with a 0.1 uf capacitor. I always get into arguments with PC engineers about this but there is a technical explanation.

Tantalum capacitors tend to become inductive at about 1 mHz or higher and would not provide filtering at those very high frequencies. Almost all chips have switching speeds of 10 nsec., or faster, which converts to frequencies of 500 mHz or more. The 0.1 uf disk in parallel with the tantalum would provide the high frequency decoupling while the tantalum helps quiet low frequency noise and crosstalk.

A PC engineer complained to me that there could be 40 of the 0.1 uf capacitors already on the motherboard so why does one need a tantalum. What the PC engineers didn't understand is that one inch of trace looks like an inductor in series at 500 mHz, so the filtering must be done at the point of entry to prevent pick up through the power input lines from other boards. Many engineers were shocked when they put a scope on the dc pin and saw those large spikes bouncing around when the boards were not being filtered properly. A good PWE would put these filter capacitors on the board automatically, whether they are needed or not. There could never be enough decoupling.

6b5. Card Guides

Card guides can be made of metal clamps that are screwed or riveted to the side walls of the chassis. This allows the front of the clamp to grip the board cooling plate. They can also be spring loaded types, which draws heat away through the springs. The metal clamp approach is more effective and is used by many companies.

One must understand, there would always be some heat losses at the card guide transitions. If the electronic is not too hot then the spring loaded guides work fine. They would hold the boards in place and provide some heat transfer from the boards to the chassis walls. Plastic card guides are alright to use for laboratory equipment but not in a military environment.

6b7. Plug-in chips.

Under no circumstance should plug-in chips be used in a military program. A Design Manager that preceded me, decided that he would use wire wrap boards with plug in chips instead of PC boards for an ATE program. The Navy allowed "best commercial practice" in the SOW because the unit would be in an air conditioned environment. The units that he built turned out to be the most troublesome design that had ever came out of our organization and unfortunately I inherited them.

The main problems with these plug in chips were in the handling the boxes by white hats. The chips would "walk" out of their sockets. The second problem was corrosion caused by dissimilar metals by the chips pins and the sockets.

There are occasions where plug in chips are justified, especially using large integrated circuits. Testing Micro-processors, Gate Arrays, or any LSIs are virtually impossible. Removing them for testing by unsoldering them would be destructive to the chips and the traces. Surface mount chips would eliminate this problem.

6b8. Shielding

Sometimes the PC board needs to be mounted into a shielded enclosure to prevent radiation or susceptibility. This is often done with modular printed circuits. If very fast chips are to be used on the board and it is in a high frequency environment, shielding is a must.

To cut down pick-up and crosstalk on the PC board, guards (shielding) may have to be used on PC boards and the mother board. On sensitive signal paths, guards are used on the traces which provide the same shielding effect as a coax cables.

On high speed boards the PWE may have to make a calculated analysis on trace resistance, guard capacitance and insertion loss, before he can start the layout. Too much shielding can deteriorate the rise time of a pulse if a trace is too long, so guards may not always be the answer.

6b9. Trace size

The trace width and thickness have to be established before a design can start. There is a MIL Spec limitation on trace width and the separation between traces. The thing that concerns the PWE on traces is clock frequency, the speeds of the chips being used, and the dc current requirements.

If the dc power for the chips are on a separate voltage plane layer, current loss may not be a problem, but there would always be driver circuits that needs a lot more power and these would require wider traces to cut down the current loss and heat.

6b10. Cooling

There are several ways to cool the electronic on a PC board. The electronics of Figure 5 does not require any special cooling techniques because most of the chips are C-MOS, which only draws micro-amps. Also, within the box there are 40 ma lamp drivers and relay drivers using transistors, but half are either turned off or saturated (turned on) and do not generate much heat in that state.

When the electronics requires high speeds, cooling would probably be required, because, as a rule, the higher the speed the hotter the chips.

Cooling through a ground plane of a PC card is sometimes sufficient to draw the heat to the sides of the chassis. If there are only a few warm chips on the circuit, they can be placed near the ends of the board close to the card guides.

A strip of raised aluminum under each hot chip that also touches the traces below may be added to insure they would be cooled as long as the bottom of the chip is touching the metal. Aluminum strips could be used under each chip in parallel to the connector and brought out to the sides of the board to draw heat away via the card guides.

Cooling using a special machined plate that is glued to the PC board is also a very popular approach. By using this method it usually doesn't matter where the chips are located because each chip would be pressed against the cooling plate and the heat would be drawn to the sides. Many electronic firms are using this technique. The only draw back to this approach is that the sheet metal must be insulated from the traces. Also, the metal covers the traces so if one wants to make a haywire change he can't cut the trace out.

The way my engineers got around this was to remove the chip, drill out the pin in question and then haywire directly to the pin. This does not always work, however, especially with fast chips. Another disadvantage to this method is when a change is required (adding a chip), not only does one have to redesign the board but the ME also has to modify the machined cooling plate which could delay the program.

Cooling by fans are still being used by many manufactures because it removes the heat quickly. In the old days the engineers would use an exhaust fan with vents located at the opposite end of the chassis. This method is no longer MIL approved. Moving air over the components is no longer allowed because of contamination and corrosion from the outside air. So now only two ways are accepted to draw the heat away using fans; by using compartmented side walls and passing the cool air through them, or a cold plate under the chassis with air flowing through it. These decisions must be made first up front before the PWE can begin laying out any PC boards.

As mentioned, there is always heat transfer loses at each transitions. When the IC is pressed against a cooling plate there would be a finite heat resistance between the chip and the plate. When the board touches the card guide and the guide touches the chassis there would be some more heat losses at each of these spots. Finally, heat resistivity in the metal itself would also have the effect of heat transfer loss. The question is, how much air flow does it take to do the job?

If the company has Thermal Imaging capability to measure heat transfer on the box and on the PC boards, one may be shocked to discover the results. The transition heat losses discovered at each junction point mentioned above would be amazingly high.

Also, I have discovered on one design, that increasing the air flow in the side walls did not reduce the temperature of a hot chip that was located in the middle of a board.

My conclusion was that there was poor heat transfer in this particular design. By adding more air flow in the side walls had little affect and no additional cooling. As a result of this test, it proved the theoretical analysis that was made by the Thermal Department was unfounded, by forcing the designers to use a very larger fan to increase the air flow. In this case, the large fan was a specially purchased item for this program and was very expensive. It also had a poor MTBF. With Thermal Imaging data I was able to get rid of that large fan and replace it with a standard more reliable military type. The Thermal Imaging test, in this case, paid for itself.

6c. PWB Design Procedure

In the procedure of designing a PC card, the EE should first develop a schematic in CADD. From here on in, everything that was reviewed about PWBs will now be discussed using a CADD system because hand layouts are practically obsolete.

Getting the EE to put his design schematic in CADD was another major obstacle for me. I had no trouble with the young electronic engineers because they were already computer oriented in schools. I tried to get the older EEs to throw away their pencils, and stop using hand sketches for schematics. I didn't get far with that approach, so I decided to have the PWE put the schematic in CADD, just for the old timers. The older Electronic engineers would eventually retire, anyway, as my ME suggested, and the CADD system would eventually be used properly by all. The recommended Printed Wire Board design steps should be done in the following manner.

6c1. Printed Wire Board Design Steps

o The EE draws the schematic on CADD

o The EE may perform a simulation test of the circuit, if the software is available.

o The PWE draws the board size in CADD. The PWE and ME have to determine the board size.

o The EE provides thermal data on the chips (hot chips may have to be close to ends of the PWB. It may not be possible at high frequencies).

o The EE selects critical trace lengths and spacing (for high currents and high frequencies).

o The EE and PWE decide the voltage and ground pins for the chips in CADD.

o The PWE places critical chips manually in CADD.

o Auto-place the remaining chips

o The EE & PWE determine the Ground and Power planes, if necessary

o The EE & PWE select guard lines (for RFI shielded lines)

o Auto-route the board (probably to a 90% level)

o The number of layers may be established automatically but may be changed for EMI

o Manually route the rest of the board

o Have all the routed layers checked by two other PWEs (check and balance) as follows:

Check point to point and to I/O's

Check for minimum feed through's

Check shielded lines

Check for shorts

Check for extended lines between layers

Check for lines that go nowhere

o The EE should also check the general routing

o The PWE produces an Assembly drawing

o The PWE produces a Fabrication drawing

o The PWE produce a Master Pattern o The PWE produces a listing or a tape routing

o Have a photo plot made

o Have the PC board made

o Design the cooling plate in CADD

o Have the cooling plate machined

o Produce a connector pin listing

When Auto-Routing first became available, it changed all of industry. I was told once, that no longer would experienced and expensive PWB engineers be required, because Auto-Routing can be done by a low paying individual. All one has to do is press the "GO" button and the board is automatically routed. Sounds easy enough.

Well, there is a lot more to it than that. Laying out a board with Auto-Routing requires an in-depth knowledge and a lot of experience.

In one electronics company that I had worked for, where the Independent Department structure was used, the PWE designed the board independently. All the EE could do was give the PWE a schematic and the EE wouldn't see the layout. After the board was assembled, only the Test Engineer would see the board.

On one occasion, the EE designer added a jumper wire to a existing 12 layer board to correct a problem. I asked the PWB supervisor how did his PWE incorporate this minor change. Did he redesign the board? The answer was a flat "No."

What they did was simply add another layer to the board. It seems that every time there is a minor design change they would just add another layer and press the "GO" button. I'm willing to bet that this 12 layer board would have only required 6 layers, if it were re-designed properly. There was a small cost savings in layout, but a new board had to made, anyway. That's the limitation of an Independent Structure. Each organization works independently and no other group has an input. "No Check and Balance."

Unfortunately, this board was being used in a TEMPEST environment, so pick-up and crosstalk became a major problem. I, eventually, left this company because of the way they did business and other reasons. One cannot achieve a decent design under these circumstances. When the EE has no say so in the board layout there would always be problems, especially at high speeds or high frequencies. TEMPEST testing (discussed later) is a killer under these conditions.

6d. Backplane Design

The different types of backplane designs used are wire wrap, PW motherboard, flex cable, and hard wired. The following is an analysis of these different approaches.

6d1. Wire wrap

As was mentioned, the wire wrap back plane approach should not be used with clock frequencies above 5 mHz. If wire wrapping is used, however, the PWE should become involved to provide the wire wrap pin-list, and he would have to work with the EE in producing this document. An Interconnecting diagram may also be required showing the connections between the PWBs and the I/O connectors.

As a former EE designer, I prefer interconnecting wire diagrams rather than a Pin Lists because it helped me in trouble shooting, and it was easier to trace the signal path. For some reason or another, the younger engineers prefer Pin Lists. I suppose whatever works is fine as long as there is control.

Wire-wrap has the advantage of being able to make changes easily. If the clock frequency in the box is above 5 mHz, then lead length and pickup becomes a problem.

My department had a design using a Manchester bus with a 6 mHz clock. The first prototype worked fine, but the second and third prototypes wouldn't work. The EE designer finally realized that the cause of the problem was that the wires in the wire-wrap cage were wired-routed differently from the first prototype. This caused different cross-talk situations which he thought he had eliminated in the first unit.

To ensure that the production units would be wired the same, Automatic Wire-Wrapping had to be specified on the drawings. Semi-automatic wiring was not good enough because the operator may run the wires over different paths. After this experience, I decided that there would be no more wire-wrap designs when the clock frequency is above 5 mHz and preferably lower.

Note: When using a Manchester Data Bus, the data frequency is actually double the 6 mHz clock frequency or 12 mHz, so the problem became even worse.

6d2. PW Motherboard

The printed wire motherboard approach is the most popular backplane at high frequencies. It is more rugged, the trace lengths are consistent, it's cheaper to produce, it's easy to assemble and repair, guard shielding for the traces are easy to design, and it provides strength to the structure. The motherboard is usually a large board and would probably be multi-layered. This is the last board that the PWE would layout. He has to wait until all the daughter boards are laid out first to establish the mother board pin assignments.

My approach to motherboard design, is to place the I/O connectors at one end of the board and locate them close to the box external I/O connectors. Hard wires are preferred from the box I/O connectors to the motherboard I/O connectors. This makes it easier to assemble, repair, and make changes. Remember, the weakest link in any electronic box is the connectors. If there is a way to make it easier to repair, the better. The motherboard should be treated as another PC board and has to meet the power and ground plane requirements stated above.

The ground and power planes could be a grid type (a bunch of Xs) if EMI is a problem. The trace length and thickness are even more important than the daughter boards because the distances they travel are much greater. This task should be assigned to an EE who understands high frequency technology.

The PWE should provide the EE with a list of questions, such as which traces contain high frequencies and require short lengths on both the daughterboard and motherboard, and which traces require guards for shielding? In addition, which traces have to be wide and heavy to handle high currents? How does the EE want to handle the voltage plane and the ground plane on the mother board? Does he want filter capacitors on the motherboard for the voltage lines and where? The PWE can help himself by asking these questions and more.

It is surprising how many EEs overlook these standards in the design. Some EEs are astonished when questions are asked. He would ask the PWE, "Why do you need this information?" Because it's important. Too many PWEs would take the schematic and layout the boards as they see fit. A poor designed motherboard can be very noisy and would become very costly to correct or redesign.

6d3. Flex Cables

Flex cables are sometimes used as the motherboard or in place of the wire harness connecting the box I/O connectors to the mother board connectors. There are some advantages in using flex cable. There may be times when there is not enough room in the enclosure for a PC back plane and the only solution is a flex cable replacing the backplane. But if the chassis were designed properly in the first place, there would have been enough room left for a PC motherboard.

As far as using I/O Flex cables is concerned, it is supposed to be cheaper than hand wiring in production, but if one has ever tried to repair or replace a connector attached to the I/O flex cable, he would become pretty irate in no time at all. This takes special talent and proper equipment.

Another advantage is that the length and routing of flex cables would always remain the same. In any event, Flex cable designing is very similar to designing another PWB. On a long production run, Flex cable is definitely much cheaper than a wire harness in assembly.

6d4. Hand Wire

Hand wired back planes are still being used in some cases. In a small box design, for example, it is much simpler and cheaper to just hand wire the I/O lines to the PC board mating connector. Why build an expensive PC mother board when a small harness to one or two daughter board connectors are sufficient.

6e. Changes

Changes are the name of the game in military electronics because most military programs are usually a short run involving possibly a hundred airplanes or less. The military would be constantly upgrading the vehicle with new electronics and new requirements, especially after production has been completed.

If a plane is twenty years old, the chances are the military would want to incorporate the latest technology. One has to be able to allow for these changes, because there are so many factors to contend with. I know of at least five instances where changes would invariably be required in the design phase, which will now be discussed as follows.

6e1. The first instance of changes would be after the first prototype or XN1 is tested in the Engineering Lab. As the EE design engineer finds problems during laboratory testing, using a special laboratory test set, he would discover changes are required. A new set of boards may have to be made to include the haywire changes on XN1 and be incorporated in XN2.

It is important to always have an XN box available in the design laboratory for continuous testing and upgrading. When XN2 is assembled it should stay in the design Lab while

XN1 is sent to the Integration Lab with the haywire changes and "dead bugs." A dead bug is a duel-in-line chip glued to the board upside down and haywired.

6e2. The second incident where changes are require are in Integration when the equipment is connected with other electronic boxes or in a simulated vehicle considered a hot bench. These changes should be cranked into XN3. When XN2 has new boards it would go to the Environmental Laboratory for shake and bake, even without the latest updated changes. The environmental test is, after all, a physical test of the equipment and not so much a performance test.

6e3. The third set of changes would come during environmental testing of XN2. If the electronics fails in the heat chamber, modification may be required to the PC boards. Depending on how drastic the changes are, they may have to be incorporated into XN2 before environmental testing can continue again. Usually, haywire changes are acceptable on prototypes as long as the wires are glued down, but eventually the changes would all have to be incorporated in the final design. EMI testing would always require some changes, almost guaranteed.

A special laboratory test fixture would be assembled in the Engineering Lab to test XN2 in the environmental laboratory. XN1 would be temporally updated with haywires and dead IC bugs.

6e4. The fourth set of change would occur when the Functional Test Procedure is being prepared using XN3 for production. While the Functional Test Procedure is being prepared, XN2 is still going through its environmental testing, so XN3 may not have all of the latest updated changes. The Functional Test Engineer would probably discover some additional problems while preparing the FTP, which may require more haywire changes to XN3. So one can see, it's a back and forth dilemma.

6e5. The next set of changes may occur during Flight Test using XN3, or after the Functional Test engineer is through with it. Now the box is being tested in its actual environment, in flight with other equipment and changes may have to be made. All changes from here on in, should be incorporate into XN1 and later into XN3.

The Integration Lab and the design Lab should always have a box in their facilities while another box is being reworked. XN2 may be a basket case after environmental testing and should only be functionally updated with haywires and dead bugs. It should never be updated as a final unit.

Note: This phase of plane testing may not be available because the plane is still in production and it would have to be done later.

6e6. The next set of changes would occur during the First Article Test (FAT) and most likely in production. For some reason when one designs an electronic box it works fine in the Lab, but when it goes into production, strange things happen. Sometimes the structure doesn't fit together or they cannot purchase the original parts that were called out. Some changes would always be required for sure. There always seems to be something in the production procedure that doesn't agree with the drawings.

Production released drawings would generally cause problems with the inspector. Especially with the parts delivery. Parts availability may continue to be a problem for a couple of years after FAT, no matter how careful the department is. It's much easier to accept this fact and make the necessary corrections as they come up rather than try to fight it. After the EE selects a particular part it may not be available in production. This problem will be discussed later.

6e7. The last set of changes would occur after the box is in the field for many years, or when ECP's are generated from the customer. Fortunately, ECP changes are funded separately and would not be part of the original contract.

The design organization should keep XN2 in the Lab forever for updating the design and to test any changes discovered in the field, or for ECP changes. If a unit is not available, a production model would have to be bailed back to engineering for updating purposes, especially if XN2 was destroyed during Qualification. Remember XN1 and XN3 were refurbished as the last production items and were shipped out.

Integration should have replaced one of their XN models with the fourth or fifth production unit, which would then remain there permanently for integrating future changes to be tested with all the associated boxes. It really doesn't matter which order the boxes are sent around as long as the Integration Lab and the EE Lab always have a working unit.

There was a time when four XN boxes were made for this reason. This number was reduced to two on one program in my organization which did delay the change cycle drastically.

The last ECP change that I made on a black box, was the result of Mr. Walker giving Cryptic secrets to the Russians. This affected a box I designed 30 years prior. The box had to be modified and retested in the Integration Lab. It eventually was tested satisfactorily on the aircraft and several hundred boxes had to be updated in the field. It seems, one's responsibilities never ends, even after 30 years.

How does one make changes on a very expensive multilayered motherboard during all of these iterations? One approach I used on an XN1 model, was to have drill-out holes laid out on XN1 motherboard next to each of the receptacle pins of the daughter boards. In this way a change could be made by simply drilling out the pin in question to disconnect it from the trace on the mother board, which then can be modified with haywires. The final design would exclude these drill out holes.

One of the early problems we discovered before we used drill-outs holes, was traces were sandwiched between layers and could not be cut out. It was very difficult to haywire. By using the drill-out hole method on the motherboard all traces to the daughter boards and the I/O connectors are accessible.

Once these initial problems are solved with haywires on XN1 during integration, these drill-out holes can be removed for XN2 or whenever the motherboard is re-laid out and incorporated in all production boxes.

I, sometimes, had to cut the lead on an IC chip to separate it from the motherboard trace and solder a haywire directly to top of the IC lead. Short jumpers between stand-offs, are another way of accomplishing these changes.

So one must understand, there are a number of things that can be done to ease the pain of making changes on the first few prototypes. Allow for it in the design to make it easy on one's self by accepting the fact that there would always be changes. After all, changes are the name of the game in the military.

7. CADD Operators

Who is a CADD Operator? He is a person with a lot of drafting experience that can manipulate mechanical drawings in CADD according to his Company's Specifications and MIL Standards. These people are worth their weight in gold because they can turn out a ton of drawings in 1/4th the time it took to do them by hand. Drafting by hand is archaic and for some reason these ex-draftsmen were able to switch to CADD with no problems. Once they started to use the CADD machines to develop drawings they immediately realized the advantages and were sold on this new technology.

One of my first experiences with CADD was when my department received a major contract to transfer many hand produced old aircraft drawings into level III CADD drawings. Level III, is basically the drawings of an electronic enclosure that are drafted in accordance to a MIL-Spec Standard, which allows the box to be manufactured by any electronic company. This gives the military the option of going to any production house to build the equipment that was designed and developed by another company. As long as the standard level III drawing procedure is followed, anyone can bid on its production run.

I had to borrowed five CADD operators from another department for this assignment. After some instructions they began transferring hand drawings into CADD. They were extremely fast in producing Level III drawings. When the program ended I retained two of them full time. As was mentioned, the drawing effort of a program is at least 1/4 the budget, so a good CADD operator is vital in saving manhours. Most companies don't like the Level III drawing approach because they could lose a major production contract if the military goes elsewhere. After all, profits are made in production and not in development.

It takes years of experience to develop a good CADD operator because he has to know the company's drafting procedures, drawing format, check procedures, the release cycle, and handling MIL-Spec requirements. These individuals do not have to be PC board designers or enclosure designers. They may be just draftsmen who can produce a drawing in CADD with minimum information.

Some of them are good enough that they could design the structure of an enclosure under the supervision of an ME. All of the CADD operators should be under the control of the ME Leadman, and the operator may be matrixed to the department, or be direct.

If one's company does not have CADD operators the MEs and PWEs would have to carry this burden. They would probably not be as fast because Mechanical Engineers usually do not know all of the drafting shortcuts and all the boring drafting practices required in a full Mil program.

Another way to supplement these tasks is to use Job Shoppers. There are many Consulting houses that supply CADD operators for short durations. Some of these operators are extremely good and some are poor. In any event, they would have to be supervised to ensure that the CADD drawings are done properly. I mentioned before that I did not approve of using Job-Shoppers, except in this case I do, as long as they are supervised.

One major problem I had with CADD Job-Shoppers was with updating drawings. When the non-standard parts engineer was trying to extract parts from the assembly drawings in CADD, he found many updated drawings were released with the same drawings number. He didn't know which drawing reflected the latest configuration. These shoppers don't know the company's procedures so they did it their own way. Unless they are instructed properly and supervised, the drawing history could get out of hand.

On one program that I was on as a Staff Engineer, I tried and failed to explain to the EDM in charge, that he should have used the revision number procedure that I mention before for prototypes, using letters after the drawings are released, and locking each drawing up in CADD as Read Only. In this way each Job-Shopper knows what to do when he is given a drawing update assignment and cannot modify an existing drawing because it is locked in Read Only. He would have to duplicate the latest revision drawing first, add an updated revision letter or number to the new drawing.

When it is released, he locks it up again as Read Only. No new drawing numbers can be assigned without the lead MEs approval. Nothing worse than having several updated drawings with the same drawing number. This revision procedure for drawing changes was mentioned many times in this book, and it always seems to crop up again on different subjects. It now applies for Job-Shoppers. That EDM, I mentioned, refused to listen to me and the drawing control documentation became a mess.

Checkers and CADD operators would be arguing with each other often because the experienced CADD operator knows the system and the proper drawing procedures, and he would not always agree with the checker's red-marked corrections. There isn't much a EDM can do about this squabbling because sometimes it's just a matter of the checker's interpretation. The best thing to do, is to incorporate the checker's corrections as he wants it, and if it turns out to be wrong, it would be picked up later at the release table or by a production inspector.

I've had situations where one checker wanted the drawing to be drawn one way and the next checker, who took over, reversed the procedure, so one has to go along with whatever the last checker wants. This approach may be a little bit costly in rework, and it may cause production delays, but the drawings must be released and delivered to production to meet schedules, and should not be delayed because of some insignificant interpretation.

Production can only work with released drawings. An inspector would buy-off an ECO to correct a drawing error to keep the production line going, as long as the correction is incorporated into the drawing and re-released within 90 days.

8. Technicians

Electronic technicians are essential in the design and development of electronic equipment. Different companies use technicians in different ways. Some use them as only wiremen, some use them as testers, some use them as mechanical sheet metal men and machine shop mechanics, and still some use them as engineering assistants. Some technicians are even used as designers.

My preference in the design world is to use them as engineering assistance. I prefer to have both electronic and mechanical types in the department with a lot of experienced and to overlap each others duties, to a certain degree.

In one company I worked for they used to have a Technicians Apprentice Program where a young high school graduate would be hired into an "on the job training program" for three years, moving him to different departments before he became a journeyman. This was a good program but then it stopped.

Most of my seasoned and experienced technicians came from other companies or from the military. The ideal technician would be a person that had some military service as a technician, or had two years of junior college in engineering or in a technical school, and had at least five years of experience in industry.

I prefer not to use a technician just as a wireman or as an assembler to do only the physical work. I would want the technician to get involved and contribute to the design and to do the electronic testing. I have seen many good technicians who could have been promoted to engineering but refused the title because they didn't like the politics or the pressure involved in engineering.

When I was a designer, I would always encourage the technician to become involved in the design. If I needed a bread board I didn't want him to just build it from my sketches. I would try to explain my design to him and have him not only build the bread board but test it first to see if it performs as the design was intended.

In many cases the technician would make some adjustments or modifications to make the breadboard work better and we would discuss it together later. An Engineer has to have a good relationship with his technicians, whether they are the electrical or mechanical types. The EE has to make them feel that they are part of the team.

The question now arises, how many technicians are needed in a department? It obviously depends on the amount of work and the type of work that is to be done. Some black boxes require just sheet metal assembly and wiring, and some require electronic assembly and testing.

As a rough estimation, a ratio of one technician for every two or three design engineers should be sufficient, which would keep the technicians continuously busy. It is essential that there is enough work to maintain a fixed working crew of technicians or they may end up being transferred or laid-off.

The technician should report directly to the EDM rather than be in a pool of technicians under another organization, and the budget should include them. In this way the EDM would know the special talents of each technician, and he would be assigned accordingly. If another department needs a technician an agreement can be reached to loan out a technician for a short period.

As an EDM, I was once forced to use a technician from a technicians pool, and on one occasion I refused to have a particular technician assigned to me because I knew he was sloppy and made many wiring mistakes. This then became a Union problem and, of course, I lost. I had to absorb that technician an give him menial work just to get another good one who could do the job. It sure hurt my budget.

The duties of a technician would be to assemble, wire and test breadboards, prototypes, and small production quantities. There are several techniques in making breadboards. On small breadboards, perforated boards were sometimes used with hand inserted terminals. These were hand wired to the components and to the Integrated Circuits.

In another technique, wire-wrap IC connectors boards were used that accepted plug in chips. These boards can be purchased and come in several lengths, and one can choose a large wire wrap board or a small one. Some companies have used these modules as their final product. Each chip is plugged in and wire-wrapped under the IC connectors. The advantage of this package for bread boards is that the chips and components can easily be removed and replaced.

One department decided to first test the unit using this wire wrap module approach and then permanently solder the chips in place later to prevent the chips from "walking out." This approach is actually more expensive than making a Printed Wire Board, because many hours would be spent doing the wire wrapping and soldering the chips, that is, if several units are required. Labor is always much more expensive than purchasing a PC board in production with automatic insertion of components. As was said, production is where the profits are made, not in development.

I have previously suggested using at least three prototypes in a development program which I called XN1, XN2, and XN3. I have also suggested that it would be more cost effective on a contract to build small orders of production units in the Engineering Laboratory and the larger quantities in production. I will now explain the advantages of this approach.

To gear up the production assembly line for a small orders of an electronic box is foolish because one has to go through inspection, release drawings, and having to buy a small number of parts with no price break. Building small quantities in the engineering laboratory makes more sense because it is easier and it helps maintain a continuous work force of technicians. There are times when there is little or no design work going on in the department and by having this small production activity as a supplement one can maintain a steady force of technicians.

I had a situation where I had an order of only one Navigation Interface unit by the Navy each year. See Figure 26, page 319. Since they were fabricated and tested in the Engineering Laboratory it helped my head count, because it meant steady work for a few technicians for six months each year. It provided work for the sheet metal development shop, for the mechanical technician, for the electronic technicians, for the purchasing department, and work for the engineer who controlled the program, ordered the parts and did the final ATP with an inspector.

To illustrate this situation further, small quantities of my ground support Test Unit of Figure 6, page 78, were being built in production. This task became a mess because production is not geared-up to handle small quantities. If the production quantities were large in the first place, then it would be understandable for production, but small quantities orders are difficult to be handled in production.

In this case, the original order was for only three units. Every year the Navy would order one or two more. Then there was a six year break and later on, the Navy ordered three more. Over a 17 year period they ordered 36 units in small quantities. After so many years, new problems became apparent of obsolete parts and ECP changes. This drove the production buyers insane, not to mention the production workers.

If the program was handled by engineering the obsolete parts problem would have been solved, the drawings would have been updated, and the units would have been delivered on schedule. This Test Unit is a perfect example of a box that should have been handled by engineering laboratory technicians. It became a nightmare in production and the last few units had to done by engineering, anyway.

Testing by a technician is an important function. He would be testing the breadboards, the prototypes, any research designs, and small production quantities. Testing breadboards are rather simple and, in most cases, all the technician needs is an Oscilloscope, a Multi-Meter, a Function Generator, and possibly a Logic Analyzer.

When he has to test a Prototype, he would first have to debug and test a specially designed Test Fixture, which he assembled in the first place which was designed by an EE.

The Test Fixture may be an automatic tester with a micro-processor in it which would be used to debug and test the newly designed prototype, or it may consist of a rack with commercial test equipment. The Test Fixture may also be used to test the unit during shake and bake for qualification.

Other duties of technicians are to support and wire the Integration Laboratory that simulates the aircraft, to support Flight Test, possibly perform the Functional Tests (FTP) on the first production built equipment, to perform Acceptance Test (ATP) on the equipment built in the engineering Lab, to repair and test laboratory test equipment, to provide field service support for maintaining company manufactured equipment, and to repair purchased capital test equipment. There is, definitely, a lot of work required for a technician.

9. Development Shops

As was mentioned, it was more economical and more efficient to have a separate engineering development shop for producing sheet metal parts and provide machine shop work. I still believe that this was the ideal arrangement to support an electronic black box packaging division. I was lucky because a development shop happened to be available for my department, which also produced special parts for other departments in the company in a short order. All it took was a memorandum with an attached sketch by the ME and signed by the EDM. Generally, the work would be done in less than a week.

This arrangement lasted for 25 years until one day the production manager convinced the president of the company that the production development shops could handle all that the specialty shops could do instead of having a duplication of effort. That small engineering development shop was then disbanded and my engineering development problems had just multiplied.

It would be understandable if the reason was because it was difficult to keep the shop busy. But that wasn't the case. They were very busy as they were supplementing their work with small orders from other departments. Altogether, they were in business for 35 years, even before we used their talents. Actually, there was no duplication of effort, because the production shop's main function was to develop production tools and production parts, and not develop prototypes.

The main point here is, one should never mix production facilities with development facilities. Production shops have to follow special procedures of inspections and released drawings, and they have to go out for bids for certain work or parts.

In order for my MEs to get work done in the production shop, they would have to first release the drawings, which could take several weeks of going through the signature mill. This scenario was discussed before, when I worked for another electronic company where they thought it would be cheaper to use Production Shops for prototype development. It is not cost effective and doesn't work well for engineering who need to be versatile.

When the design is in development many different configuration may be tried out. It is sometimes difficult to visualize the machined part on a drawing board as compared to when the part is made. Trial and error sometimes is cheaper in the long run than a paper design. I have seen it happen too often, where a mechanical design of a part did not fit after it was made and had to be scraped.

As an example, the shop made a sheet metal part from a drawing and it was cut too short because the engineer didn't allow for all the tolerances. If it were made in Production, they would have produced hundreds of sheet metal parts before it was detected.

Many times there were errors on the drawing with improper dimensions. If it can happen it will. If one had to go through the production development shop to make these parts and had to correct the errors constantly, the delays of drawing release and the signature mill would have killed the program.

10. Liaison Engineer

A Liaison Engineer (LE) is a person whose assignments are to interface with many organizations. He interfaces with the support organizations, the design engineers, product support, the environmental Lab, the customer, the integration Lab, Flight Test and especially with Production.

During the design phase, he would be ordering and tracking prototype parts, he would provide the list of non-standard parts, the advanced parts list (AMO's) for production, and participate in all design reviews. He would be interfacing with all outside vendors that are required, such as PC boards or sheet metal work. He would be the go between engineering and support. He would accumulate all of the breadboard and prototype test data for the final report. Many of these functions are done by the lead ME if no LE is assigned to the program.

During integration the LE would interface with the Integration Engineers and monitor the initial tests. He would accumulate all of the integration test data for the final report. He would monitor the qualification testing and accumulate all the environmental test data for the final report.

As was mentioned, Integration test problems should be made available to engineering as soon as they are discovered. The LE is the go-between to get that data to the EDM quickly rather than wait for the Integration Lab's final report, which may not be available for months or because it may be too late to institute the changes without a major redesign.

He may be required to prepare the Functional Test Requirements (FTR) for production. This is a task that I prefer the Equipment Systems Engineer to do, but the LE may be stuck with it. The LE would also help review the Functional Test Procedure (FTP) that was prepared by a Production Test Engineers. The LE would monitor the first FTP using a Prototype and record the test data to see that it agrees with the engineering data.

The LE would have to accumulate the various test data and inputs from the Engineers to be given to publications and maintenance. These would consist of Descriptions, Theory of Operation, Flow diagrams, Wiring diagrams, Software programs, and Test procedures.

The LE would be responsible for accumulating and submitting all the data submittal (CDRL's) to the customer. This is a major task in itself. If the program is large more than one LE may be required.

The most important function of the LE is to interface with Production. He would monitor all of the ECO's and parts, and if there are problems he should try to resolve them by himself first. If he cannot, he must report back to the EDM. He has to make sure that all of the outstanding ECOs are incorporated into the released drawings.

He should be a capable engineer and be given the authority for making minor production decisions. He should, however, always provide the EDM with a status report on all of his activities. A summary of this report should always be given to the EPM by the EDM.

There may be a production representative assigned to the program, so the LE may have to work directly through him and not with the production supervisor. Many of these duties that I just mentioned, I had to do by myself as an EPM, which should have been assigned to an equivalent LE.

The final act of the LE would be to prepare the Final Report on the whole program.

10a. Final Report consists of many of the following:

History of the electronic unit

Description of the unit and it's functions

Theoretical analysis and calculations

Design information

Thermal calculations and test data

Mean Time Before Failure calculations Bread board test data

Prototype test data

Integration test data

Qualification test data

Functional Test (FT) data

First Article Test (FAT) data

Photographs of box and sub-assemblies

Block diagrams

Assemble and repair information

Interface wiring or Pin list

Schedule and Budget information

Hours quoted vs hours used tables

And anything else the EDM includes

This report may be the only chance that an EDM would have the opportunity to "blow his own horn" so to speak, and it should be done properly. The customer may not require the submittal of a Final Report but I think it's good practice anyway, and one would be surprised how often this report would save the department's hide many years later.

If the report is being generated while the program is in progress, it is surprising how little effort there is left to complete it at the end. Treat it like a diary, and don't wait till the end of the program to start it because it then becomes a major task and everyone forgets what they did.

As one can see, the Liaison Engineer has many major responsibility. In most companies, no one is ever assigned this task. Yet, someone has to do it. It usually ends up being done by many different engineers or not being done at all.

All of the tasks and recommendations expressed in this book are for an ideal managed electronic department. I understand that costs, schedules and capabilities may be the dominating factors, so the ideal situation may not always be obtained. Very few companies, for example, would prepare a final report or even assign a Liaison engineer to the job. The EPM may be stuck with most of these additional tasks which he generally never has the time to do it, either.

11. Competition

The Electronic Design Manager has to know his competitors and some of the political maneuvers associated with competition. It was my company's position that there shall be no competitive bidding against an out-side supplier. In this way out-side suppliers cannot accuse my company of stealing their proposal and underbid them. My company did not want them to have this impression because these suppliers would never want to participate and quote again as subcontractors for new programs.

However, some of these immoral practices still went on but in reverse order, and I became the victim. In one case, a super secret division asked me to quote a job that had to do with voice synthesis. This was a complicated new program and, basically, had to be invented. My engineers spent countless hours on this quote and actually invented a system that would do the job.

After the quote was submitted with our design approach, this secret organization selected an out-side company to do the job. They actually gave my department's invention to them, and used my financial quote to bargain them down. My company's policy was ignored by this organization. I did learn a lesson, though, not to give away inventions again.

In another instance, I was asked to provide a quote on two black boxes for a different experimental secret aircraft. One had to do with Hydraulic Pump controls, and the other had to do with Automatic Steering controls.

After the EPM, on that program, saw my costs for development and fabrication of three units, he told me I was 10 times too high and that he would go out-side for bids. That EPM, obviously, didn't know what a fully qualified, full Mil box should cost, so I let him go out-side without a struggle. I couldn't do it any cheaper.

After one year, that EPM came back to me with his tail between his legs, and said he had no out-side response to the RFQ for those boxes and would I still be interested in doing the job for my original quote. I had him by the tail and almost turned him down, but I couldn't because we both work for the same company. I was also tempted to raise my quote but I didn't.

Out-side suppliers are not interested in doing small quantities for a complicated research program, because there won't be a production run and it would tie up several of their design engineers. There was no profit to be made on the above proposal and it could even result in losses. As was said, suppliers make their money on production, not design.

It turned out that these boxes went through five major redesigns after we were awarded the job. This happened because my engineers didn't have the proper clearance to work on their program, so we had to rely on second hand information with hand waving directions, which, sometimes, turned out to be incorrect. Had these boxes gone out-side the cost could have tripled my price because of the changes. We were able to make these changes without a new contract, but we did need a some additional budget, which they gladly supplied.

Understanding competitors is very important. Putting bids out to any electronic house can be disastrous because there are many companies that don't have the foggiest idea on what they are bidding on, but they would turn in the lowest bid and win the job.

I was involved in many of these situations and I was asked to evaluate their bids. As was mentioned, the box of Figure 5 was one of those programs that initially went out for bids. Many companies didn't have the slightest idea how complicated it was, but they bid on it anyway.

This is why I recommended earlier that our design organization should at least be used as consultants to the Project engineers in the proposal of special boxes, and that we should be involved in the "make or buy" decisions. Project engineers are not hardware oriented and cannot always make a proper evaluation of an out-side proposal.

Another competitor that the Electronic Design Manager has to be aware of is the Military itself. The military over the years have developed research divisions with civil service personnel. These research divisions have, eventual, increased their capabilities and are now actually producing military equipment. First of all, this should be illegal, and is certainly unethical. They are designing to their own SOWs and ignoring the "check and balance." They are violating the government's basic philosophy of competition. I have lost several programs to the military because they decided to do it themselves.

The worst offence they made was when they specified an RFQ for a specific design. I bid the job in my normal fashion, but after the military saw my prices they decided to do it themselves. This practice has been going on for 20 years and no one wants to challenge them for it.

They used my Quote from that RFQ to prove to their superiors that they can do it cheaper. Naturally, they can. They don't have the cost of overhead, they don't have to pay taxes for their facilities, and they don't have to pay rent or mortgages for their development buildings because it's Government property. They don't have to show a profit to their stockholders. How can any company compete against them. It's one of the most unethical practices that the government can be involved in, but it is continually being abused, even today.

The question then arises is, would the military be doing a good job in producing a design? Who is checking on them? Are they designing to the full Mil requirements and qualifying their own boxes? The chances are, they are not.

When I had the chance to examine one of their designs I could tell it was being done by amateurs. The worst thing about a military design is, there would be many changes and schedule slippage. They don't have to meet any deadlines and they can work on a design at their own leisure.

We had to interface with one of their boxes on one program I was on, but we could never test it together because they spent years in development. We had a deadline to meet but they dragged their design out forever. Who's checking on them?

With one company that I worked for, I had to bid on the Control Functions for a Global Positioning System (GPS) which interfaced with two Navigation Interface boxes that were designed, built, and produced by the military. The military had already worked on these boxes for several years and they were still not finished. They were changing the design weekly. When I called and asked them when would these Navigation boxes be available they just laugh and said, "Some day." There is just no way that industry can out-bid the military? It's actually a monopoly created within the government.

In another instance, my company was asked to modify our reconnaissance airplane to carry people and cargo, which we did. After the first aircraft was modified, the military asked for all of the modification drawings which, of course, they owned from the contract. With these drawings they decided to modify the next 6 airplanes themselves because they could do it cheaper at a Naval base. They didn't even go out for bids. But, of course, when they had problems we had to send our engineers to their facilities to bail them out. Such is the life in a design department.

Chapter II

OTHER ORGANIZATIONS

Throughout this text I have often mentioned Support Groups. These are other organizations within the company whose job is to insure that the final product is of the highest quality and meets all of the required military documents. These organizations would not only support the design but would also monitor it throughout the program.

They have two major responsibilities to support and monitor in-house designs and to monitor supplier designs. For supplier designs, they would act more like policemen to insure that the commodity received from sub-contractors are of the highest quality and meet the Equipment Specification and military standards. They would work with their equivalent departments within the sub-contractor's organization.

As was mentioned, some vendors believed that we had a conflict of interest when these organizations supported our in-house design. Some suppliers felt that when these support groups were making recommendations to a design, they were monitoring their own suggestions, which, of course, would be unethical. But that was not the case.

When the support organization make a recommendation in the design, they don't tell the designer how to implement it in the design, they merely provide information on the weakness of the design, and it's up to the designer to solve or correct that weakness. So, when these organizations monitor an in-house design they are observing how the designer executes their recommendations.

It is very important that these organizations be present during most design reviews. They don't all have to be present at every review because the agenda may not be covering their particular expertise. They all should be present, however, at the first "kick off" meeting to establish the program requirements and to introduce the players with the different departments.

There are certain support organizations that should be present at all design reviews. These are; Testability, Reliability, Maintainability, Quality Assurance, Master Scheduling, and Production. The other departments would be invited only on an "as needed" basis. In this way the EDM can cover those subjects when it's appropriate as they become involved in the design. It makes no sense to have Human Factors present, for example, at all meetings.

As was mentioned, on a program I was on, the EPM actually excluded most of these organizations from the program to save budget. This was a major mistake because the civil service engineer would only discuss a subject with his counterpart in our company. Military Human Factors personnel, for example, would only collaborate with my company's Human Factors department. They wouldn't even discuss any Human Factors requirements with the EDM or EPM. I know, because I tried to deal with them directly with no success. They wouldn't even talk to me as the EDM.

On an ATE program that I was involved in as a Staff Engineer, the EPM decided not to provide budget to our Non-Standard Parts Department. I asked, "Who would handle this task?" He said engineering.

Now engineering, in this department, had never submitted a non-standard parts list directly to the government before, and they didn't know the first thing about parts data submittal. Even when I was a EDM I didn't know what forms to use or who in the government to submit the Non-Standards Parts List to.

I relied completely on the Parts Department to do this task and to provide me with an approved substitute parts. One is only fooling himself if he thinks he can bypass all of these support organization.

Sometimes, by-passing some of these organizations is acceptable if it is not stated in the SOW. Remember, the SOW provides the contract requirements between the customer and the prime contractor, and if a particular support subject is included in the SOW, such as non-standard parts, the EPM had better have that organization included in the program because the government won't discuss that issue with engineering.

The following is a list of all of the different support organizations that the EPM and his Design Department may have to interface with:

- o Human Factors
- o Safety
- o Non-Standards Parts
- o Testability
- o Production
- o Reliability
- o Quality Assurance
- o EMI
- o Maintainability
- o Weights
- o Thermal
- o Electrical
- o Stress
- o Producibility
- o Integration Lab.
- o Flight Test
- o Functional Test
- o Check
- o Drafting Dept.
- o Qualification
- o Drawing Release
- o Inspection
- o Material
- o Purchasing
- o Computer Service
- o Finance
- o Master Scheduling
- o Contracts
- o Administrations
- o Product Support
- o ATE
- o Manuals
- o Training
- o Field Services
- o Logistics

That is quite a list. The following discussions concerning the responsibilities of these support organizations are not listed in any particular order of importance.

1. HUMAN FACTORS

Some engineers consider Human Factors to be insignificant. I was told once that all a Human Factor Engineer does is round off the square corners of a box or square off the round corners. This shows how ignorant some engineers are. Human Factors personnel are not always engineers and many are graduate Psychiatrists who may have a Doctorate degree. They have to understand the needs of the operators for all equipment. The Dash Board of an automobile, for instance, had to be approved by a Human Factor Engineer.

If I were designing a piece of test equipment that is to be used on a test station, a Human Factor person has to be able to determine accessibility of the test leads, of viewing meters, maneuverability of the operator, etc. He has to analyze the equipment to determine if it would operate on the station without interference from other equipment.

He would look at the general appearance, the handling, and the installation. The carrying weight must be considered for a female that is 5 ft tall and weighing 105 lbs. Should it be a one person carry or more. This all has to be done early in the program. Many a bad design required major modifications later because a Human Factor Engineer was ignored and not brought in the at beginning of the program.

At the "kick off" meeting three dimensional drawings of the enclosure should be viewed by all who are present, and a copy must be given to the Human Factor Engineer to study it at his own leisure. His comments would be noted at this first meeting but no action would be taken until the next design review is presented, or later on in the program when the Human Factor Engineer would be asked to introduce his own recommendations, if any.

Also, at that first kick off meeting, a cardboard or styrofoam mock-up of the box should be provided so that everyone can visualize what the enclosure would look like. The Human Factor Engineer would appreciate this because, even as a Doctor of Psychology, he may have trouble visualizing the enclosure from a drawing.

After he accepts this cardboard configuration and possibly provides recommendations he may not be invited again until the first prototype is built. By using the prototype, the Humans Factor Engineer would simulate the operator's functions on the station to make sure the enclosure fits properly and can be operated with simplicity. He would try lifting the unit to determine if it is unbalanced and difficult to handle. Does the unit require handles and how many? Does it have a nice appearance? It's surprising how important the appearance of a unit is to the military.

Should the corners be rounded to avoid hurting the technician? Would the technician mash his fingers when installing the unit? He may ask the operator to demonstrate a normal operational procedure to see if there are interferences.

Hopefully, during his evaluation, most of his questions would have been answered and the demonstrations are only to verify that his recommendations have been incorporated. After this simulation, the Human Factor person would write his report. The LE should receive a copy of this report and include it in the Design Department's final report on that unit.

2. SAFETY

The Safety Engineer provides a small but important function on any program. He would examine the equipment to insure that there would be no electrical shock to military personnel. He is mostly concerned with power inputs, power outputs and proper personnel protection. In most boxes the power input could be 110 to 220 volts ac or 270 volts dc.

He would want to know if the power is properly grounded to the chassis, and is the point of contact at the bottom of the chassis free of paint so that it would make a good electrical ground contact to the installation rack? He would examine the box to see that there are no cutting edges on the enclosure or inside the enclosure. He would want to know if there are any components that could explored, such as electrolytic capacitors, if they are being used under rated and where they are located within the box. Are the capacitors properly confined to protect the operator. He would also examine the complete enclosure for allergic or toxic materials.

3. NON-STANDARD PARTS

The Non-Standards Parts department provides three services to the design department. These services are; (a) processing and submitting the non-standard parts list to the military in a proper format and following it up, (b) help prepare and monitoring Specification Control Drawings (SCD) for sub-contractors, and (c) Support the design engineers on selecting approved military components or recommend replacement parts.

3a. Non-Standard Parts List

Many companies have a Parts Selection Standards Manual. This manual has a list of components, including metal parts, nuts and screws, etc., that the company recommends be used for production. The parts may be stored in the company's warehouse and are already on the qualified military approved parts list.

If and engineer can use these parts, then no special parts approval is required from the military. It is also cheaper because the company would buy them in large quantities at reduced rates, and they are tried and proven reliable parts. This inventory of parts sometimes limits the engineers in his design, but that's what Non-Standard Part submittal is all about. To give the engineer some latitude to be able to use in-house parts or the latest chips on the market. However, if a new part is used, engineering must prove to the military that they meet at least one of four categories, (1) no other standard part can do the job, (2) there is a dual source for that new part, (3) there is a substantial cost savings, and (4) there is a technological improvement using that part.

The military has added another parts requirement in the SOW which is called the Program Parts Selection List (PPSL) for selecting components. Each program would have its own individual PPSL list assigned to it. This list does hamper the designer and that's where the fun begins.

Using this document, it seems, an engineer would need a Law degree to be able to select parts from this list. There are so many parts that are superseded by superseded parts and, worst of all, there may be errors in this document. It could sometimes drive an engineer batty.

When a resistor is selected from the company's standards, for example, the part may still be allowed to be used even if it is not in the PPSL, that is, if one knows how to read the document. There is a clause in the PPSL that states, "If the part has a higher reliable notation than the part in the PPSL, then it is an acceptable part to be used." This comparison alone takes a substantial amount of research.

If one's company doesn't have an approved parts standards manual then there would be more paper work required. Each part selected that is not listed in the PPSL would have to be submitted for approval to the military. This includes screws, nuts and material. The problem is, there is no guarantee that the non-standard part that one has selected would be approved by the military.

I've had situations where the Navy did not approve a part till after production has started, which resulted in a PC board modification to include an approved chip at my company's expense. The unfortunate way of handling some non-standard parts is to submit and resubmit over and over again. It may take six months to a year before approval of a part is resolved.

This is why a company needs a Non-Standards department. They know how to handle the submittal of components. They know the proper procedures, how to fill out the required format, and what government agency in Washington to submit the documents to. They also know ahead of time which parts have a good chance of being approved and would give the designer a preliminary go ahead to use it.

3b. Specification Control Drawing

A Specification Control Drawing (SCD) has to be prepared on a part that is to be manufactured by a subcontractor such as a transformer or filter. Parts like these are not on the PPSL because they are specially made by an outside supplier for the company's program. These parts may still be considered non-standard parts and don't always require approval.

The requirements on these drawings would be generated by the Equipment Systems Engineer in conjunction with the designer's drawings, which would contain all the pertinent data to build the part. Dual sources of the part are preferred but are not always required because they were sub-contracted out. Usually, only one source is available because it is a specially made part.

An engineer, who worked for me, once used a sole source special electronic module that was potted by a supplier and was submitted to the military as a non-standard part. While the approval stages were still in process a year later, the manufacturer discontinued producing that module. This engineer had to go to another manufacturer to design a replacement part. While this was going on he received the approval for the original part from the Navy. Now the engineer had to re-submit the replacement part all over again. The paper work is sometimes insane.

Note: parts that are made in-house are covered by the contract as part of the company's drawing submittals, and do not require non-standard parts approval. When the box is approved for production, every part in that box that the company has manufactured is automatically approved. In other words, the company is responsible for producing that part.

The SCD is not just an ordering document, it could be a formal drawing that specifies all the parameters required for that part.

Sometimes an Equipment Specification (ES) is used as the SCD. It depends on the complication of the part, the contract or the unit itself. This drawing, or SCD document, is handled the same way as any other drawing in the program and has to be released and signed-off in the same manner.

Note: the EDM and EPM have to deal with the Contracts Department directly for SCD parts. The PM is not involved in this selection. He only deals with a customer. A SCD can cover such complicated parts as a plasma display, a special power supply or a crystal oscillator, for example. It includes any purchased part that would be enclosed in the unit that is not considered a single standard component, and has to be specially manufactured to the company's SCD.

3c. Supporting the Design Engineer

When an engineer is in the process of a design, he spends an awful lot of time just looking up parts in catalogs or in the company's Standards Manual. He may have to go to the company's library, call component manufacturers, look in magazines, or look in the department's catalog files which are usually out of order or out of date. Finding the right part that would do the job takes a special talent and plenty of experience. After the engineer goes through all of this hassle, he still doesn't know if it is an approved military part.

One engineer selected a power transistor that our department had been using for years on many military programs only to find out later that the part is no longer on the Qualified Parts List (QPL). It had been superseded by a higher reliability letter on the part's specification number in the PPSL. I had ordered several hundred of these parts for this engineer to use on the prototype and even paid for the pre-screening costs. It was later discovered they were superseded and we could no longer use them.

A lot of this looking up and using the wrong part could have been avoided had the Non-Standards Department been utilized. I have preached to my engineers many times, that they should submit the Non-standards parts list to the company's Parts Department as soon as possible to avoid this problem. But on that transistor, even I was fooled because my department had been using that component for many years. It was a good part then, why not now?

One of the main reasons it pays to work with the Non-Standard Department is, if an engineer selects a part and submits it to them right away, they can evaluate that part immediately. They have access to many files in their computer which are being updated continuously in the MIL-Specs. If the part selected is not on the Qualified Parts List (QPL), the Parts Department can quickly find a substitute and save the engineer many hours of searching. To make sure an engineer is covered, I suggest that he submit the complete parts list to the Parts and Equipment Department. They then can tell the engineer that some of the old parts are no longer on the PPSL list, and can recommend a substitute.

In another situation, an engineer had ordered a 50 mHz crystal oscillator in a TO-5 can from a supplier who said he had plenty of them. I signed the order for 20 pieces for our prototype and 12 were received right away. My engineer waited a month and then five more showed up. It was delayed because the order went from Boston through Canada and back to the States again. The supplier finally told our buyer that the manufacturer stopped production of these components and could not fill the order for the remaining three parts.

Had this engineer worked with the Non-Standards Department, as he finally did, they would have told him right away that that part was no longer in production, and a replacement part would have been given. The Parts people knew about it months before he selected it, but no one asked them about it.

This happened because my upper-manager, on that program, refused to deal with the Parts and Equipment Department to save budget. It backfired. A supplier would lie and tell an engineer that a part is being produced to get rid of stored parts, but he would never say that to a Parts and Equipment engineer or a production buyer.

4. Testability

The Testability department should be involved in the major portion of the program, especially in the early stages of design. They would monitor the test methods using the ground support equipment or by using ATE. Their job is to see to it that the design engineer implements the testability requirements that were set forth in the SOW.

4a. Diagnostic Test

A Diagnostic Test is a vehicle test on all WRAs, which is known as the "O" level test. There is two parts to this test. One is a readiness test which tells the vehicle operator that all systems are "GO" or if a Weapons Replaceable Assembly (WRA) is not working properly. This test is controlled and performed using the on-board computer. There would be a display on board that specifies which WRA or black box is out of order in a particular sub-system.

The second part of the test is to determine what particular function is bad and, in some cases, what Sub Replaceable Assemble (SRA) in a WRA has failed. A PC card is considered an SRA and a replacement on board may be feasible.

The diagnostic test is sometimes performed periodically and automatically while the vehicle is in operation or in flight. This test would provide a systems readiness alert throughout the flight so that a mission failure can be avoided by either on board repairs or by aborting the mission.

Nothing worse than, after doing a ground test on an aircraft, to find out later in flight that a sub-system had failed. The mission would have to be scraped and wasted. A diagnostic test would help avoid this problem because the system is being tested automatically before and during the flight.

I had been involved in an EW flight test program where the aircraft took two hours to fly to its test site but had to return again because of a failed unit. The pre-flight ground test was fine, but because this aircraft did not have diagnostic test capability there was no way of knowing if all the EW units were mission ready until the plane reached the site and began its maneuvers. Similar missions had to be aborted may times on this program because of failures in flight, which cost a bundle.

4b. Built In Test

Built In Test (BIT), is basically a self test of a WRA or black box. After the Diagnostic Test is performed on the aircraft and a faulty WRA has been identified, the operator may perform a BIT test while still on board. BIT should be capable of locating a fault within the enclosure to an SRA or module. The WRA can either be repaired by replacing the SRA and retesting it, or the black box can be replaced by another known good WRA.

If BIT is required in the SOW, the Testability Engineer would ask the design engineer how he intends to meet the requirement at the beginning of the design program. If there is a micro-processor in the unit, he would want to know how the designer intends to use BIT using this processor.

BIT usually requires that a self test program be capable of locating a fault to an SRA such as a PC board and in some cases to a group of components on that board. If the WRA is removed from the vehicle it would then go to the "I" level maintenance. BIT would again be performed at the "I" level to confirm the "O" level data by using more detailed testing, and repairs could take place there. If repairs are made, the WRA should be retested again with the Automatic Test Equipment (ATE) to insure it would operate in the vehicle. BIT is only a limited internal test of the electronics. It cannot make an overall performance test to check out I/O connectors, for example, which the ATE can do.

The Testability Engineer would always ask for more test points on PC cards, which are used to aid the technician to trouble shoot and repair the card. The philosophy of the design engineer should be; the design should be made in such a way to facilitate easy testing and to be able to arrive at a faulty component quickly.

For example, if a 2 input NAND gate is part of the design in a series of logic circuits, the engineer may use a 3 input NAND gate instead so that the extra input pin can be used for test purposes. This can be accomplished by using a pulled-up resistor on that pin and a lead can be brought out to the connector to provide an input signal from the ATE when testing the SRA. In other words, the designer may have to add circuit elements to make it more testable. It's going to raise the cost and reduce reliability somewhat but the SRA would then be much easier to trouble shoot.

The BIT test may find a fault to a group of chips or to one chip, depending on the SOW. As was stated before, if the design engineer assumes that he would be the field tester, why not make it easier on himself and provide better test capabilities.

There are several ways to provide test points. Usually they are located on the top of a card so that they can be tested while the card is in its normal slot. Small individual pin jack test points are available that can be put along the top of the board. These are pretty good because one can attach a test lead to it. Terminals are also good to be used because one can clip a test probe to it.

As was mentioned, I prefer to use a connector on the top of a card as test points. In this way I can either probe it or insert a mating connector into it and automatically test the board in production, or using ATE while the card is in the WRA or on a test bench.

Some test points may be required to be placed in the middle of the card. This would allow the technician to be able to probe directly on the card while using an extender board in a WRA. A military card is conformal coated for environmental protection, and probing within the card may mean breaking the coated hard seal which would expose the board to contamination, if it is not re-coated. If test points are available, probing the card and breaking the coating may not be necessary.

The ATE engineer has to provide the software program to be able to test a WRA to detect a fault to an SRA. He also has to be able to test an SRA separately to detect a fault to a component or a small group of components.

For example, in an "I" level test, if the ATE card test displays a fault, it could be any one of four chips and the military technician would replace all four chips. This can be costly and time consuming, so if the designer can arrange his design to be able to test the board to a single component, it would simplify the test and minimize the repair time and cost. It's not always that easy to do.

These considerations are the kinds of things that the Testability Engineer would impose on the design engineer. The EE, ME, and PWE would sometimes resent this additional effort of using extra components, but it may have to be done.

The Reliability Engineer would also squawk at these additions, because adding more components reduces MTBF. The problem with many designs is that the engineer would try to squeeze as much as he can onto one board, and would sometimes ignore test requirements.

On the last ATE program I worked on, the ATE engineers had to provide the software to test a very complicated dual board combination (two boards on one connector) that had components on both sides of each board which had no test points. This board was previously designed by a major firm who did as they pleased in the design, and completely saturated the two board with components. They had a monopoly in the EW field so the government let them have their way.

This is one of the weaknesses with many government contracts. The government would overlook their own testability requirements in the SOW in order to purchase that firm's equipment. There was no consideration in their design to support any ATE. To this day I don't know how that firm tests the cards in production because the same problem must exist with them.

Our ATE engineers could not meet the requirement of getting to a small number of components on several of these dual boards. This meant that many good components would have to be replaced when there is a single component failure. It is sometimes very wasteful in this field.

5. Production

Production interface, is one of the most important requirements there is. Too many programs go through major designs without consulting production personnel. On one program that I was on, the PM decided not to use our company's local production facilities. They selected another production facility in our company that was located in another state, but he did it after the design was completed. This was a bad decision because no production personnel were involved during the design of the equipment.

It became a major problem because our engineers now had to learn how to work with this new production organization after development. As a matter of fact, my company was reprimanded by the Navy for not having a production representative present at all design reviews. The PM had decided to save budget by not inviting our local company's production personnel to the design reviews, because he knew he was not going to use them for production. No one knew about this decision but him. So our production personnel were not invited. This is another case again where non-engineers are making engineering decisions. If I had to go into all the disruptions and problems that followed with this arrangement, it would require another book.

5a. Production Interface

During the design most engineers ignore production problems. Their main objective is to get the design done within budget and on schedule. This is important of course, but to disregard production problems is inexcusable. There are several things that an engineer must consider in his design to assist production. Testability is one consideration as discussed above. Not only does the design engineer have to provide testability for maintenance in the field, but also for production.

The designer has to know what type of test equipment the production facility are using. On blank PC boards, for example, does production do a bed of nails test first for continuity tests and test for shorts? Do they have an automatic wiring tester to check wire-wrap wiring and any hand wiring in the enclosure? Do they use a computer to functionally test the box and what kind? Do they use flow soldering for the PC boards? These questions and more, are the reasons why the designer has to make provisions in his design for production, and he most know their capability. This is where the Liaison Engineer comes into play.

5b. Production Wiring

The ME has to route the wiring in such a fashion to help simplify the wiring in production. He should create a pictorial drawing showing the wiring harness routing. It can't be left up to manufacturing as to how to run the wire harness. This harness is usually part of the I/O requirements in the design and the ME must allow room for the harness in the enclosure, especially in a high frequency environment.

The wiring pin list drawing is another problem. Remember, production would take the engineering wire pin list and recreate it on a separate table for the girls in production. Every time this transfer is done there could be errors. My suggestion is that engineering should automatically create the production wiring tables for them in CADD using their format. In this way there would be no human errors in transition.

Transferring data from one piece of paper to another by hand would almost always cause human errors. The biggest problem that I had in dealing with production, was in making changes in the wiring. My engineer would provide ECOs to incorporate the changes into the next box coming down the line, and he would correct the drawings according to the release cycle. But the problem was, the production manager would red mark the girls tables and the Inspector would buy it off for that one box with the ECOs. But the tables were not updated in production. Unfortunately, after production ended, the marked wire tables weren't destroyed.

A year later, when a new production order came in, the production wire tables did not have the latest revisions incorporated and the girls would wire the box using the old red marked tables. What a mess. If the changes were incorporated automatically on CADD as I am now suggesting, these errors would not have occurred.

5c. Enclosure

When the ME is designing the enclosure, the first thought on his mind should be production assembly. I have seen too many amateurs design enclosures that cause nothing but aggravation in production. In one box that was mentioned before, the whole box was put together by hundreds of screws. Every time one had to repair a PC card in production it took a half an hour just to remove the cover.

In another poor box design the electronics was installed in compartments like a two story home. To get to and inside compartment, all the upstairs components had to be completely removed. Packaging can make or break a company in wasted costs and time when it is poorly designed.

The best approach is to think accessibility. The Production

Interface Engineer has to make sure that the design engineer provides this accessibility. He should attend all design review meetings and try to influence the design for ease of production assembly. It's surprising how little effort it takes for a designer to implement production accessibility.

The example box shown on Figure 5, is what the I consider to be an ideal packaging approach for production assembly and accessibility. In this package all of the internal parts are removable and can be assembled separately. The wire wrap cage is on a separate chassis. It can be wired separately on a Gardener Denver machine and is installed from the bottom.

The I/O wiring and connectors are on a separate panel and can be wired off line. The front panel is on hinges and flips open for access. The two Power Supplies are removable from the top, and all the PC boards are removed from the top. The top cover is held down by nine quick-release Zeus fasteners. The box is welded together by Dip Brazed aluminum sheet metal.

This particular package had the highest compliment from the production manager and from the military as far as assembly, test, and repair. The automatic test equipment of Figure 6, was designed to locate a problem to a PC card on the box of Figure 5, and to a chip in most cases. My company produced over 1,000 of the WRAs of Figure 5 in the 1970's and are still flying today.

5d. Drawings

Drawing procedures and drawing release are the most difficult transition from engineering to production. For some reason or another when the drawings are produced to the latest MIL specification, and have been approved by the check organization, there would always be additional abnormalities in production. Engineering would receive countless calls from production because of drawing problems. They don't fit, or they can't buy the parts.

One major problem is in the development stages. Engineering builds and tests the unit under different conditions. The parts are purchase by engineering buyers in small quantities and are usually available for them in a reasonable time. Not so for production. They have to buy pre-screened parts through production buyers, which means they have to go through shake and bake before delivery. Also, production cannot buy from any outside supplier or machine shop they please, especially if the part has to be made. They generally have to go out for bids.

At one time, I had a Band Pass filter made by a small filter house which I used for development, but later, this company lost the production bid to a lower bidder. The house that finally won the bid ended up not making the same part electrically or physically. They produced hundreds of them before it was discovered that the replacement part was no good.

By the time they redesigned the filters the production schedule was completely out of sight. To solve this dilemma and meet the first production lot, production accepted a set of filters made from the original supplier at a premium price and later would use the new filters from the new company.

There was another situation where there was a drawing dimension error on pilot holes that are used to line the box up for installation on the aircraft. Production caught this drawing error and corrected it without informing engineering. For ten years this went on, until one day that production person, who was involved and knew about the error, retired. The next man, of cause, made the box to print and sure enough it didn't fit. Talk about embarrassment. I'm glad I wasn't the EDM on that job.

Engineering needs feed-back when there is a production problem. A production interface engineer or LE may have caught this and reported it back to engineering, but no one had been assigned. It's not known how that individual passed the box through inspection with that error over all those years. It's scary. I wonder how many other errors slipped through over the years?

Production has many more different requirements to consider than engineering. They are governed by so many manufacturing procedures and rules it is sometimes a wonder that anything can be produced. This is why a PE (Production Engineer) working with a LE should be involved in the development stages and to attend all design reviews. However, this production interface engineer must understand that the final design decisions are up to engineering, and he may have to give in a little in some instances. One can't design a box for production requirements alone. There must be a lot of give and take.

There are other considerations to be made, such as vehicle installation, ATE, and operations. As was said before, each department sees only their own problems, and do not know the effects his inputs would have on another support department's requirements. The EDM has to referee this confrontation between departments and he has to make the final decision or use the best compromise.

Note: Sometimes the LE and the PE could be the same person.

6. Quality Assurance

Quality Assurance provides a major task in industry. There are several special classes that may be taken in this field. This subject could even be a major course at a University, that is if all of the parameters were known. Unfortunately, like so may other engineering subjects, it is not taught in either under graduate or graduate classes. It is a specialized field and most of the people in this field have learned this trade on the job.

The object of Quality Assurance is to ensure that the product is delivered with the finest quality. On one program that I was on, Quality

Assurance was intentionally not invited to attend the design reviews by the EPM to save, again, budget. The Navy later reprimanded our company for excluding them. The Quality Assurance Chief engineer read the riot act to the PM and the EPM as a result of this exclusion. It is sometimes mind boggling how these PMs or EPMs get away with these bad decisions.

To ensure that the company delivers the finest quality equipment, Quality Assurance would go over the drawings to make sure engineering has selected the highest quality material and components. They would examine the workmanship and test procedures. They may want to be present during environmental testing and may ask for the qualification report.

6a. Inspection

Quality Assurance is also responsible for inspection of the unit for performance. Unless it is not specified in the SOW, Quality Assurance would use their own plant inspectors or from any other facilities to inspect Functional Test. Usually the company inspector works together with a government inspector on the first production box and then he would turn over the responsibility to the company's inspector.

The following is an example of what may be required during inspection:

6a1. QA Standard Test Conditions

The standard condition for the performance tests are usually at room ambient temperature conditions of:

o Temperature: +23 degrees C
o Altitude: 0 to 5,000 feet
o Humidity: 50% RH

6a2. Visual Inspection

o Cable Set
o Test Fixture
o Electrical
o Complexity
o Optimization
o Expansion
o Accessibility
o Drip Proof
o Handles
o External Connector
o Strain Relief
o Design and Construction
o Contacts
o Parts and Material
o Coding
o Electrical Connectors
o Spacing
o Internal Connectors
o Quantity
o Protective Caps
o Circuit Cards
o Connector Keying
o Finishes
o Terminal Blocks
o Color
o Buffering Devices
o Lacquer
o Interchangeability
o Safety
o Protective finish
o Grounding
o Human Engineering
o Equipment

6a3. Inspection by Analysis

o Performance Characteristics
o Performance Testing
o Fault Isolation Testing
o Test point Interface
o Complexity
o Useful Life
o Operational Service Life
o Storage Life
o Mean Time Between Failure
o Maintainability
o Mean Time To Repair (MTTR)
o Transportability

6a4. Inspection by Tests

o Definition
o Performance Characteristics
o Performance by Interface
o Electrical Incompatibilities
o Self Test (BIT)

6b. Environmental Tests (Typical)

6b1. Low Temp. test -55 degrees C

6b2. High Temperature test +55 degrees C and +71 degrees C for one half hour.

6b3. Altitude - 40,000 feet

6b4. Vibration - Sinusoidal 5 G's

6b5. Humidity & Salt Spray

6c. Quality Conformance Tests

An acceptance test shall be performed on each deliverable equipment at a level sufficient to assure that the equipment satisfies the Functional Test requirements.

6d. Manufacturing Screening

Each equipment shall be subjected to manufacturing screening consisting of random vibration and burn-in as specified in the SOW. The equipment shall be subjected to functional tests as required.

6e. Components Re-screening

All semiconductors, micro-circuits and non-standard passive components shall be subjected to re-screening inspection in accordance to the SOW, before they are installed in the equipment.

7. Reliability

Reliability is a very important function in the design of equipment. The Reliability Engineer should be present at all design reviews to provide his expertise, and ensure that the equipment would meet or exceed all reliability requirements stated in the SOW. When the program begins, the Reliability Engineer has to implement and submit a reliability program to the customer to meet the specifications called out in the SOW and MIL-STD-785 or other applicable Standards. Typical Reliability Requirements are as follows:

7a. Reliability Program Plan

The contractor shall prepare and submit for approval a reliability program plan in accordance with Task 101 of MIL-STD-785 as modified by the SOW.

7b. Reliability Indoctrination and Training.

The contractor shall conduct appropriate indoctrination and training for personnel assigned to this program to ensure a good understanding of how important the achievements of the reliability requirements are to the success of the equipment in the field.

7c. Records and Reports

The contractor shall maintain verifiable objective evidence that all program tasks have been performed in accordance with the specified requirements.

Contractor internal records and reports such as: engineering logbooks; charts, material and processes selection criteria; trade-off data, failure data/reports, etc., shall be maintained in the Contractor's format and made available for review upon request.

7d. Fabrication Techniques

All parts and circuitry used in the fabrication of the equipment shall be derated in accordance with the requirements of AS-4613. All non-compliances to this derating criteria shall require written approval.

7e. Reliability Analysis

The contractor shall prepare and update as necessary, reliability prediction reports for the equipment in accordance with MIL-STD-785 and MIL-STD-756. The Contractor shall conduct the reliability analysis/prediction at the piece-part level, assuming continuous operating at worst case conditions. The reliability prediction shall be used by the Contractor as a tool to improve the overall reliability of the equipment, by using it to direct efforts to simplify/reduce component count and circuitry to obtain maximum reliability. Failure rates shall be approved by the procuring activity prior to use. Failure rates submitted for approval shall be fully substantiated by the Contractor with back-up data.

Reliability is an another example that illustrates the problem of how one support department affects the requirements of another department. The Reliability Department always wants to reduce the number of components for a higher reliability, and yet, the Testability Department wants more circuits for BIT and test points for ease of testing. It's a battle that only the EDM can resolve. If "push comes to shove" the reliability person would win, especially if the design exceeds the reliability criteria.

7f. Prediction Report

The prediction report shall provide failure rate predictions at the piece-part, assembly, and end item levels. It shall also document the methodology in performing the prediction, describing the approach used to calculate the temperature, stress ratio's, etc. A sample failure rate calculation shall be included.

7g. Component Re-screening

A Functional check shall be performed on all active micro-electronic or semiconductor devices at +125 C, -54 C and at +25 C.

7h. Program Reviews

Reliability review activities shall be combined with the regularly scheduled Program Reviews as specified in the SOW. Reliability status at the reviews shall be in accordance with MIL-STD-785.

7i. TAAF

A Test Analysis And Fix (TAAF) philosophy shall be instituted at the earliest stage of fabrication. It shall be the Contractor's responsibility to identify and analyze all deficiencies, including self-test anomalies, and to implement corrective action to improve the overall reliability of the equipment. During the formal testing requirements and the informal testing planned, the Contractor shall keep a log of all failure-discrepancies, analysis findings, and corrective action. Plans for implementing the TAAF philosophy shall be detailed in the Reliability Program Plan.

7j. FRACAS

Failure Reporting And Corrective Action System (FRACAS) - All failures occurring during pre-production and manufacturing shall be analyzed and corrective action as appropriate shall be instituted to achieve maximum reliability and self-test effectiveness. Corrective action shall consist of modifications to the design, changes in parts, material, manufacturing processes, etc., in order to reduce deficiencies and the probability of failure occurrences. The Contractor shall institute a FRACAS in accordance with MIL-STD-785.

7k. Soldering

The Contractor shall implement soldering methods, a training program, and an overall soldering system in accordance with WS-6536.

7l. Printed Wiring Boards

The requirements of MIL-P-55110 and MIL-STD-275 apply.

7m. Manufacturing Screening

During production, all end item equipment shall be subjected to manufacturing screening in accordance with the guidelines established in NAVMAT P-9492 and as specified below. These test have nothing to do with Qualification. They are simply production screening tests to eliminate weak components and manufacturing errors using the following two "Shake and Bake" requirements.

7m1. Temperature Cycling

The equipment is temperature cycled to detect weak components in manufacturing of a box before delivery. It is better to find out if there is a weak component ahead of time in the production stages, rather than in the field. The temperature shall be cycled from -54 degrees C to +55 C. The rate of change shall not be less than 10 degrees C per minute. The equipment shall be energized, and operated during temperature cycling, except the equipment should be turned off during chamber cool-down to permit internal parts to become cold. The test shall be completed after accumulating a minimum of 10 temperature cycles, the last four being failure free.

7m2. Random Vibration

Random Vibration in production is used to determine if there are any loose solder joints in the equipment. Even though the unit was inspected, the naked eye cannot always detect a poor soldering joint. So, by shaking the unit vigorously poor soldering joints and connections can be minimized before production delivery. The Random Vibration shall be at least 6 G's RMS and in accordance with the random vibration spectrum called out in the SOW. The duration of random vibration shall be at least ten (10) minutes perpendicular to the plane of the circuit boards.

It must be noted that these burn in-tests have now subjected the unit to hash treatment which would reduce the MTBF. The military, however, is willing to take that risk.

7n. MTBF

Mean Time Between Failure (MTBF) is a requirement imposed on the Contractor that the equipment shall have a warranty period and any repairs made during this period would not be charged to the customer. Each unit shall have a minimum MTBF of 2500 hours (typical) and shall be calculated in accordance with MIL-HDBK-217.

7o. Useful Life

The equipment shall have a useful life of ten (10) years under any combination of operating and storage conditions without exceeding the operating service life. Note: Some of the boxes that were designed in my organization are close to 40 years old and are still operating. I retired in 1990 and many of our avionic boxes are still flying on Lockheed's planes including the P3 and S3 aircraft.

7p. Service Life

The equipment shall have an operating service life of ten thousand (10,000) hours under the operating environmental conditions specified in the SOW with normal maintenance.

7q. Storage Life

The equipment shall have a storage life of thirty (30) months under the environmental storage conditions as specified in the SOW without parts replacement or unscheduled maintenance.

7r. Electronic Parts Derating

The Contractor's circuits and equipment designs shall provide stress derating of all electronic parts. This derating shall encompass the voltages, current, fan-out, power, temperature, mechanical stress, and duty cycles. The Contractor shall conduct a part level stress analysis to verify derating compliance. This analysis shall verify that all parameter deratings are compliant under the conditions of continuous operation. These derating criteria shall be at least as restrictive as the example specified in Table 2 that follows.

For example, the working voltage of a Ceramic Capacitor shall be derated by 52% of that parts maximum rating at +25 C. The equipment maximum operation temperature is +55 C. Derating Factors are in accordance with AS-4613. RADC is just another derating guide. The following Table 2 illustrates component derating factors.

TABLE 2 COMPONENT DERATING FACTORS

PART TYPE	MIL-SPEC	PARAMETER	DERATING FACTOR	MAX PART RATING
		CAPACITORS		
Ceramic	MIL-C39014	Work Voltage	52%	at 25 C
Tantalum Solid	MIL-C39003	Work Voltage	43%	at 85 C
		Ser Resistance		Minimum 3 ohm/V
		Rev Voltage		1 V Max
	MIL-STD198	Ripple Voltage		MIL-STD
Mica	MIL-C5 MIL-C39001	Work Voltage	48%	at 25 C
Variable	MIL-C14409	Work Voltage		RADC
Tantalum Non-Solid	MIL-C39006 /22	Work Voltage	42%	at 85 C
Metal Plastic Film	MIL-C83421	Work Voltage		RADC
		RESISTORS		
Carbon	MIL-R39008	Power	50%	at 70 C
Wirewound	MIL-R39007	Power	35%	at 25 C
Wirewound	MIL-R39015	Power	45%	at 85 C
Accurate Wirewound	MIL-R39005	Power	33%	at 25 C
Insulated Film	MIL-R39017	Power	60%	at 70 C
Metal Film	MIL-R55182	Power	60%	at 70 C
Variable	MIL-R94 MIL-R39002	Power	30%	RADC at 45 C
Composite	MIL-R39035	Power	32%	at 85 C

TABLE 2 (CONTINUED)

PART TYPE	MIL-SPEC	PARAMETER	DERATING FACTOR	MAX PART RATING
TRANSISTORS				
Silicon Bipolar	MIL-S-19500	Power Current Voltage Junction Temp	35% 75% 60% Derate Max	at 25 C at 25C of Vceo 110 C 179 C
Silicon Power	MIL-S-19500	Power Current Voltage Junction Temp	35% 75% 60% Derate Max	at 25 C at 25 C of Vceo 110 C 179 C
Mosfet	MIL-S-19500	Power Junction Temp	35%	of 25 C 110 C
THYRISTORS				
SCR TRIAC	MIL-S-19500	Forw Current Gate Current Voltage	33% 50% 60%	RADC
DIODS				
Silicon General	MIL-S-19500	Junction Temp Current Rev Voltage	 35% 60%	110 C
Silicon Power	MIL-S-19500	Power Rev Voltage Forw Current	35% 60%	
Zenor	Mil-S-19500	Junction Tenp Current	 39%	110 C
Voltage Reference	Mil-S-19500	Junction Temp Current Power	 39%	110 C RADC
Varactor	Mil-S-19500	Junction Temp Rev Voltagfe	 60%	110 C
CONNECTORS				
Circular Rectangle	Mil-C-38999 Mil-C-15305	Contact Cur. Operat Voltage	50% 50%	of Rated

TABLE 2 (CONTINUED)

PART TYPE	MIL-SPEC	PARAMETER	DERATING FACTOR	MAX PART RATING
INTERGRATED CIRCUIT				
Dig TTL	MIL-M-38510	Supply Voltage Output Current Sink Current	+5 to -5 70% 70%	Fanout
Dig CMOS	MIL-M-38510	Supply Voltage Output Current Sink Current	-2.5V +1.5V 80% 80%	of Max of Min Sink Drive
Linear Amplifier	MIL-M-38510	Supply Voltage Diff In Volt	80% 60%	at 25 C
Linear Regulator	MIL-M-38510	Line Voltage I/O Dif Volt	80% 150%	 of Min
RELAYS & SWITCHES				
All	MIL-R-39016	Coil Voltage Capacitive/ Contac Current Inductive/ Contac Current Motor Contact Current Filament Cont Current	10% 75% 40% 20% 10%	of Norm
COILS, CHOKES, AND TRANSFORMERS				
Coil, RF	MIL-C-39010	For all transformers and coils use Max. allow formula for Hot Spot Temperatures HSP = derated -40 C		
Wirewound	MIL-R39007			
Inductors Saturable Reactor	WS-1612			
Transform Audio/Pwr	MIL-T-27			
Transform RF	MIL-T-6531			
Transform Pulse	MIL-T-21035			

8. Maintainability

Maintainability is another important requirement that is imposed on Contractor's equipment. Testability is somewhat related to Maintainability. To maintain a piece of equipment it should be tested at the "O" level first, then at the "I" level, and finally at the Depot or at the manufacturer's plant.

Although the Maintenance Engineer is concerned with the testing, he is also concerned with how to repair the unit. He wants to know how the unit is installed and removed within the vehicle. Are the PC boards accessible and easily replaced at the "O" level? When the unit is at the "I" level can it be disassembled, repaired, and assembled again easily and within a specified period of time? He would want to know if bad connectors and wiring can also be easily replaced.

His job is to make sure that the design engineer incorporates the maintenance philosophy in accordance with the SOW. Because of this responsibility, the Maintenance Engineer should attend all design reviews or provide a substitute. Sometimes, the Testability Engineer and the Maintenance Engineer are one and the same person, depending on the size of the program and the company's structure. The following examples illustrates what is included in the overall maintenance program:

8a. Maintainability Program Plan

The Contractor shall implement a maintainability program to meet the requirements of MIL-STD-2084. He shall prepare a plan in accordance with MIL-STD-470 and as modified by the SOW.

8b. Maintainability Indoctrination And Training (MIAT)

The Contractor shall address and conduct appropriate MIAT for personnel assigned to this program to ensure a good understanding of how important maintainability is to the success of this equipment in the field.

8c. Maintainability Predictions

The Contractor shall prepare, and update maintainability predictions in accordance with MIL-STD-470.

8d. Program Reviews

Maintainability review activities shall be combined with regularly scheduled Program reviews as specified in the SOW in order to be more efficient. Maintainability status presented at the reviews shall be in accordance with MIL-STD-470.

8e. Subcontractor Control

The Contractor shall monitor and control subcontractors in accordance with MIL-STD-785.

8f. Physical Design

Interchangeability and accessibility shall be in accordance

with MIL-STD-2084 and MIL-I-8500. The equipment shall be designed of MIL-STD-2084, MIL-STD-2076, "Type I Maintainability" of MIL-T-28800, and "Design for maintainability" of MIL-STD-1472. The equipment shall be designed such that calibration and scheduled maintenance is not required. All circuit cards shall be identical in size.

8g. Fault Isolation

The equipment shall be capable of fault isolation with BIT to the following:

8g1. 3 or less SRAs for 100% of the faults

8g2. 2 or less SRAs for 95 % of the faults

8g3. 1 SRA for 90% of the faults

8h. Mean Time TO Repair (MTTR)

The MTTR of each unit, including time to run BIT and repair time, shall not exceed thirty (30) minutes.

9. Electromagnetic Interference (EMI)

The EMI organization is responsible for several fields; Electromagnetic Emission, Susceptibility, TEMPEST, Ground Loops, Oscillations, Atomic Armament, and Electro-Static Discharge (ESD).

The EMI engineer is not only responsible for the equipment's EMI but the EMI of the complete vehicle. Most companies have their own screen room to perform EMI testing of the equipment. If the company does not have this capability it would have to subcontract the test to be done by a qualified EMI house.

EMI is part of qualification, but is separated here because it requires special test equipment to be tested under RF conditions. It is not considered an environmental test which is more of a physical test.

The EMI engineer is a special type of a person. He has to have had a lot of experience in communications in RF and micro-wave technology. This is one field that I feel is not a science in the practical sense, but is more of an art. It's a mystical field, where no one knows where the interference is coming from but it takes very sensitive test equipment to detect it. How does one design in a fourth dimension?

On my last job our EPM's upper-manager asked me if it was wise to pay for an EMI test on a modified box and do I think it would pass. I told him flat out, it probably wouldn't pass because I have never passed a box through EMI on the first try. A design engineer may think he has done a fine job on his enclosure for EMI considerations and, no matter how careful he is, sure enough it fails EMI testing on the first try and maybe even the second.

When an engineer starts a new program, one of the first things he should do is make contact with the EMI organization even before the first "kick off" meeting. Get his inputs as soon as possible. The packaging is so dependent on EMI filtering, gaskets, etc. I have seen engineers that would `over' design the enclosure for EMI purposes and to find out later that all his special compartments and EMI gaskets were not necessary. Then again, I have seen the opposite approach, where an engineer refused to provide special EMI filtering in the power lines and later had to completely modify the enclosure after he had failed EMI testing.

First of all, one must know from the SOW what the EMI requirements are and where the equipment is to be used. If the box is in a fighter, for example, one can bet it would have a high EMI requirement. If the box is for test equipment, the EMI requirements may not be as stringent. But, if that equipment is to sit on the deck of a Aircraft Carrier and would be subjected to the ships RADAR, the box is in trouble from radiation from the RADAR before the design begins.

If the unit is required to handle secure data or voice, the engineer is really going to have his hands full. This is where TEMPEST comes into play. How does one meet TEMPEST is one of the most difficult EMI tests to pass? Also, if the engineer is dealing with Atomic Armament he had better know his business or look for another job. The whole design may be based around EMI as the primary requirement.

When one is dealing with EMI the "glitch" word is often used. A "glitch" by definition, is caused by a high speed pulse that is generated by a circuit and picked up by another and may cause the other circuit to be enabled or disabled. Cross-talk is similar, except it is information that is pick up from one channel to another channel. The "glitch" and crosstalk are usually a design engineer's nightmare and he would have to solve them himself. The EMI engineer is mostly concerned with radiation, susceptibility, and ground loops that may cause oscillations and not so much with `glitches' or crosstalk. He may make recommendations, however.

9a. Electromagnetic Emission

Electromagnetic Emission is the radiation of the Contractor's equipment to such a level that it degrades the operation of adjacent equipment or causes the vehicle to exceed it's EMI limit.

The equipment shall operate satisfactorily in a Government active environment, and shall not be degraded nor degrade the performance of other equipment installed close by. The equipment shall be designed to comply with MIL-STD-461 and the SOW.

9b. Susceptibility

Susceptibility is the interference from another source that would degrade the performance of the contractors equipment. Susceptibility can be radiation from another box, from a Radar, or through the power lines.

9c. TEMPEST

TEMPEST is a special type of interference that has to do with security. Voice and digital data can be picked up by the enemy through telephone lines, from radio transmissions, or from vehicle radiation. This is a very difficult test to pass. If the equipment is part of the communication system and is used in a vital environmental situation, TEMPEST requirements may be imposed on the design.

In general TEMPEST is more of a crosstalk or pick up problem. If the pilot is transmitting secret information or an officer is telephoning on a secure red line, then the other lines, or black lines in the system should be clear or it cannot be detected below a set level. The lower test limit that is usually imposed on the equipment is that the black channel shall not be able to detect the information from the red line by at least -120 db down or no greater than X db above the noise threshold in the black line. The X level has to be established in the SOW.

I don't agree with this -120 db number in an airplane because it would be in the air during red transmissions and any enemy intercepter wouldn't be close enough to the aircraft to be able to pick it up. But, that's how the TEMPEST Specification reads. There are other variation to these requirements but this should provide the reader with the gist of what TEMPEST is mostly about.

One ridicules requirement that was specified in an SOW was, we had to meet TEMPEST up to 1 gHz for voice. This is another example of a "boiler plate" specification that didn't apply to our box and should have been negotiated out of the SOW in the beginning. After the fact, it took a lot of time and traveling to Washington to convince the civil service people that this requirement didn't apply to our box. It was difficult to convince them because we were dealing again with non-engineering types.

One way TEMPEST is measured on an aircraft when it is on the ground is, red data that was radio transmitted to a secure receiver is measured at the output of the black transmitter located a short distance away from the vehicle using a sensitive radio receiver.

In one design I was involved with, we couldn't get the TEMPEST level below the -120 db requirement even in the Lab at our plant. The best we could do in the Lab was -110 db. But when it was measured on the aircraft, the measured noise level of the equipment was already at -90 db which was above our TEMPEST signal level, so the red signal was undetectable by their radio receiver and we passed.

I jokingly told the PM, who was with me at Pax River, that if we didn't pass the test, I would introduce a noise generator into the black channel to muffle out the red signal. That would have been a little underhanded of me but it certainly would have worked. Generally, black voice or data that is 40 db above the noise is acceptable so why meet a -120 db TEMPEST level? I could have set my noise generator at -60 db and no one would be the wiser. I was told that I sound like a typical designer, always looking for an advantage.

9d. Ground Loops

Ground Loop oscillations are another one of those engineering philosophies that are not taught in class and one has to learn it from practice. A simple explanation of what a ground loop is, is oscillations that occur in the box when there is more than one ground path for the signal to return back to the source. It's a known fact that ground loops do cause unwanted oscillations. It is preferred that there is only one ground return path. This usually applies to the wiring external to the box but it can also happen within a box.

The philosophy that I have applied to minimize ground loops is to consider the grounding system as the spokes of a wheel without a rim. The hub is the one point ground and the spokes provide the ground connection to each PC card or to black boxes in a vehicle that seem to be floating in space. Only one ground pin of each PC card or box is returned back to the hub.

The motherboard, for example, in a box could be considered as the hub. In this way there is no way a signal can be looped back to the hub through an extra ground lead which could cause oscillations. RF signals, however, are treated differently where every circuit goes directly to a ground plane or shield to prevent radiation. The designer has to be aware of all of these problems and know how to handle them.

9e. Atomic Armament

Atomic Armament only applies to specific aircraft that are to be used in this fashion. The testing of the equipment for this application is secret, but as a designer, one must be aware of the consequences in this environment. A single part failure shouldn't cause an accidental launching of a missile, for example.

This is when the designer has to ensure that it would take at least two or more independent component failures within his unit and not on the same board before there is an inadvertent firing of a missile or the release of a bomb. Also, the inadvertent firing shall not be caused by EMI or a "glitch." The electronics has to be fool proof with plenty of back up safety features. Generally, for additional assurance, more than one red switch is required to be thrown by different operators before the sensitive electronics is even activated.

9f. Electro Static Discharge (ESD)

ESD is generally considered as part of the handling of Integrated Circuits, but it also applies to bench testing and chassis grounding. ESD is no different than the static discharge that one gets from walking on a rug which creates a spark when that person touches an object or someone else.

The test technician must be grounded to the frame of the test bench by using wrist bands that are connected to the bench. The equipment must be properly grounding so that when a module is removed it would not cause an ESD spark in the box which could blow out several chips. The EMI engineer would recommend certain types of ESD cable covers designed to control ESD and fool proof grounding techniques before power is applied to the equipment.

9g. EMI Packaging

Packaging of the equipment for EMI is essential for reducing interference. There are several basic applications that should always be applied to the design whether there is an EMI problem or not.

o All input power lines should have EMI filters.

o The enclosure should have no open ports or screens.

o Covers should make electrical contact with the chassis.

o One point grounding to the chassis.

o No ground loops.

o Proper filtering on each PC card and the motherboard.

o Proper filtering within the Power Supply.

When EMI becomes a major problem then additional considerations must be implemented. EMI gaskets may be required on all loose covers and on screwed down plate assemblies. EMI filter capacitors may be required between modules. The modules may have to be enclosed in metal shielded containers. PC boards and motherboards may require guard shielding. Compartments may have to be installed between Red and Black electronics with double walls. Internal harnesses may have to be coax or shielded wiring. The I/O connector may require EMI back-shells. Coax cables with proper connectors may be required.

Double relays isolation techniques and special optical isolators may be required in the electronics. Each relay provides 40 db of isolation between contacts at audio frequencies. The double contact approach using two relays would improve it to 60 db. Optical isolators can provide over 100 db of isolation. Fiber optics may be the best solution to prevent EMI, especially on the cables between boxes. Encoded Bus systems may provide security because they can be scrambled.

There a lot of options that the EMI and Design Engineer can employ for EMI, but they should work together as soon as possible to establish a plan of attack. Unfortunately, most of the EMI problems are resolved by trial and error.

The designer should go as far as he can with the above recommendations, so that later, when the EMI engineer tests the unit in a screen room he can finesse the design and recommend modifications to meet the EMI requirements.

Like I said, "EMI is more of an art than a science." When one considers EMI he might visualize an enclosure having internal gases trying to escape threw the cover, like a creeping crud. That's the way RF may seep in or out of a box.

If fans are used for cooling, the designer cannot use DC fans with commutators because they arch and produce a lot of interference. A DC fan with field effect commutation, however, is acceptable. AC fans with slip rings are, of cause, acceptable.

10. Thermal

The Thermal Department has one of the highest priorities in supporting the design of military electronics. Heat especially, causes more component failures than for any other reason. As was mentioned, the accumulated reliability data of military electronic over the last 30 years has shown that a component's life can be predicted by its heat dissipation. The cooler the device is the longer it would last.

The problem today is, new systems require higher and higher speeds which are mounted in smaller and smaller IC packages. The higher the speed, in general, the hotter the chip would be. The more compact the ICs are, the hotter the internal parts would be. Surface Mount ICs with Large Scale Integration (LSI) would generate a lot of heat.

The Thermal Department should be involved with the program right from the start and a Thermal Engineer should attend many of the design reviews, especially the early ones. How much involvement he requires depends on the application of the equipment, the speed of the electronic chips and the clock frequency.

The various packaging techniques I discussed in Chapter I, to help reduce heat, won't be repeated here. One important fact that was mentioned, was that the thermal approach should not come before the electronic packaging approach. There are many ways to tackle the thermal design. The design engineer should not be straddled with aircraft thermal ideologies before he even begins his design or he may use it as a `crutch.' The Thermal engineer should first oversee the designers approach and then provide his inputs and requirements.

What was discussed before was a situation where the thermal approach, for the electronics on an airplane, was to use cold plate cooling with air being pumped through it from a central source. All the suppliers took advantage of this cold plate and reduced the cost of their designs, but now they had to depend on air flow through the cold plate at all times.

What resulted in the end was a series of boxes that could not operate without the aircraft powered up and the blowers going. This became a problem in ground testing and trouble shooting where ground power was not always available. If one gives a supplier a thermal crutch up front, he would certainly take advantage of it.

Indeed, there are some cases where a thermal approach has to be established before the design begins. One example that illustrates this is in the Advanced Tactical Fighter program. In this case, each supplier module is now considered a WRA which are mounted together in a special rack. The initial cooling requirements of this system was that the junction temperatures of VISIC Integrated Circuits cannot be greater than +60 C in a +55 C environment.

The only way that this temperature ceiling could be met on a chip, is by using refrigeration cooling in the rack for each sub-assembly. So some sort of cooling approach had to be established first. I believe the Air Force finally reduced the temperature requirement because refrigeration would be too heavy.

After the Thermal Engineer reads and interprets the SOW he has to get together with the design organization to see how they plan to design the package. This interface could even come before the first "kick off" meeting to make sure the designer and the Thermal Engineer agree on the initial thermal philosophy.

After this meeting and a few more design reviews the Thermal Engineer would prepare his theoretical thermal analysis report which would be presented to the military during the customer's design review.

This report would consist of junction temperatures on all semiconductors and ICs that are in operation. It would contain power dissipations of all components. It would recommend the amount of air flow if fans are to be used. The report would show the heat transfer analysis from radiation and conduction.

At this point, when all parties agree on the thermal philosophy, the Thermal Engineer may not be invited again to the design reviews until sometime after the prototype hardware is built and qualified.

A Thermal Imaging should be performed on the first prototype (XN1) box to determine if there are any hot spots on the PC cards and how close the thermal analysis compares to the actual conditions. The actual data would be analyzed by engineering and the Thermal Department. If modifications are required they would be incorporated in the next XN2 box. Hopefully, if the EE and the Thermal Engineer have done their homework, modifications would not be necessary.

As was mentioned earlier, theoretical analysis doesn't always agree with the actual data because there may be losses in the heat transfer on a PWB and transition points. One can only go so far with theory because there are so many unknown parameters in trying to predict the losses in heat transfer.

Junction Temperature, on semiconductors and integrated circuits, is the only measured parameter that is accepted throughout industry as the way to measure the life expectancy of these components. One cannot use the Case Temperatures even though some IC components have little losses between case and junction but some have quit a bit. The case to junction delta variation could be 0.3 C to 20 C.

The example shown on Table 3 on page 141, illustrates the actual measured values of Junction

Temperatures of a PC card that is still being used in the field today. The temperatures are based on an ambient temperature of 55 degrees C and the PC card was air cooled in its enclosure. The Maximum junction temperature allowed or derated value is 110 C. The Maximum manufactures Junction Temperature is 179 C.

COMPONENT IC's	DELTA T DEGREE C CASE TO JUNCTION	COMPONENT-TEMPERATURE CASE	JUNCTION
54F00	2.5	83.4 C	85.9 C
54F04	3.7	83.4	87.1
54LS21	0.3	83.4	83.7
54F112	5.0	83.4	88.4
54F253	9.1	86.4	95.5
MC10504	5.8	95.4	101.2
MC10524	13.5	98.4	111.9
MC10525	13.2	98.4	111.6
MC10541	11.7	98.4	110.1

TABLE 3. IC JUNCTION TEMPERATURES

As shown in Table 3, there are two chips that are slightly above the derated Junction Temperature requirements of 110 C; the MC10524 and MC10525, but they are well below the chip manufacturers maximum Junction Temperature of 179 C (not shown). These temperatures are based on the unit being in an environment of +55 C ambient continuously. So, the reliability of these two chips, as shown, are only considered fair. It is not known what the temperatures would be with air flowing in it's normal enclosure.

The main purpose of this table is to illustrate the procedure by which a Thermal Engineer can monitor and rate the ICs in the design. The design engineer should also take these parameters into consideration during his PC card thermal design. A Thermal Image was run on a card mentioned before at room temperature and it was found that there was a temperature reduction of 13 degrees C from no air to with air. This indicates that this unit must have air flow when used under it's normal environmental conditions to meet MIL-Specs.

The following Stress Factor (my own approach) may be used as a guide in determine the reliability of a Chip:

STRESS JUNCTION FACTOR TEMPERATURES

None Below to 59 C

Low 60 C to 75 C

Medium 76 C to 90 C

High 91 C to 110 C

Very High 110 C & Above

My recommendation is, try to keep the Junction temperatures of the Chips in the Medium range area or below to maintain the highest reliable box. Even if the SOW derated Junction Temperature is 110 C, I still recommend using a lower derating temperature factor when designing electronics to ensure even the highest possible reliability.

As was mentioned several times, (MTBF) is the most important factor in military designs. Design cost is probably the fifth most important factor, but then a reliable box would reduce the maintenance costs in the field and the overall savings would be less. Also, one cannot put a price tag on continuous and guaranteed operation, especially in a combat environment. For clarification, I use the word Chip with reference to both Semiconductors and Integrated Circuits.

11. Weights

My company spent two years on the internal design of a special aircraft using sophisticated subcontracting equipment. The aircraft was a very large plane and, for some reason, no one considered making a weight analysis during the design phase. The Weights Department were not included in the early part of the program to save budget. But then it was too late. When the Weights Department were finally asked to do a weight analysis, their conclusion was that the plane could not fly effectively because the equipment on board exceeded the aircraft pay load. It was now panic time to eliminate many black boxes.

I was the Project Engineer for the mission recorders to be used on this experimental aircraft. When I was given the responsibility for the ships recorders, I was told to use 3 ship-borne recorders weighing 200 lb each because there were many available in the Navy's inventory. I couldn't understand this because the same company that made the ship's recorder also made a miniature flight version weighing 40 lbs for the Air Force with somewhat less capability.

There are three reasons this recorder was selected. First the system's engineer wanted more capability. Second the recorder would be GFE so the cost would not be charged to the program, and finally using an Air Force recorder in a Navy program is taboo. Military politics again...? So in the end, only one ship-borne recorder was used and we had to sacrifice capability in the program to help reduce the weight.

The point I'm making is, the program was designed by the "gun shot" method rather than properly planned. This is what happens when an organization is not set up properly and many support departments are left off the program to save costs. Had the Weights Department been involved early, even on a part time basis, they would have caught this overweight problem immediately. The program was eventually scrapped by the Navy for this and other reasons, including cost overruns.

I am using this example to illustrate again, how the support organizations can influences a program. One does not eliminate a support department from the program simply to reduce the cost. The Program Manager and the Engineering Project Manager in conjunction with the System Engineer must establish who the players are and present it to the Vice President and the Director of Engineering for approval. It cannot be just a one man decision.

On ship-borne electronic equipment, weight is usually not a factor. As a matter of fact, ruggedness is the most important factor on a sea going ship. The Weights Department are usually not included on a ship-borne program for this reason.

As far as lifting the equipment by an individual is concerned, it is the responsibility of Human Factors Department. A 200 pound recorder would even be too much for two little girls to carry.

On airborne equipment, weight is critical. Every pound of weight that can be reduced means more fuel or more payload. On any aircraft development program the five most important goals are Performance, Weight, Cost, Quality, and Schedule, and not necessarily in that order. These goals are all interrelated and any one of these functions can adversely affect the other.

Improving the Quality may cause the weight to go up, for example. As was mentioned, when I designed the box of Figure 5, I had a conflict between the Weight Engineer and the Stress Engineer for the bottom cover. The Stress people wanted it beefed up and weights wanted it lighter. Stress won out because, on the vibration table, the cover "oiled canned" and shorted out the pins of the wire-wrap connectors.

Never the less, these types of conflicts must go on. Each organization has experts in their own field and they have to submit their inputs for evaluation. I would rather have my design criticized by all departments, rather than have a passive engineer who is afraid to "rock the boat." We can always come to a satisfactory solution.

In a package design, the ME has to consider how to keep the weight down. My recommendation before, was to use a Dipped Braze enclosure with a Motherboard backplane similar to that was used in Figure 5. Cast Aluminum can be pretty heavy and costly. Rivets and screws for sheet metal enclosures add weight to the box, not to mention increased production costs. I also preferred to use round connectors as shown in Figure 5 versus DPX types. Round connectors are easier to install and they are much lighter than the DPX type. DPX connectors have problems mating together unless they are screwed down properly at each corner.

Some people prefer to use Flex Cable for the mother board which connects directly to the I/O DPX or round connector. This does save weight but other parameters must be considered; such as assembly, repairability, and cost. Every decision must be negotiated to produce the best and the highest quality box.

In the kick off meeting a Weight Engineer should be invited. Later on, at the time the enclosure is in mock-up the Weight Engineer should be invited again at that design review.

Before the first box is assembled, the individual piece parts should be weighed. The weight for the metal is easily calculated and added to the weight of all the parts. This information should be forwarded to the Weights Engineer early. After the box is assembled the box should be completely weighed. The Weights Engineer would again be invited to the next design review and probably for the last time if he is satisfied.

When one considers the minimum manhours that the Weight Engineer is needed for his support, it doesn't pay to leave him off the program. An EPM once said to me that the Weight's Department charged too much in their quote. If this was truly the case, he should have challenged them about their quote and not just cancel them. It sometimes takes a little bargaining to get the right price, but that's what the EPM's job is, to make sure that all support departments don't over bid.

Weight Engineering are important in the design of airborne

equipment. They have to consider the weight of the entire aircraft and all it's WRAs so they would know how the weight of each electronic box affects the aircraft payload and also its balance. On a Fighter, this analysis is essential because even an ounce is considered very important.

12. Electrical Department

The Electrical department of any vehicle provides the design for the overall vehicle wiring and the design of the internal structures for equipment. They have to design the power distribution rack of the vehicle which includes Control Panels, fuses or circuit breaker panels, and relay panels.

The duties and responsibilities of the Electrical department are as follows:

o Electronic Interface
o Power Distribution Design
o Circuit Breaker Panel Design
o All Control Panel Designs
o Relay Panel Design
o Junction Boxes design
o RF Microwave Interface Design
o Vehicle Wiring Design
o Vehicle Dash or Cockpit Wiring
o Vehicle Rack Designs
o Vehicle Drawings and Release
o Antenna interface

The Electrical Department is a major organization and has many responsibilities. In order to provide the above services, the Electrical Department should be composed of Electrical Engineers, Electronic Engineers, Mechanical Engineers, and CADD Operators or Draftsmen. The following is a breakdown of these services:

12a. Electronic Interface

In order to interconnect the different equipment together in a vehicle the Electronic Engineer in the Electrical Department has to know the interface requirements for all the boxes on board. On a large reconnaissance aircraft, this could amount to hundreds of boxes and thousands of wires.

There may be Computer Buss wiring such as 1553, Manchester, or RS-232. Coax cabling and twisted shielded wiring may be required. There would be micro-wave connections and plumbing to be established. There may be Fiber Optic wiring on board. Many stranded hard wires would definitely be required.

In order to accomplish all this wiring the Electronic Engineer must investigate each unit on board the vehicle. He would have to read each unit's Equipment Specification and understand their interface requirements. He would have to study all the I/O drawings of each box and possibly influence the I/O design requirements of each box. He would then have to design the interface wiring diagram to connect these boxes together. He would also have to establish the bulkhead connections and has to make sure there are no ground loops in the vehicle which can cause oscillations.

There are many sub-systems that have to be considered in the Interconnecting Diagrams. Some of these are:

Electrical Interconnecting Diagrams

- o Communications
- o Navigation
- o Intercommunications
- o RADAR
- o ASW
- o EW
- o Infra Red
- o Armament
- o Flight Electronics
- o Mission Record
- o Antennas
- o Ships Computers
- o Missile Guidance
- o Video Systems
- o Teletype Systems
- o Power Functions
- o Cockpit Instrumentation
- o Flight Controls

The personnel in the Electrical Department have to have an in-depth knowledge of all of these systems in order to provide the proper interface. In an example I provided earlier, I designed a Teletype Interface Box that had to interface with a High Speed Printer. As was mentioned, the problem was the ES for the I/O of my box was different from that of the HSP. In that case, the Electrical Department did not pick up the discrepancy because the wiring was supposed to be the same for the HSP and a LSP, but the wiring was not the same. The LSP required plain hard wires with bi-level signals but the HSP required RS-232 interface which uses several wires.

Had the Electrical Engineer studied both ESs, he would have caught this discrepancy, but he didn't. The fault is not really his because it is the responsibility of the Equipment System's Engineer in the first place who created the ES for the HSP.

Had there been a proper check and balance procedure in the Electrical Department, which I always recommend, this discrepancy might have been caught before it ran into a major cost modification to correct the problem. If the Electrical department had been set up properly with each engineer informed of their duties and responsibilities many other anomalies such as this one could have been avoided.

The problem was that the Electrical Department for this program was composed of mainly CADD operators, some MEs, and only a few EEs. This Department needs Electrical, Mechanical and Electronic Engineers with a lot of experience.

A set of Interface Guidelines should have been created for the department such as the following:

Electrical Interface Guidelines

12a1 Examine all Equipment Specifications

12a2 Examine all I/O Drawings

12a3 Design the Interface Wiring Schematic

12a4 Design the Harness and Vehicle Routing

12a5 Attend all Design Reviews by Suppliers and Internal Departments

12a6 Interface with the EPMs and PMs

12a7 Interface with Equipment EPMs

12a8 Interface with overall System Engineers

12a9 Interface with Equipment System Engineer

12a10 Interface with all Design Organizations

12a11 Interface with all Support Organizations

12a12 Interface with Production

12a13 Provide Interconnect Drawings.

12b Power Distribution Design

In any Plant or Factory, a power distribution analysis has to be made of all outlets to operate machinery, lights, and air-conditioning. It is no different on any military vehicle, especially on a large aircraft.

The Electrical Engineer has to establish the electrical distribution requirements throughout the aircraft. The generators from the engines provide AC and DC electricity for all of the electronic boxes and for the flight operations of the aircraft. The General Distribution Panel is usually located in the center of the aircraft to minimize the length of the power lines. Power has to be provided to the Galley, to the Cockpit, to the electrical Bomb Bay doors, to Flight Controls, and to the avionic distribution panel.

In addition, the Auxiliary Power Unit (APU) power distribution panel has to be designed. An APU may be required to provide power to only essential equipment in an emergency or during ground testing.

The Electrical design would consist of an investigation of power requirements for all electronics (WRAs) and where they are located. The Power Distribution Panel and Avionic Distribution Panel would have to be designed structurally in such a way as not to block the normal operation of the aircraft and rigid enough to hold equipment.

12c. Circuit Breaker Panels Design

The circuit breaker panel, located near the Distribution Panel, would consist of the Main Circuit Breakers that provide circuit protection for a specific area of avionic boxes. Twenty Main Circuit Breakers are not uncommon for a large aircraft. In addition to the main circuit breakers, there would be many individual small circuit breakers on different panels which could be located at different parts of the aircraft. These small breakers would be in the 1 to 10 amp range and be used for individual electronic boxes or A/C functions.

It is unfortunate, but many of these smaller circuit breakers are often used by the crew as On/Off switches for the electronic boxes. A power switch is usually provided for several boxes, but they prefer to power up or shut down a small section or a single unit with a circuit breaker.

To set the requirements correctly, the circuit breaker is to protect the ships wiring and not to protect the electronic box. In the old days each box had a fuse in it, which was suppose to protect the vacuum tube electronics.

In this new age of semiconductors no fuse is fast enough to protect a chip, and besides, the military hates fuses because they get lost and they don't like having to keep an inventory of all the different fuses required. As a result fuses are not allowed anywhere on the plane.

To get around not using fuses, I cheated once by using a fusible resistor (fusistors) to protect an expensive circuit on a PC board. It worked great. Some individuals have used thin traces or very fine jumper wire on the PC board as fuses. The things engineers have to do to protect their designs and still meet MIL-Specs.

Our Power Supply engineer used the "crow bar" approach with an SCR power transistor, so that any peculiar failure would force the one amp circuit breaker on the box to blow. This was done on the box of Figure 5 to protect the electronics. Worked fine until a new EDM decided to remove the circuit breaker from the box. The reason he did it was that the aircraft also had a circuit breaker on its panel, so why have two breakers in series was his reasoning.

But a problem arose when bench testing the box. He didn't know that the Military's Test Benches used 100 amp circuit breakers and when this crow bar failure occurred on the Navy test bench all the wire-wrap wires in the box melted and fused together, but not before all the Chips in the box were blown. It was a costly change which was discovered years later.

The crow bar was a quick fix until the 1 amp circuit breaker was removed from the box. This is a fault that usually happens when there is a change in command. The new EDM thought he could save production cost if he removed the breaker. My approach when working with another man's design is to leave it alone unless it becomes necessary or at least to contact the original designer, if available.

One rule of thumb that I have always followed was, when one examines an existing design and see's something stupid, don't be in a hurry to change it because one doesn't know what the original designer had in mind. He probably had a good reason at the time. Unless one knows why he designed it that way and the box is still working fine, leave it alone!

If Circuit Breakers (CB) have to be designed into the Distribution Panel for individual boxes they should be mounted on a separate panel located close to the electronics. The Main CBs are in the order of 20 to 50 amps and they may be AC or DC. Most large aircraft carry 220 and 110 volts 400 Hz AC power, and 28 volts of DC power. Some of the new large aircraft use 270 volts DC to power the boxes from converted 220 volts AC. On smaller aircraft, Fighters and Helicopters, they may still be using only 28 volts DC to power all boxes.

12d. Control Panel Designs

Throughout the Vehicle there are many requirements for Control Panels. These panels would contain such items as switches, meters, potentiometers, relays, lamps, digital keyboards, and digital displays. The need for these panels usually comes after the fact or after the sub-systems have been established.

It is very difficult to determine how many and what the requirements of these panels would be at the start of a development program for an aircraft. One panel may provide the controls for more than one sub-system which was not defined at the start of a program. These sub-systems have to be defined first and then they have to go out for bids. This may take a year or more before the electrical department knows what the complete panel requirements are.

On an aircraft, the panels are mounted on upright rails that have standard 1/4 inch holes separated by 3/8 of an inch for Zeus Fasteners. These dimension are governed by the ARINC ATR sizes. The smallest panel would be 2" high by 4" wide, with a minimum of 4 Zeus Fasteners. The width and height could go up to 20 inches if need be. These panels may be located in any part of the vehicle but they should be close to an operator. Many would be located at the cockpit. The cockpit control panels would usually be known in the beginning of the program because they are part of the cockpit design which must be established early.

Small panels that do not have active electronics in them, do not have to be qualified because they already meet the standard ATR configuration. Some control panels may be loaded with relays and can be very heavy and generate a lot of heat. These large panels may have to be subjected to environmental testing but in most cases they are not. They are usually considered as part of the aircraft structure.

12e. Power Relay Panel Design

On a large Vehicle many power functions that are required are in the order of 10 to 100 amps. These functions may be controlled by miniature switches or by electronics. In order to control these large relays and solenoids, sometimes smaller intermediate relay are used. It's all a matter of power requirements and power dissipation. For example, a relay requiring 30 amps through its contacts would require a 1 amp drive current at 28 volts.

If the designer were to use solid state switching to provide this 1 amp drive current, he would have to consider several limitations. First of all, the relay coil is considered an inductive load so the initial turn on current is actually 10 times the continuous required current, or 10 amps.

In a +55 C environment the voltage across the solid state switch could be as high as 2.5 volts. So now this switch would have to dissipate 25 watts for a short duration. Time enough to blow out a 5 watt transistor. If there were 50 of these solid state switches in a box, the power dissipation in the box, momentarily, would be 1,250 watts. Imagine trying to cool a box with that much power dissipation (PD), especially if the switches are continuously going on and off.

It is much easier to use small relays to drive large relays. In this way the power dissipation is much lower because the voltage drop across the contacts of the small relay is 0.05 volts or less and the momentary PD is 0.5 watts in comparison to 25 watts for the solid state switch. This is why a power relay panel is required. It would be composed of many small relays from 1 to 10 amps to drive the larger relays.

The example just discussed, actually happened to me. It was over the box I discussed earlier that went out for bids with exactly the above requirements where large relays were driven by many solid state switches. That design requirement was replaced with the box shown in Figure 5, where no solid state relays are used.

The original Equipment Specification on the requirements, called for many solid state relays to be used and installed in the box with the electronics of Figure 5. Thirteen companies bid on the job and they all said they could implement the design in that same size box.

What a joke. My calculations showed that the box that they were all proposing would have to dissipate over 2,000 watts which would require an enormous heat sink. Where would they put the electronics in the box? As anyone can see, the box is already crowded with electronics and no room for solid state relays.

It was a bad ES, but what was worse, the Suppliers bidding on the job did not acknowledge or even mention there was a problem. They didn't want to be eliminated because of not being responsive. They would wait till after they were awarded the contract and then they would point out the deficiencies by asking for more money for a new design. When uppermanagement reviewed my report on this box, as I mentioned, they immediately canceled all bids and awarded my department the design. However, the first thing I insisted upon was a separate relay panel, and to design the panel the old fashion way using small relays to drive large relays and solenoids. It is sometimes hard to beat a relay.

The Power Relay Panel can be located anywhere on the vehicle but it should be accessible for repair. The Panel may or may not be removable as long as the relays can easily be replaced. Designing this panel is the responsibility of the Electrical Department.

12f. Junction Boxes Design

Junction Boxes are another one of those items that come into play after the Supplier designs are completed. The Electrical Engineer already knows that several Junction boxes would be required and tries to allow for it.

A Junction Box does several things; it provides a terminal strip or a linkage to terminate a group of wires and feed them to different destinations, it may contain relays for signals that are controlled by the electronics or by a panel, and it may contain resistive networks for terminations and for other reasons. Unfortunately, many Project Engineers use these Junction Boxes as "catch-all" boxes and add PC cards with electronics in them. Junction Boxes were originally intended to be passive so that they can be designed by the Electrical Department and they don't require qualifications. There was no check and balance here, again.

Because of this added electronics in junction boxes, my organization had to design many of them. Some of them were so large that it became a two man pick-up (and not two girls weighing 105 pounds, either). They just grew and grew.

It's always more economical to add to an existing box than to produce a new one. It is less inventory for the Military, less parts, less ATE testing, etc. So because of this restraint, the Project Engineer would rather make the Junction Box larger and save the cost of designing a new box. Sounds reasonable to him? Sounds more like the old saying, "There's more than one way to skin a cat."

While reading this section one must wonder how all this ties in with the EDM. I was deeply involved in Junction Boxes because we had to design the enclosure and before I knew it we were designing electronics in the Junction Box. Also, my organization depends heavily on the Electrical Department on such assignment as the wiring of the Integration Laboratory and other simulated facilities or a hot bench using their drawings and our technicians.

The EDM has to know more than just black box design. He and his organization has to know what equipment goes into the vehicle, how are all of his boxes are used, where they would be located and how they are powered up.

12g. RF Microwave Interface Design

The Electrical Department is responsible for the installation of Microwave Waveguides, Flex and Rigid cables. The RF design should be the responsibility of an RF System Engineer or the Antenna Department, if there is one. On a Reconnaissance aircraft there could be as many as 100 antennas. Some antennas would be large dishes, some would be small rotating dishes (spinners), and many could be just whips or stubs.

To interface RF transmitters and receivers with antennas, waveguide plumbing or Rigid cables are required. There are also many small microwave components that they have to interface with such as mixers, couplers, attenuators, amplifiers, splitters, switching matrix, etc. This field is a specialized field and one must know microwave theory to be able to implement the interfacing technology.

One major problem with this interface is signal loss. Every time a coupler or splitter is used there would be a signal lose. The front end of a receiver should be as close to the antenna as possible or the transmission line losses may cause a reduction of signal sensitivity. The Electrical Engineer has to have enough knowledge to be able to discuss the interface requirements with the RF System Engineer so that he can route Rigid cables, Flex cables, and install the waveguides properly in the aircraft.

12h. Vehicle Wiring Design

The vehicle wiring is more than just providing wires. It provides the special wiring require for all of the above systems. If coax is used the bulkheads would require special connectors as feed-throughs to insure the coax impedance is maintained with proper shielding. This is also true for twisted shielded pairs.

Terminal blocks are used throughout the vehicle for fanning out signal paths. Fiber Optic cables have to be installed properly through the bulkheads.

Service loops have to be provided for many black boxes so that the equipment can be removed and installed. If DPX connectors are required, special DPX mating frames have to be designed to withstand the human force of inserting the black boxes.

On one aircraft my company developed, all the Electronic boxes required DPX connectors. After a few flights in the field it was discovered later that the mating DPX connector pins were not making very good contact. When a white hat (sailor) installed a black box into the mounts he would slam them in with such force the frame would bend. To this day I detest DPX connectors because there is no normal way of knowing if the connector is properly engaged and they require too much force to make them engage properly.

I got around some of these problems by using a trick to ensure my box was engaged. By running a 28v dc line through all four corner pins of the connector in series to a test light on my box front panel I was able to detect when a DPX was fully engaged. If the light remained off after it was engaged it meant that either the connector was not properly engaged or the light was burned out.

Four pins had to be given up for this detector but it was worth it. The round MIL-Std connectors are still the most reliable, easier to engage, and easier to wire especially if EMI backshells are required.

Unfortunately, the Program Systems Engineer on that aircraft insisted that there wasn't enough room for round connectors. I don't buy it. If there is room for a bulky DPX mating frame there would always be room for round connectors. On a Fighter or helicopter I can see the advantages of using DPX connectors because the avionics would be installed from outside of the plane and no room for service loop cables.

The wiring diagram on the aircraft has to be designed showing the overall interconnection of each black box with the pins called out. This could simply be a pin listing on 8 x 11 inch sheets. I still prefer a drawing because it is easier to trouble shoot and to follow the signal paths.

I remember pasting many sheets of drawings on the walls to trace the paths of the signals on the aircraft. It took the length of a 100 foot room and passed over peoples desks. The pin listing approach, on the other hand, may be written in a book with 200 or more pages. To follow the signal path means going back and forth throughout the book. It could be a nightmare, but some engineers prefer it that way, which is okay.

As was mentioned before, Ground Loops are one of the major problems in the aircraft wiring, and one has to be very careful whenever shielded wiring is used. There are special EMI backshells connectors with rings in them for the purpose of connecting all the shields together. It is preferable to use twisted shielded wires for signals that have frequencies below 1 mHz. The twist reduces the noise and the shield reduces pick up.

By the way, it is against all MIL-STD's and dangerous to personnel to put power on shielded wires.

12i. Vehicle Dash Board or Cockpit Wiring

The wiring of a Vehicle dash board or Cockpit is a specialty by itself. The Electrical Department is responsible for designing and mounting the pilot's and copilot's instrumentation. Special Instrumentation Engineers would design the cockpit system, but the wiring is usually the responsibility of the Electrical Department.

In addition, as was mentioned, several cockpit control panels would have to be designed by the Electrical Department. The main problem with control panels are limited space in the cockpit. The cockpit is always overcrowded and if one has to add another switch or knob, it could sometimes be baffling. As mentioned, Instrumentation Engineers are usually ex-pilots who know cockpit requirements.

12j. Vehicle Rack Designs

To design racks in an aircraft, the Electrical Department has to interface with the Aircraft Structures Engineer and the Stress Department. As was mentioned, ATR sizes are used to determine the amount of real estate required for all the electronics, including panel mounting. In a rack, most of the electronic boxes would be mounted on a shelf for extra support. In the rack design, one has to consider shock mounts, cooling, service loops, and the wiring. Space is always a problem, especially if the black box is an after-thought. There would always be several "oops I forgot" boxes or an ECP or GFE that the military wants to add later on in the program.

On cold plate cooling that was mentioned before, it becomes the responsibility of the Electrical Department. Tubes would be installed as part of the rack. They have to provide the ducting for airflow passages to and from the black boxes to the blower. Remember, it is no longer allowed by the Military to cool the electronics by passing air over the components. Indirect cold plate or side wall cooling has to be used. In the Advanced Tactical Fighter rack design mentioned before, the electronics may have to be cooled by refrigeration through each modular duct. This rack design could be a major undertaken.

On panels that have read-outs and switches, for example, they do not have to be mounted on a shelf. The zeus fasteners along the sides should be sufficient to hold these panels.

12k. Vehicle Drawing and Release

All drawings are controlled by the Technical Data Department or the company's Library. What we are now considering is the Engineering Control and signature mill. In order to have control of the drawing changes it is best to have one organization responsible. Otherwise, changes would be lost or not updated.

If all changes are controlled by the Electrical Department, they should sign off each drawing, see to it that they are updated to a higher revision number or letter, get the necessary signatures, and submit them to the release desk or department. Technical Data would release and distribute the updated drawings to the appropriate departments.

When I was an EPM for an Electronic company, their procedure was to have the EPM be responsible for getting the mechanical department to create the ECOs, check the ECOs for the changes, getting the signatures, see that the changes are incorporated, and having the changes tested and inspected. It was a one man show and a nightmare. This was over and above all of the other responsibilities that I had.

First of all, the EPM is not a draftsman nor is he a checker, so to have him be responsible for drafting practices is utterly ridiculous and then to have me go around obtaining signatures is actually degrading because I had to practically beg to get them to sign. It's called Empire Syndrome. "I won't sign until the other department signs them first", or "I don't agree with the mechanical engineer, so I won't sign." This was definitely a poor way to control drawings.

The reason I prefer that the Electrical Department be responsible for Drawing Control, is that they would be creating thousand's of drawings for the vehicle by themselves, anyway. They would have the drawing procedures down "Pat" already, because they would have done it so many times and it would be strictly routine for them. If all of the other departments submitted their drawings to the Electrical department, the signature and release cycle would go much smoother. As was mention before, two of the three most difficult things to accomplish in engineering is Drawing Release and Name Plate approval. Non-Standard Parts is the third.

On one program that I was on as a Staff engineer, the EPM on that program could not understand why the drawings had to go through a release department, and why shouldn't engineering handle the drawings themselves. This man could be forgiven because he was software oriented, so I had to explain to him how the drawing system works for military programs.

His experience was in the ATE world and before this new program began he never had to release a drawing before. In his previous programs the drawings that were delivered were to the "Best Commercial Practice." That is a completely different ball game. Drawing check and signatures are not required for best commercial practice.

In this new program, the contract called for full Mil-Specs and Category III Drawings. If engineering had attempted to release the drawings themselves the Government would never have accepted them and they would not be to Mil Specs. Inexperience like this could cause the loss of a contract or the company may have to eat the over-runs.

As one can see, the Electrical Department has many tasks and they have to meet the same requirements as the Electronic Design Team does. They are responsible for all the Junction boxes and Panel requirements, and in some rare cases, they have to qualify their Junction boxes. The only difference between the two organizations is that the Electrical Department should not design black boxes that contain active electronic circuits in them. It is quite possible that the Electrical Department and the Electronic Design Department could be a combined division, but then again, there would be no "check and balance."

One of the main duties of the Electrical Department is to monitor the interface and external wiring of all boxes, but if they designed the electronics themselves, they would essentially be monitoring and checking their own design. So they really should be a separate organizations to maintain "check and balance."

13. Stress

Engineering also interfaces with the Stress Department to ensure that the black box is strong enough to withstand the vibration and shock that is required in the Mil-Spec. The responsibility of the Stress Engineer is also to ensure that the black box is mounted properly in the vehicle and would not cause any damage to other black boxes or to the vehicle itself.

The Stress Engineer should be invited to the "kick off" meeting. He should be given the preliminary mechanical drawings for evaluation and the final set of release drawings. He should be invited to some design reviews, before and after fabrication.

He should be informed when the start of the environmental testing begins and is provided a copy of the vibration and shock test results. He has to make sure that the black box doesn't resonate at the same point that the ship resonates, which could cause the ship to shake itself to pieces. The ME lead engineer should work very closely with the Stress engineer.

14. Producibility

Producibility could be another Production Engineer's function to monitor. In a large company this would be done by a separate department, especially if many black boxes are to be developed by the design department. The Producibility Engineer's job is to make sure that all production considerations have been included in the design of the black box. Each company has their own way of doing business and they may have limited production equipment for assembly and test. This would influence the design because it has to be tailored to meet the capabilities of the production equipment. If production has automatic component insertion equipment, for example, the PC boards should be designed accordingly. If they are hand inserted, the PC designer may take a different approach to his design. This analogy applies to all phases of the design.

The Producibility Engineer should be invited to many design reviews as the design is in progress. He and the Production representative should be working together or may even be the same person. I have seen too many black boxes poorly designed for production capabilities.

On one aircraft program that I was on, a black box was designed by another department. It took twice as long to fabricate and assemble it in comparison to fabricating an equivalent black box that my organization designed. Production would never let them forget that poor design. This was the result of a Project Engineer's ego, who convinced his upper-management that he could design that box for 1/4 the manhours that were quoted on by my department. He ended up using twice the manhours that I quoted.

This particular subject was discussed before when the Project engineer thought he was a designer, but had never design any hardware before. Not only was it a bad design for production, but it was very unreliable and it became a "hanger queen."

15. Integration Laboratory

The responsibility of the Integration Laboratory is to test the operational functions of all electronic equipment that would be on board the vehicle before they are implemented on the aircraft, and to retest the equipment when there are changes made. Their objective is to stimulate and exercise the Unit Under Test (UUT) to determine if it would interface properly, and work with its associated equipment. This task is a major responsibility which requires special Test Engineers. See Figure 35 on page 328.

Many aircraft or vehicle companies don't see the need for an Integration mock-up Lab. So they end up using the aircraft or the vehicle as their test bench. Very costly in the end. Many times a simple bench test with the associated equipment is enough to test the box. Several times, I've gone to Pax River in Maryland to test our box using the aircraft as the test vehicle only to find it didn't communicate properly with other equipment. What a waste.

I had the same problem on the U-2 bird where we tested vendor EW black boxes using a flight recorder. After a flight exercise I would bring the record back to our facilities to analyze the data and discovered an equipment failure. If a test bench had been available some of these problems could have been caught ahead of time and not have to be tested in flight.

The Integration Laboratory Engineer prepares a special test program under a prescribed format which he has to submit to the customer for approval. No box design would be accepted by the customer until it has passed all Integration testing. Every box, whether it is made in-house or by a sub-contractor, would be subjected to Integration testing.

When my organization designs a box, I usually provide the Integration Lab with a prototype (XN1 model) to do preliminary testing. They would first prepare a test program and then test the unit with its associated equipment. In some cases the tests may involve several separate electronic boxes just to test one black box.

The main objective of the Integration facility is to ensure that all electronic boxes performs its functional requirements in a mock-up airplane. It is sort of a pre-flight test to catch any bugs before production and before a flight test. Because of this, the design engineer or LE has to interface with the Integration Department and work closely with their engineers.

As was mentioned, if the Integration testing is done in a formal manner, the design engineer may not receive the test data until after production has already started, which is too late. If there were any major problems discovered in the Lab and not reported immediately, the design would have to be changed later which could kill the schedule and be very costly.

That is why there must be a close relationship between the designer and the Integration Engineer, so that changes can be incorporate in the proper sequence in the updating cycle using XN boxes, and discover errors early. Integration would be the first time the black box is tested under actual conditions and it is very crucial in the development phase of the program to perform these tests with XN boxes. Preliminary data from Integration would save a lot of time and money later.

Many times, in my company, the aircraft was used as the test station, because they by-passed the cost of using our Integration Facility. This, eventually, became very costly in the end, because it is very difficult to evaluate the equipment while flying. An upper-manger felt that it was cheaper not to use the Integration Laboratory (which was available) and to use the aircraft as the test bench. All he could see up front was the enormous cost savings in manhours by not testing the unit in the Integration Laboratory.

If there were no Integration mock-up, in this situation, I could understand his decision, but the Integration Facility was available and working. Unfortunately, it is sometimes the customer who refuses to pay for the support of an Integration Facility. It's their money so who could argue with them.

Consider what the cost would be if a test has to be done on an eight hour flight and the unit doesn't work, especially if the aircraft is a Fighter. Worse yet, consider the cost of reworking many production units when a major design error is discovered.

As was mentioned, I was on the U-2 black program, and my job was to analyze the operation of new equipment that was installed in the aircraft which had to be tested in flight. There were all kinds of sensors and test equipment on board to monitor the operation of that new equipment. That aircraft had to fly over and over again because there was always an associated equipment failure which were not related to the new unit.

If there had been a Integration Lab, these kinds of problems could have been resolved in days. It sometimes took months to resolve a problem by using the aircraft as the integration test platform while flying. Of course, the final test has to be in flight.

Even a simulated hot test bench could have provided some help in this case. Some problems can be solved during ground testing using the aircraft, but when the aircraft is used as the test bench, the box has to be a qualified production box. An XN1 box may not be allowed to be used for flight test.

If the aircraft is a newly developed plane, it is possible that the electronic box could be in production before the first aircraft was assembled and there was no way of testing the box. This situation has happened. On one particular contract, many boxes were produced and were sitting on a shelf waiting to be tested on the aircraft under their actual environment. Rework was indeed very costly.

The Integration Lab usually remains within the company's facility for many years, to provide a test base for any ECP changes and for new or modified equipment. Some military airplanes would have different missions. Some may be used for reconnaissance and different version of the same aircraft may be used for attack purposes. New missions are always being generated by the military which would require changes to a box and tested. This may mean updating the wiring in the Integration mock-up.

I worked on one type of aircraft that was used for ASW, EW, Attack, Transport, and as a fuel tanker. Each mission required several different types of electronic equipment. If the customer had paid for updating the mock-up, which was available, the Integration Lab could have easily tested these different configurations, but because they refused to pay for integration on some configurations, the aircraft became the test bench. What a mess.

When I worked for another company they had a black box (separate contract) that was designed to replace another box on the aircraft of the company I had just left. (That was why they hired me in the first place). I tried to convince their managers that there was an Integration Lab available in my old company but they wouldn't go for it. They continued to use the aircraft at the Navy base as their test platform. This meant flying back and forth to Pax River in Maryland over and over again to retest errors before any changes to the box are made. I had a hell of a time passing TEMPEST, for example, and that box went back and forth to California several times before we solved all the problems.

They also had a temperature problem. The box wouldn't work on the plane in a cold hanger for at least an hour after it was turned on. This problem went on for years. When I was hired, I made a technician put the unit in a cold chamber. Sure enough it acted the same way. We finally fixed the problem. I was told that the box had already passed low temperature environment and there was no need to put it into a chamber again. But, many changes had been made to the box since then, which caused the box to have problems.

Because there was no test bench available the Navy flew the aircraft to Greenland and left it exposed to the freezing weather outdoors overnight. The next day, after they started the aircraft, the box passed with flying colors. It was a pretty costly flying test station, but I was proud of my fix.

16. Flight Test

Vehicle or Flight Testing is the next step in the normal sequence of evaluating an electronic box. It is the final test that would be performed on an electronic box that's to be installed in an aircraft or in another type vehicle. This final test is done to ensure that it meets its predicted functions while in flight or on maneuvers.

If the unit is on an aircraft, Flight Test should be done with the latest updated XN model and then again using a production model. Sometimes a new aircraft is still in the development stages and is not ready for the XN model, so it would have to be done later using a production model. Flight Test is a test that is performed by the aircraft manufacturer at their facility, as apposed to military testing at a military base such as the Navy Test Center at Pax River, Maryland.

The EE design engineer or LE has to interface with the Flight Test Engineer during these situations because there would always be some problems that were not discovered in the Integration Lab.

The Flight Test Engineer would develop a flight test program for the aircraft similar to the one that was created by the Integration Engineer and his program would have to be submitted to the customer in a particular format for approval.

If a major problem is discovered during Flight Test, the box may not get a chance to be retested again by Flight Test personnel because the aircraft has now been delivered to the customer and is not available to Flight Test Engineers any longer. This just means the final testing would have to be done at a military base.

Flight Testing on the plane is usually done after the development stage of the unit is completed, and at the company's facilities. Thereafter, any modifications that are made on the black box, would have to tested later, at a military base, as was discussed. At some point in time during the development of the aircraft, there would be a transition where military pilots would take over the airplane but Flight Test Engineers would still be responsible for the equipment. When the aircraft leaves the company's facility the Flight Test Engineers may still have to support the unit in the field.

The Government is forever trying new equipment in the aircraft. On one aircraft that I worked on as a Flight Test Engineer, the plane was already 30 years old and it seemed that engineering was always testing the latest versions of EW equipment.

My company had to modify the installation and the wiring on the aircraft for the many newly developed units. As I mentioned before, I had to analyze the performance data using a mission recorder over those units while they were in flight. As long as the military has an aircraft in operation it has to be supported by the company and its FT Engineers. I spent countless hours analyzing the data using the mission recorder, but that was my job.

17. Functional Test and Acceptance Test

The Functional Test Procedure (FTP) is a test procedure for production testing. The procedure is generated by the Functional Test Department. When a black box is designed in-house, the Functional Test Engineers would provide a test procedure to sell off a production box to the customer.

If the box is to be sold-off by engineering, an Acceptance Test Procedure (ATP) is used. The difference between the two procedures is that the FTP is geared around Production test equipment while the ATP is written around Engineering Lab test equipment. The main objective of both procedures is to provide some sort of written test procedure to meet the requirements of the ES so that an Inspector can evaluate the performance of the box before he puts his stamp of approval on it. As was mentioned, the ES is the bible which the inspector depends on to corroborate the FTP or ATP test procedure.

The steps required in preparing an FTP is taken from the System Engineer's Functional Test Requirements (FTR) document. The FTR has to be approved by the EDM and the EPM. Then it is submitted to the Functional Test Department.

When the System Engineer prepared the FTR, he usually takes the requirements from the Equipment Specification (ES) which he probably prepared in the first place and the SOW. The FTR tells the Functional Test Engineer how the box should perform under certain stimulations showing tolerance limits. The System Engineer does not tell the Functional Test Engineer how to perform the tests as long as he covers all the parameters specified in the FTR.

The Functional Test Engineer uses the FTR to generate his FTP. The FTP is then submitted back to the System Engineer, the EPM, and the EDM for approval.

On the electronic design of Figure 5, the Functional Test Engineer convinced his upper-management that he didn't need an FTR or an EE approval, so he by-passed the normal procedures and generated his own FTP to test that box using a sophisticated HP automatic test computer. After the production boxes were installed in the aircraft there was a steady stream of boxes going from production to the aircraft and back again to production for rework and repairs. The problem was, Production was not testing the box properly, so they were continually installing bad boxes in the aircraft. When the aircraft went through a flight test many flight functions did not work properly. It took two years and many arguments between Production and Engineering before they realized that their sophisticated computer was not programmed correctly.

The box shown on Figure 5, is considered a life supporting equipment, and without that box the plane could not fly except under emergency conditions. Fortunately, there was enough redundancy in the design (two boxes) that if a circuit failed in one box the function would still work from the other box, and the plane was still able to fly under some of the flight maneuvers. After the production Functional Test Procedure was corrected by using my Test Box of Figure 6, the steady stream of boxes finally stopped.

Anyway, one can see what happens when an individual tries to go it alone. In this case, again, their was no "check and balance." If engineering were allowed to reviewed his FTP from production, this problem would have never happened. Think of the wasted cost that was done by the Functional Test Engineer who thought he knew it all. After that goof-up that Functional Test Engineer quit the company. He had tested the box by functions only, so if one circuit had functioned properly the redundant circuit was never tested. He did not test for specific time delay's, and for signal level tolerances. This program was a perfect example of what not to do.

Engineering and the Functional Test Department should have a close relationship. They should be invited to almost all design reviews. After the FTP is approved and the test equipment is ready, one of the XN prototype boxes should be given to the Functional Test Engineer to check out his own FTP. After he receives his first production box he can then return the XN box back to engineering.

18. Check Department

This department was discussed before and is now expanded further. Check is always taken so lightly in engineering. Upper-management and non-engineers cannot understand why engineering has to waste so much budget on another department just to check drawings. Why can't the draftsmen do the drawings right in the first place? What's so special about a stupid drawing? All a Checker does is dot the "I"s and cross the "T"s. This is the kind of nonsense I've had to listen to for 40 years in this business.

My Division Engineer told me once to forget about checkers because checking would be done by engineering. It was obvious that this man had never delivered a box before. First of all, a drawing cannot be released without a Checkers signature. It's obvious to me, that an engineer does not know all of the drawing practices and drafting procedures to properly provide "check and balance." An independent Checker is definitely required for all military drawings.

On another occasion, an EPM did not allow any budget in my quote for Checkers. When I asked why, he said, "It's a waste of time and they just eat up the budget." It seems that everyone is always trying to by-pass the system and look for an easier or cheaper way to do business. In the end, the checkers had to be brought in, anyway, because they cannot release the drawings, and the clean-up after the fact ended up costing twice as much as the original checker would have cost.

A Checker is a special type of a person. I for one, could never be a Checker. I would be totally bored stiff in that position. If one watches a checker in operation (he looks as though he is in slow motion all the time) he would check every line and every dimension according to the drawing procedure to ensure that all the Mil-Specs are met, and that all of the company's drawing practices have been followed. He has to know all the proper drawing practices. Many times he would call the Chief Draftsman's office for advice on a questionable item.

The Checker would not sign a drawing until he is completely satisfied with it. A drawing would probably go through a check cycle of rework and back-check many times before it is approved. It's a boring job and somewhat costly, but it has to be done.

If engineering makes one change after the checker has signed the drawing and the drawing was released, a revision letter has to added to the drawing to reissue it. This is when the cost goes up. My philosophy is not to bring the Checkers into the picture until after Integration and Environmental testing is completed. These tests would, invariably, require many changes, and to save costs it is best to delay the check and recheck cycle until the program is down stream. The only problem with this delay is that production cannot sell-off a box until the drawings have been issued and signed off. That's why we had a special drawing release procedure such as mock-up release or engineering release which are used to get production started. Inspection would buy off the first few production boxes using these special drawings but not the follow-on units.

Checkers should not be invited to any design reviews, but should be part of the program. One problem with checking is, a Checker may start the check procedure on a drawing and later a different Checker may be used. He may even disagree with the first checker on what was approved in the first place. Then the arguments starts.

My approach to this, as was mentioned before, is to just go along with the new checker. He is going to get his way anyway, so why fight it. Change the drawing to the way he wants it, and get on with the design and not waste time and budget trying to find out who was right and who was wrong as long as the final product is right.

It's surprising how an engineer thinks he is following the proper drawing practices until a Checker reviews his drawings. On top of it all, many years later, the drawing practice procedures may change. Engineers are usually not aware of these changes so if an engineer is used as a checker, the mistakes may be compounded. It's probably good practice that engineers check one another in the beginning, anyway, for technical reasons but not to be used as checkers over drawing practices.

19. Drafting Department

Each company has its own method of providing drafting. Some companies use the mechanical department as draftsman, some have a separate department that does all the drafting, and some have CADD operators that are loaned to engineering to do the drafting in CADD. Some companies don't use checkers.

My position is that engineering should do their own drawings in CADD, and checkers should be brought in to do the checking. Some companies use a separate department for hand drafting. Then engineering has to first sketch the drawings by hand and then give it to the drafting department to redraw. It's just a duplication of effort and more chances for errors.

Now the engineer has to check the drawings from the draftsman and when he finds errors, he has to send it back. These drawings may go back and forth several times. After this maneuver, it still has to go through check and back-check and back to the Draftsman again.

My philosophy is, the fewer people in the loop the fewer chances of making mistakes. If a separate drafting department is used they do not have to be invited to any design reviews.

20. Drawing Release

As was said before, Drawing Release is one of the most difficult and frustrating operations in this business. One of the reasons for this is, it requires so many signatures. Many engineers suffer from "signatureites." This is a disease that assumes one would lose his job if he signs a bad document. So what do they do, they sit on it. Maybe it would go away, I don't know.

One of my old bosses deliberately delayed signing a purchase order for the same reason. One time I had given him a purchase order of $ 1,000 which he delayed for two months. He did this often, so he had a pile of back orders on his desk. On that $1,000 order, it was later discovered there was a design flaw and that purchase order had to be canceled.

My boss gloated over it by saying, "See, I saved the company $1,000 by delaying that order." He now had justification for his delaying tactics, and he reminded me many times thereafter whenever he was being pushed to sign an order. He could not be convinced that the $1,000 he saved was peanuts in comparison to the number of manhours lost in delays of assembly and test time. When I took over his job as EDM I used to say, "I'll sign anything to keep things moving as long as it isn't a blank check."

I discovered that the release signature disease "signatureites" can never be cured. I have tried everything imaginable to speed up the drawing release signature mill. I finally gave up. I'm convinced it's an incurable disease.

What most companies do, is have a table or a room set aside just for signatures. Department Managers or Inspectors would drift by this table occasionally and see if they have a drawing that has to be signed. If they see something wrong in the drawing they won't sign it and just walk away. No one would be informed that he wasn't satisfied or there is an error, so the drawing just sits there waiting for signatures. Some drawings don't get released until three months after they were given to the Release Department. There is no excuse for this. But, as I said, this sickness is incurable.

The reason drawingings go through this release cycle is because the customer requires project approval to ensure that the drawings meet the Military requirements and, in production, the Inspector has to have a way of knowing that the drawings are official and that they are up to the latest revision. It is the only way to have some sort of drawing control. If the proper signatures are left out, the system falls apart and there is no "check and balance."

The following are just some of the Departments that may be required to sign the drawings before they are released:

Design Department
Check Department
Quality/Assurance
Material
Electrical Department
EPM
Release Department

The EPM selects the departments that must sign the Drawing Release. One weakness with this is, the EPM may purposely leave off some of the critical departments to save budget or to speed up the release cycle. This has been done time and again. As part of the EPM Project Plan he should at least include those departments listed above as a minimum so that if there are any disagreements they may be challenged by the PM. Saving budget is no excuse for poor efficiency.

Without drawing release control the purchasing department might be ordering wrong parts in advance, and production may be fabricating to the wrong drawing revision. Even with control, production would find many errors when they go into fabrication for the first time. Inspection would allow ECOs to be used temporarily, but all ECOs must be incorporated into the final drawings and re-released. It is important that an Inspector should not allow an ECO to be left on the books for more than 90 days. I have seen boxes being assembled with ECOs that were over a year old and units were being assembled with obsolete ECOs.

A proper Release program is essential in this business. I don't have the solution to speed up the signature cycle, but the Design Department should not deliver any drawings for release until production actually needs them.

Everyone in the loop from design to production is geared to working with the Released drawing system and as soon as one tries to by-pass the procedure, everything stops. This is why production wants the drawings released as soon as possible. Except, it may cause chaos if the drawings are released too soon because engineering is still making changes by the day.

Master scheduling should be aware of this situation and should provide a proper period of time for drawing release in his schedule. Production can work with a mock-up or engineering release drawings to get started, but they can't sell off follow-on boxes without the official drawing release. If the program is handled properly everything should fall into place and the first released drawing should be clean of all revision letter or numbers to start with.

On one program I was on, as the Staff Engineer, the released drawings were up to the letter "m" and were still being updated before the first production box was ever fabricated because the EPM had insisted that the drawings be released as soon as possible. That way he met a "carrot" on his schedule and would look good to his upper-managers with his weekly report that states, "20 drawings were released this week." But, they were not correct to the latest configuration. His schedule fell apart in the end.

21. Inspection

Inspection, in some companies, is usually part of Quality Assurance Division. The Inspector has to be a well rounded individual. He has to understand electronics, structures, materials, parts, and testing. He has to be able to read blue prints, schematics, and has to know Mil-Specification and company documentation. Most Inspectors come up through the ranks as technicians, machinists, sheet metal men, and mechanics. Some just specialize in one category of inspection, such as electronics or vehicle assembly, and some have all around capabilities.

The important thing about an inspector is credibility and product assurance. He has the final say so in the sale of a product. His stamp of approval is the only way of knowing that a product has been tested properly and that all production procedures have been followed.

The way it works is, the Customer or Government Inspector would inspect the first or second production unit and witness the First Article Test (FAT) together with the Company's Inspector. When the unit passes this first test and is accepted by the Government Inspector, he would then authorize the Company's Inspector to be the Official Production Inspector for all follow-on units. Even though he works for the company, his inspection authority cannot be overridden by anyone, even by the president of the company. He is now <u>bonded</u> and could lose his job over an improper inspection.

Working with Inspectors is not all that difficult. My approach is to treat him like he's a God. Be kind and patient with him, and above all never get into a conflict with him. One can yell at an Umpire in baseball but to yell at an Inspector would cost one dearly in the end.

One of my engineers got into an argument with an Inspector once during an ATP of a box, and the Inspector just quietly walked away and said, "Call me when you get it fixed."

When the problem was fixed my engineer called the Inspector who said he would be there in an hour, but he never showed up. The inspector did this for two weeks to teach my engineer a lesson. Then he would nit-pick the unit to death. I finally replaced that engineer with a likable old timer, and the unit passed without a hitch in an hour. The Inspector got his point across and was very happy to pass that unit for the old timer as long as it passed the test.

When one is ready to sell off a box, be sure that all the paper work is in order for the Inspector. Have a duplicate ATP ready, make sure the ATP can be read easily. Provide the Inspector with a comfortable chair, and then buy him a cup of coffee. Before one calls the Inspector make a few dry test runs first with another engineer. No surprises, please. Be courteous and be humble. After all, the Inspector is really "God."

22. Qualification

Qualification is composed of two tests, Environmental Testing and Electromagnetic Interference (EMI). To Qualify a box the unit may be required to run through a performance test while it is being subjected to extreme environmental conditions. It depends on the SOW. When a box is Qualified it assures the customer that the unit meets or exceeds military environmental condition and EMI requirements in accordance to the MIL-Spec called out in the SOW.

These tests can be done on any of the prototype units (XN) or on a production unit. If a prototype unit is used it must resemble the finished product in structure and function. It cannot be a non operational box and must run through a performance test. The electronics in the prototype may be altered later without re-qualifying it as long as there are no major structural changes. Environmental testing and EMI testing do not have to done on the same box.

Environmental testing is done to prove to the customer that the equipment design would meet the extreme military environmental conditions set forth in the SOW. These tests include vibration, shock, temperature, salt spray, humidity, altitude, etc. The SOW may not require all of these test. EMI tests should be run by a separate independent organization, either within the company or by an outside house that specializes in this field. It is important that a separate organization do the tests independently to ensure "check and balance."

I have seen the results of some out-side vendor's reports, where the designer also did his own environmental testing. When I later looked at the package and the design, I could see there was no way that his box could have passed environmental tests. On one PC board, for example, a vendor used plug-in TO-5 transistors which would easily have fallen out on a vibration table. It's too easy and also illegal to doctor-up the results if the same person who designed the unit does the environmental testing.

Note: this testing is different from that of Burn-in testing which is required for production. Burn-in, or shake and bake, is done on all production units while Qualification is done only once, if it passes.

When performing the environmental tests, an engineer or LE should be present to ensure that the tests are being run properly and if any problems arises he would be present to witness and record them. The LE would be a good candidate for this task.

EMI testing consists of Radiation and Susceptibility. TEMPEST and ESD may also be the responsibility of the EMI Engineer. The EMI tests are done independently and tested in a Screen room using sophisticated RF equipment. The object is to ensure that the unit would perform its normal functions and would not fail or degrade when subjected to RF radiation from a RADAR transmitter or other sources in accordance with the SOW. Also, the unit shall not radiate radio signals at specified levels at RF frequencies that would interfere the normal operation of other equipment in close proximity.

TEMPEST is not normally a qualification requirement but is handled separately and probably performed by the customer at his facility such as Pax River. ESD is also not a qualification requirement and usually requires just a visual inspection of the unit and of the drawings.

There isn't much interfacing needed with the Environmental Test Engineers before the tests. They are not required to be present at any design reviews but should be present at the kick-off meeting because they may have to design a holding fixture for Qualification and be able to quote the fixture.

It is always a good idea to first run a preliminary sinusoidal vibration search on the box at low Gs to see where the resonances of the box peaks at. In this way the ME can modify his structure and beef it up in certain areas, if need be, to dampen the resonance curve and lower the peaks.

The most important consideration is to make sure that the unit does not resonate at the same frequency as the vehicle does. If it did, it would multiply the vehicle's resonant frequency and could contribute to the destruction of the vehicle, especially if it's an aircraft. Every vehicle has a peculiar resonant frequency associated with it, and it becomes very important in a light Fighter.

An environmental report and an EMI report is required, which would be part of the CDRL data submittal to the customer. All data taken during these tests should be recorded and saved for later reference. The results of qualification should be included in the final report of the unit.

Most companies do not require a final report but it is good practice as reference because it provides vital information and the history of the design. It is not known how many times our boxes, which were over 15 years old, had to be updated for ECPs and if we had had a final report on each box the modifications would have been simple, but since there weren't any reports written on those old boxes, many weeks were spent just analyzing the old design before it could be changed. The history of a design is vitally important.

23. Material

When the Design Organization orders a part, the Material group would first make a search on that part to locate a distributor. He would first see if the part is in the company's stores. If so, then the part can be retrieved in a day or so. Most companies would purchase large quantities of parts that are used frequently in their product line such as diodes, transistors, screws, resistors, etc. In this way production would obtain a very substantial price break.

The material person would then record the ordered in his computer with all the details such as cost, date of order, delivery date, etc. In this way the engineer can track the delivery of his parts and have a summary of all the parts that were ordered in the development phase of the program. Material would assign a purchase order to that part an submit it to the Purchasing Department.

I prefer that engineering should have a separate small Materials group to handle development parts. Parts are better controlled by engineering when only four or five pieces are ordered rather than a large production run.

If the parts are ordered through the overall company's Material Department, then other situations could slow the process. The parts may now have to be pre-screened, or it may have to go out for bids.

Drawings would have to be released, SCD may be required, and they must be pre-inspected. All it does is hamper the Design Engineer.

I worked for an electronic company that had only one ordering department, whether it was for a prototypes or production unit. The problem was they required formal released drawings before any parts can be released by Material to Purchasing, even for the prototype design. This caused many delays in the program.

Material personnel do not have to be invited to any of the design reviews but should be invited to the kick-off meeting. It's always a good idea to have all of the people on the program to be aware of what's going on in the beginning. It helps in the end. It gives the material people a better understanding of what engineering is trying to accomplished and it makes them feel they are part of the team.

24. Purchasing

After the Purchasing Department receives an order from Material with an order number, the buyer would dispatch a purchasing order to a supplier. If the order is for production, the buyer would have to solicit the part for bids from the different supply houses.

If the parts are for PC boards, for example, the engineer may have used a specific PC house to developed his first prototype, but this doesn't matter to production buyers. When PWBs are ready to be ordered for production, the buyer has to go out for bids for the PC Cards, usually from the companies preferred PC supplier's list. Because of delays in bidding, and to save time, the design engineer should have selected a PC supplier from the company's list in the beginning of the development stage so that the tooling costs are paid only once. There is still no guarantee the buyer would select that PC house but it could influence him.

The Preferred Parts Vendor List applies to other items as well, even sheet metal parts. The reason this list was created in the first place was because many small outfits work out of their garages and they are not reliable. To ensure a good product and lengthy production line for the future, the company must investigate all the bidders and eliminate the fly-by-night outfits. The reputable suppliers would then be put on a Preferred Parts Vendor List and only they would be allowed to bid.

Some suppliers would start out great doing a fine job and, later in years, their quality degenerates. This may cause them to be removed from the preferred list of suppliers. Many suppliers that were once removed from the list were later reinstated after they proved they had improved their quality.

The list is actually a form of Black Ball, but it is justified when dealing with military programs, and, in some cases, with life supporting equipment. Quality of parts have to be controlled. The company, however, has to be cautious that there is no payback to key employees to put a poor vendor on this list. This does happen.

Some of the considerations a buyer has to make is the cost of the part and availability. For some reason, different distributors have different prices for the same military item. Integrated Circuits, for example, can be made by several IC manufactures and their base price may different, or the overhead cost for each distributer may be different. For whatever reason, the cost is important especially on large production orders.

Availability is the next major issue. As was mentioned, suppliers would lie and tell the engineer that a particular part would be in production for some time, so the engineer goes ahead and uses the part in his design. The distributer then provides the engineer with a few parts for his prototype from his "dead parts" pile which he thought he couldn't get rid of.

When the production buyer finally places a large order he finds out that the item that the engineer selected is no longer available and out of production. There are some bad suppliers out there and one must be aware of all of their shenanigans.

One problem that I had was, I selected a part for development from a large supplier company but we couldn't use them in production. I found out later, that our Government Contract specifically stated that we must only deal with small supply houses to help the small businesses in our community. Because of this policy we had to pay much more for the part in production. What a joke! Whatever happened to competition?

My Division engineer had the habit of finding new and untried chips in a magazine and made the design engineers use them in a research device. That was fine for research, but when the device was to be used on a military contract the Buyer would say no-way. It wasn't an approved part and there wasn't a second source. Always be aware of new devices.

A word about the cost of purchasing a hammer which received so much political fan fair in the media. When Lockheed purchases tools for an aircraft it buys them in bulk, possibly one thousand kits of various tools at a very low cost. After twenty years in operation, a navy personnel somehow lost a hammer on one our aircraft so he order another one through Lockheed. In order for Lockheed to buy that hammer, which they no longer have in stock, is to go through Material, Purchasing, Distribution and Field Services. The price would be very high by that time with all these organizations charging to the program. It would have been much cheaper for the Navy to purchased the hammer at a local store. Lockheed or any other Aeronautical company cannot purchase single items at a low coat. They are not geared for single items.

A word about toilet seats. The toilet seat on a P3 aircraft, has to meet full Mil requirements and be qualified. If the plane, for example, has to make a crash landing the navy doesn't want the toilet seat breaking loose flying around the aircraft decapitating the crew. In addition, it has to meet jungle environments and extreme cold temperatures such as in Alaska or Greenland. Of course an aircraft toilet seat would be expensive under these conditions. Enough said.

25. Computer Services

Some companies have a computer service department that provides services for all departments. They handle payroll, provide budget records and material cost records. For engineering they can provide software services for such items as connector pin-listings, non-standard parts listing, special memory loading programs, and provide the program and service required for the company's Computer Aid Design (CADD) machine. Lockheed's CADD machine was called CADAM.

In today's world, every manufacturing company should have some type of a CADD system to be able to compete. There are some personal CADD systems that do not need the aid of Computer Services but they are not as extensive as a main frame unit. Also, how does one lock in drawings in ROM, and consider the how long it would take to send a drawing to another CADD station through a modem. It takes forever. By using a main frame with fiber-optic cables, transferring drawings from building to building is pretty fast. Computer Services could also be used to provide the software for the Budget Tracking System I've proposed in the text.

When I worked for an electronic company several years ago, they were still doing drawings by hand. When I asked them when are they going to install a CADD system, they said it would be too expensive, and they don't feel it would save any time or money. Yet, they had personnel computers all over the plant to assist engineering to track schedules and costs.

The reason they were behind the times was because that company was run by finance people. They understand the advantages of using personnel computers for administrative purposes, but they cannot envision the advantages and the cost savings of a CADD system to produce drawings. CADD systems, would be considered overhead costs and not direct charges to a contract, which they didn't like.

My rebuttal was, it would take less engineering hours to do a drafting job. Guess who won that argument? Finance people are just number crunches. They do not understand design and development problems. I don't know of any engineer who is capable of creating a sophisticated design by the numbers.

Some day this will happen and, as a matter of fact, I do have a design approach (not included in this book) for an automatic design technique which, one might say, would be a method of designing by the numbers. I tried to implement this approach in our CADD system but I couldn't get my upper-management to spend the research money to develop it. It would have saved us a lot of time and money on future programs, but I was turned down. Oh well! I had the same rejection on my standard enclosure design.

If the company does have a separate Computer Services Department, then engineering would have to interface with them. They are usually not invited to any design reviews or even to the kick off meeting, unless they play a part in the design. This department is not part of the Engineering Software Department that was covered, but simply a service organization to the company and to engineering.

26. Finance

In most companies, Finance plays a major role in the company's operation. Over-budget programs could be canceled because of engineering costs, so their budget progress must be tracked. When a job is quoted, it is the Finance Department that puts the finishing touches on the quote.

Usually, the EDM does not interface with Finance directly on in-house program. Finance would only interface with engineering directly when engineering is soliciting sub-contract work out to a vendor, being used as a sub-contractor themselves for another company, or is working on a military Contracted Research program (CRAD).

When work is being sub-contracted out to a vendor by engineering, Finance has to get involved with the bidding and engineering must submit an Equipment Specification to the vendor.

Over the years my company has sub-contracted sheet metal fabrication, PC boards, Flex Cables, Antenna design, RF Electronics, Displays, and some special software designs. On those programs, Finance would have to interface with the Electronic Design Department. In a complete aircraft program, Engineering would be sub-contracting possibly 70% of the electronics, and only Project Engineers (EPM) would be interfacing with Finance. The PM would not be involved because there is no customer to deal with.

When the Design Department is doing sub-contract work for another company or is contracted directly by the military, Financing would be required to interface with engineering. This is when the Design Manager is needed to follow some sort of a quoting procedures as I outlined in Chapter I.

CRADs require even more support from the Finance Department because Engineering deals directly with the military and the work may be independent of the company's product line.

I had done a lot of CRAD work directly for the government and upper-management had little control, which was a blessing. The players involved in a CRAD program would be the Design organization, Administration, Finance, and Contracts. What a pleasure it was. Little interference from upper-management and no politics.

With CRADs I didn't have to answer to an EPM or a PM as a EDM. I didn't have to solicit the help of any support organizations. I didn't have to deal with Production, Drawing Release, and Non-Standard parts. Not even Master Scheduling was involved. My organization would simply do their job, prove the capability of the design and submit research reports to the government as required. As long as I stayed within budget and met the schedule, Finance was happy. CRAD is just another research program to design and build a unit to prove a theory or a proposal.

In one situation, my Director of Engineering didn't even know I was working on a CRAD program until it was completed. His favorite expression was, as I mentioned before, "Just keep me out of trouble."

Now that's what happens when a Director is an Aeronautical Engineer who could care less about electronics. My organization didn't have a legitimate home within my company at the time, so I was put under this Director's jurisdiction. I kept him out of trouble alright, so he left me alone. I didn't agree with this arrangement, but that's the way it was.

27. Contracts

In general, what ever applied to Finance above also applies to Contacts so this subject won't be expanded any further.

28. Master Scheduling

Master Scheduling plays a major role in an electronic design. For some companies they are sometimes used as the coordinator of a program. Someone has to put all the pieces together on a planned schedule to tell when each event would take place throughout the program, so that all of the departments know when they would need to become involved.

When a program begins the Master Scheduler starts out with the Design Organization's Schedule and then branches out to all of the other departments. The following is some of the players that should be included on the Master Schedule Plan and not necessarily in the order shown:

28a. Master Schedule Plan

Program Plan, Level I & II Schedules (Proposal)

Engineering Project Plan, Level III Schedule

Kick Off Meeting
Equipment Specification
Design Department Schedule
Level IV Schedules
Overall Master Schedule
Begin Design
Bread Board
Design Reviews
Fabricate Prototypes (3 XN's)
Engineering Drawings
CDRL's Submittal
Support Departments
Electrical Department
Customer Reviews
Non-Standard Parts
Environmental Search
Qualification Testing
Environmental Tests
EMI Test
Integration Lab Tests
Product Support
Manuals
Training

Field Service
Automatic Test Equipment
Logistics
Functional Test
FTR
FTP
Drawing Check
Flight Test
Production
AMO's (Long Lead Items)
Preliminary Parts List
Mock-up Drawings Release
Complete Parts List
Design Verification Demonstration
Preproduction Demonstration
Production Test Demonstration
First Article Test (FAT)
Drawing Release
Deliveries of Production Units
Customer Field Tests

All of these functions and departments have to be placed on the Master Schedule. The Master Scheduler should attend all design reviews. He should get involved in the program but only from a scheduling point of view.

Unfortunately, on one program I was on as a Staff Engineer, the Master scheduler was given full authority of the program, representing the Program Manager's office. There was no way that this person can call the shots in engineering. His main concern was to meet schedules regardless of the condition of the product. The program was already 2 years over-due because the electronics had to be redesigned over and over again. My hands were tied on that program because the Master Scheduler ran the program.

Whenever non-engineering tries to manage an engineering function it would eventually be a disaster. Conversely, if an Engineer were to manage the business end of a program it could also be a disaster. This over-due program, I'm referring to, was not all caused by this Master Scheduler but he did contribute a great deal to the confusion to the program which in the end caused many delays.

The Master Scheduler must be on top of the program and not be just a yes person. He must express his position and make sure that the program schedule doesn't get out of hand. He should not be put in charge of the engineering portion of the program. That's what the EPM's position is for. Every time a company tries to by-pass a function it always gets into trouble.

The Master Scheduler coordinates with all other organizations, such as Production and Field Service. Slippage in engineering can greatly affect the production schedule and final delivery. When there is a slippage in design it could cause a domino effect for other departments. In many cases, the Master Scheduler has the option to manipulate the schedule in such a manner so that the end date does not change.

Manipulating schedules may be compared to the greatest juggling act in a circus. It has to be done and done often to avoid a major slippage in the final delivery. Remember, there would always be a budget penalty when there is a delivery schedule slippage. Manipulating the schedule doesn't mean cheating. It means rearranging the schedule to make up for the internal delays in the program so that the final delivery date doesn't change much.

29. Administrations

The Business Administrator on a black box program, provides an important function. Engineers, first of all, are poor business people and someone has to do these tasks. It all starts out with the RFQ. Engineering would provide their quote to an Administrator, but it is usually in a poor format to be submitted to a potential customer. The Administrator would put the total quote together after he receives the Quotes from all the different engineering organizations (Design, Qual, Integration, Shops, etc). He would format the proposal properly and add whatever overhead charges are required.

When the contract is awarded, the Administrator would again ask all departments to re-quote the program and provide the final hours for in-house reviews. The contract may be awarded 5 years after the initial quotation, which could cause the costs to be much higher than the original quote due to inflation, higher production wages, etc. Also, a different Administrator may have been assigned to the job. Therefore, it is important that a re-quotation be done before the program begins.

My predecessor, as a Design Manager, quoted a job four years before I took over his position. After I became the EDM and my company received the contract, I could see that it was very badly under-quoted. My re-quote became twice the original quote. When upper-management saw this I was called into a meeting to explain the differences. When I convinced them that the first quote was a bad quote, they still refused to give me the additional budget I needed, and told me to get started on the design, anyway.

So, my engineers went to work and proceeded to design the enclosure until the budget was close to running out. I then wrote a memorandum stating that all work would stop on this program at the end of the month, and I would have to transfer people that are doing the design because of lack of budget.

I was told by my management to continue to work on it and just show the budget as a over-run, so I did. That was a mistake because, from then on, I gained the reputation of being an overrunning manager, and this particular program was always used to substantiate it.

Has anyone ever tried to convince his managers that it was someone else's fault, for any reason? Don't try it. No one would ever believe it, anyway. The program was finally completed but, from that moment on, word went through the grape-vine that I was an overrunning manager who charged too much to do an in-house design. I was used as the escape goat. Like I said before, it comes with the territory.

After the final quote is re-done the Administrator and the Design

Manager have to get together and produce Level IV schedules, and send it to Master Scheduling. A Tracking Program is then set up with the Administrator. Manhour Spread curves have to be generated, such as those shown in Figure 3, for the overall program and for any individual tasks. A Tracking Chart (Figure 2) has to be generated which would be used to track the manhours used by each task. These hours would be accumulated each month to assist the EDM to determine when or where a task is falling behind in the schedule.

Internal weekly department meetings should be conducted by the EDM with his engineers and with the Administrator to discuss these schedules. The Administrator would then combine these schedules and manhour spreads for the EDM's presentation in the weekly EPM's design review. The Administrator should be at all meetings and may be the one to take the minutes.

30. Product Support

Product support covers many organizations. It covers Automatic Test Equipment (ATE), Ground Support Equipment (GSE), Manuals, Training, Field Services, and Logistics. Some companies separate some of these organizations because they consider ATE and GSE as an engineering function since they produce and design deliverable equipment. In a way this is true, because Product Support is a service division whose basic charter is to support the company's equipment in the field.

My recommendation is that the ATE and GSE organizations stay under Product Support, but only act as the Prime Contractors to engineering. This organization would prepare an equivalent SOW or some sort of a Requirements Document for engineering to design all of the required Product Support Equipment. They should have the authority to control and monitor the program from a customers point of view.

ATE and GSE should have a representative at all the design reviews. He would essentially assume the Program Manager's function. By using this recommended approach, ATE and GSE would still be under Product Support, but they should not do the designing themselves in order to maintain "check and balance." Doing the design to their own requirements is not a good idea.

The following paragraphs assume ATE and GSE would be as recommended above:

30a. Automatic Test Equipment

Automatic Test Equipment (ATE) is a special division of Product Support because it provides automatic testing software for a Test Station (TS). The Unit Under Test (UUT) may or may not be the company's product line and no hardware designed boxes may be necessary from engineering. However, in most cases, an Interface Device (ID) is required to interface the UUT with the TS. The Test

Station is usually composed of several test equipment and a patch panel that somehow has to be connected to the UUT. This is where an ID design is required.

The ID may be simply passive, where connections are made to the UUT through special connectors, or it may be active which provides I/O signals to the UUT. In any event, the ID has to be designed to suit the requirements to test the UUT. The ATE engineer would provide these requirements, like an ES, so that the Design Department can develop a prototype ID unit. The ATE engineer would use the prototype ID unit to develop and test his software with the actual UUT at the Test Station.

The ID design should be treated the same way as any military electronic box. It would probably have to be qualified and be produced under the same conditions as any other box, except it would be under the document control of MIL-T-28800 for Test Equipment which is not as stringent as aircraft equipment.

My company decided that the ATE department should be completely under Product Support and that they should do their own designs. This introduced a mountain of problems and, unfortunately, I became involved in it as a Staff Engineer.

First of all, the ATE Department did not have the design capabilities and personnel so they had to hire many Job Shoppers. These shoppers didn't know any of the design procedures that were required by my company. They couldn't understand why checkers were needed and why the drawings had to be released. They were used to working with a contract allowing the design to be done to the best commercial practice.

The main problem is, what happens when the program ends? What does the ATE department do with the designers? If they are released, who would support the equipment for ECPs and problems in the field? Can the ATE department maintain a continuous talented hardware design work force? These questions and many more have to be resolved first before ATE can have its own design department. Needless to say, that program was way over budget, very late, and had many technical problems.

Figure 7 on page 179, is an example of an Interface Device (ID) that was used to interconnect several UUTs with the Test Station. This particular ID is used for testing supplier's Printed Wire Board UUTs or SRAs. The ID sits on a shelf of the Test Station and is plugged into a patch panel which has over 400 connections.

This ID provides special electronics located in the card cage as shown, to interface the TS with the UUT. There is a very large mother board in the base of the ID (not shown) that provides the interconnections from the patch panel to the cage and to the UUT. In this application, air flows over the back of the UUT as it does in its original environment. As one can see the ID is very large and bulky.

It is my opinion, with a little ingenuity, this ID could have been replaced by a separate box or GSE using a micro-processor, and the GSE would be no larger than this ID shown. In other words, the Test Station would not be required at all. If the mating patch panel were removed an awful lot of electronics could be provide in the same area, plus the card cage could have been expanded. The ATE system did not get rid of all of those loose test boxes as the military wanted, it just replaced them with many Interface Devises. Again, this is just my opinion.

Some companies that specialize in ATE do have their own design department, but they have a continuous work load in ATE. In that case it is understandable. Most electronic companies, however, have a main product line, and a separate Product Support Department. In another company I worked for, the ATE department was a separate company, and the UUT company didn't know how the ATE engineers tested their box. It just so happened that that arrangement seemed to have worked, so it all depends on the structure of the company and if the Product Support department can support a full time design group.

Remember, every design needs the support of many other organizations so when a design is done by a separate ATE department, it is unknown if they would have these same support departments available to them. If they are a separate company even within the corporation, then these support departments may have to be duplicated by both companies. It could present a problem.

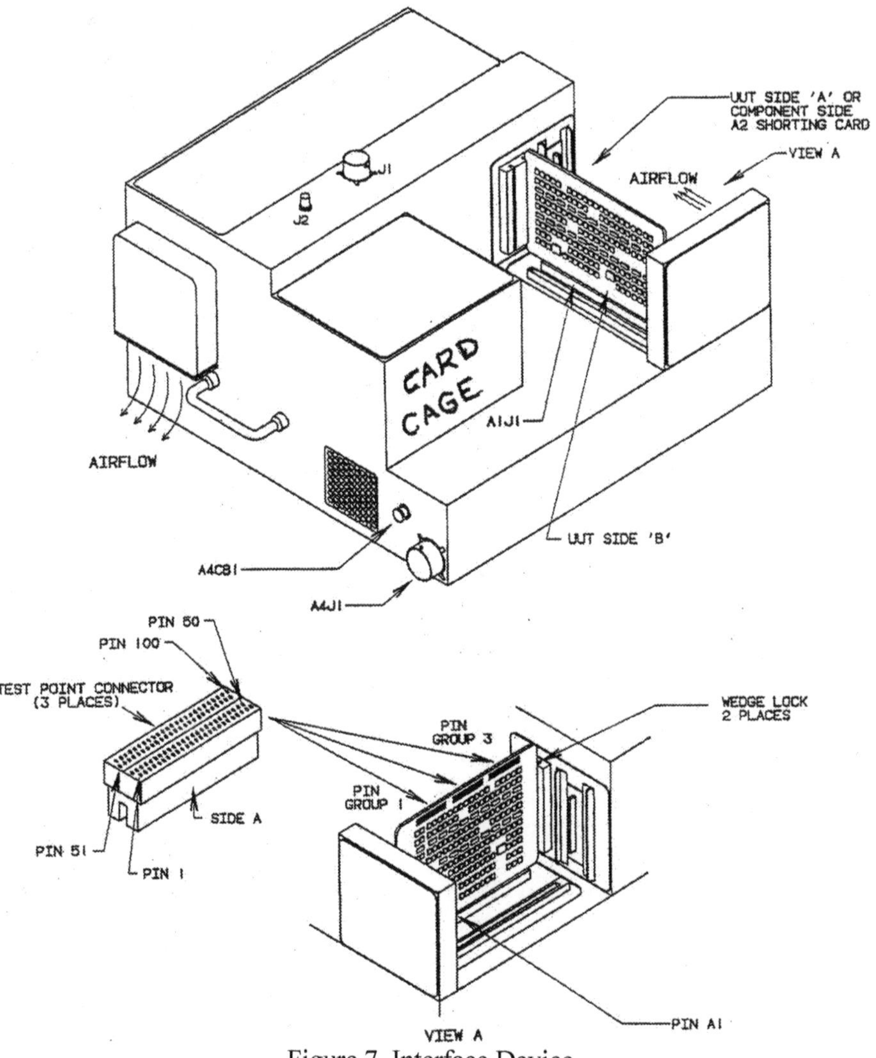

Figure 7. Interface Device

30b. Ground Support Equipment

Ground Support Equipment (GSE) has the same basic requirements that the Automatic Test Equipment (ATE) has. The difference is that GSE is usually a separate test box that is carried to the vehicle or installed in a repair shop to service the UUT. Not all UUTs contain electronic. The GSE may be a hydraulic tester or even an engine tester. In these cases, the UUT does not lend itself to being tested by an ATE.

In the old days, there were no ATEs. Everything was done with GSE. As was mentioned, the military was getting tired of having so many small GSE boxes stored in bins at the repair stations, so they were sold on the idea that an ATE was the solution to reduce the number of GSE.

Having one or two automatic test station at a site sounds great, especially to Congress. The problem is, they now have replaced the GSE in the storage bins with many IDs instead. Note: This decision was made before micro-processors were invented and may have been justified, but, today, the testing could be simplify using automatic GSEs with micro-processors.

My argument, again, is that they haven't accomplished anything, and may even have made things worse. The government is now paying for a very expensive ATE, with complicated Test Software, but they still need many loose IDs. These IDs in the field are large and many require a two person lift.

The worst of it is, the ATEs being used today are over 20 years old and they are required to test advanced equipment on modern vehicles which contain the latest sophisticated high speed components. There is no way that the ATE evaluation can keep up with the advancement of new vehicles. It takes years to develop new ATE Station, and when it is ready to be deployed it is already ten years old and outdated.

My approach is to go back to the old way of supporting the field equipment because in today's world of micro-processors, the GSE would probably be smaller than the ID now being used. This would eliminate the need for a very large ATE Station.

The second advantage of having separate GSE for each UUT, is that the designer of the UUT should also design the GSE. He knows what his UUT is capable of doing, and he knows how to test it. He would have tested the UUT in development, in qualification, and in production. The design of the GSE then would be relatively simple as the production test software and test procedures can easily be adapted to the GSE. The chances are his in-house tester would be initially designed with GSE in mind. Again, all this is just my opinion.

The design of GSE is handled the same way as airborne equipment or any other vehicle equipment. There are Mil-Specs to contend with, and all of the headaches associated with any military design still apply. The one difference may be that the

GSE department (under Product Support) would prepare the SOW instead of the military. The GSE Engineer provides the same function as a Program Manager would and should attend all of the design reviews.

Figure 6, Power Control Test Set, is an example of a GSE that my department designed to support the UUT of Figure 5. This GSE was designed before micro-processors were invented, so my engineers had to invent their own special purpose mini-computer using discrete components. As one can see it is a large box and had today's micro-processors been available then, that GSE would have been half the size. This GSE is no longer called a Black Box but, instead, it is a Yellow Box, signifying ground support equipment.

The point is, had the UUT of Figure 5 been tested on an ATE station, a special ID would still have to be designed which would be just as large as the GSE of Figure 6. So the government is paying twice the cost; for the ATE and the ID. It all could have been replaced with one GSE, as Figure 6 verifies.

Since my department designed the UUT box of Figure 5, it was relatively simple for us to design the GSE box of Figure 6. Both of these boxes are still in operation after 30 years. It seems only logical that the company that designs the UUT should also designs the GSE, instead of using a separate company to try to provide the test software on an ATE and design an ID. This other company doesn't know the UUT, so it would be difficult for them to prepare the test software of someone else's design.

30c. Logistics

Logistics provides the services required in the field. They provide the spare parts to repair and maintain the equipment. They would handle the delivery of any Ground Support Equipment required. It is their responsibility to make sure that all of the vehicles in the field are adequately supplied with parts, special tools, and equipment to keep them running. If there are any changes to the design they must be made aware of it to support the vehicle in the field.

In any new design, the Logistics department should be invited to the kick off meeting and to some of the design reviews. They need to have a copy of the parts list as soon as possible for their own quoting purposes and to get an early start on ordering long lead items for the field.

Any UUT that is in the field that requires repair in the company's factory or by engineering, is handled through Logistics. In some programs the company that developed the UUT may be considered the DEPOT because there was no test GSE in the field.

30d. Publications

The Publications Department provides the Manuals for the equipment being manufactured for the customer. In most cases Publications can prepare the manuals on their own. On the more complicated equipment, such as the GSE of Figure 6, Publication had to request assistance from engineering. The information they needed from engineering was descriptions, theory of operation, flow diagrams, software programs, test procedures, and final drawings. The Publications department would prepare the manuals in the proper format as specified in the SOW and the customer. They would then ask engineering to examine the manual for errors and technical acceptance.

30e. Field Service

Field Service represents the company in the field. The Field Service Engineer would provide failure reports on any equipment and would have the UUT repaired in the field, if possible. They would report back to engineering of any weakness in the design, and would do some field testing on new equipment. During the first delivery of the equipment, there would be many problems in the field and engineering needs their feedback so that they can make the necessary changes. Even though the unit was tested in the Integration Lab and in Flight Test, it seems there is always a problem after the first production box is delivered.

Sometime, years later in the field, the box starts acting up because of new parts problems. Parts do become obsolete and the replacement parts may not be doing the job. There is always something that has to be done to the equipment even after 30 years of service.

30f. Training

The Training department would teach the Field Service Engineer how to test certain equipment and how to operate the electronics in the vehicle. They would also train military personnel. They would have special training classes set up within the company's facilities and in the field. Training may require the assistance of engineering, but most of the time they do not. On the GSE of Figure 6, the Government had sent a Civil Service technician to my company's facilities and my engineers provided the training. The technician remained with my company for a couple of months and then he went back to train other military technicians. Most of the time Training can get the required information out of the Manuals and set up their training classes, but sometimes they require the assistance from engineering.

40. Management Course Explanation

The Management Course that is outlined later may be used to teach students, engineers, electronic companies, or anyone associated with military programs on the duties of the Electronic Design Manager and his subordinates such as; EE, ME, and System Engineers. It discusses the responsibilities of other support organizations, such as Maintainability and Thermal, and how they contribute to a program. It covers the Electronic Integration and other test departments.

The course also covers the responsibilities of non-technical departments, such as Administration, Master Scheduling and Finance. It finally ends with the Product Support Department who provide Automatic Test Equipment, Ground Support Equipment, and who supports the program with Logistics, Field Service, Publications and Training.

It teaches future Electronic Managers how to track programs, and how to prepare a quote for a new program. The course covers all aspects and responsibilities of an Electronic Department, from Quotation, design and eventually delivery.

Note: The Management Course , as illustrated on page 331, requires this book be used as the text.

CHAPTER III
COMPANY MANAGEMENT

1. DIFFERENT ORGANIZATIONS

There are many philosophies on the subject of how to run an engineering company. It really depends on the contract, the type of work, the manpower available, the production facilities, the support organizations and what works best for one's company. There are five basic structures that I know of that are used to set up an engineering branch. These are Matrix, Projectize, Service Type, Independent Groups and combinations of these types. Of these structures I favor only two, Projectize and the Service Type. There are advantages and disadvantages with all of these methods which are covered in the following paragraphs. Every company is structured differently and, for the sake of simplicity, the following company chain of command would be used:

President
Vice President
Engineering Director
Division Engineer
Department Engineer
Design Manager (Supervisor)

The analysis that follows is based on the above chain of command which can be used in any of the above structures.

1a. Matrix

A Matrix organization is made up of specialized people drawn from other organizations, like a pool, to support and work on a particular project. When a new contract is awarded, the Project Manager (PM) would develop the organizational structure by drawing key people from the various organization and build from there. As the project increases new people may be hired and more individuals may be drawn from other organizations.

One drawback with a Matrix organization is that the departments that the people are being drawn from are reluctant to give up their key personnel because they are working on other projects. This would leave them strapped with a shortage of manpower. Also, no one wants to give up a good person. Consequently, they would rather give up their poorest engineers. This results in a new project that would be starting up with mediocre or poor engineers instead of sharp and experienced engineers.

The second drawback is trying to get a Department Managers to release a person for the new program within a reasonable time frame. The Department Manager would delay the transfer as long as he can because this would cause a reduction in his manpower which may cause him to lose his empire.

A Department Manager needs at least two sub-groups to maintain his position. He doesn't know how long these people would be matrixed to the new program. It may be weeks, months or years, so the reluctancy sometimes is justified because it could jeopardize an existing program.

I was an Engineering Project Manager (EPM) for several years in a matrix situation, and it was difficult trying to get these departments to give up their people. One thing these Department Managers would never say, however, is that they would not cooperate with the EPM because that would be violating company policy. Although they would assure me that somehow they would support the program it's easier said than done. Meanwhile, the start of the program would have to be delayed because all the players have not been established.

Another disadvantage of Matrix is that the individual that is on loan does not get a proper evaluation on raises and promotions. The way it's suppose to work is the home department would call the new project organization and inquire about how well his engineer is doing, or request a written letter on his accomplishments.

I was also involved in this situation, when one of my engineers was loaned to another military program. The matrixed department manager would not tell me what my engineer was doing for security reasons, yet unsubstantially, he indicated that the engineer was not doing a good job. This man was one of my best men and I could not believe this statement to be true. So I called my engineer directly and was told that his basic functions consisted of pushing paper, a task any mediocre engineer could easily accomplish. I tried to get the engineer back, but was told that a transfer back to his original organization, at that time, would disrupt his whole program. Politics again. As a result he was deprived of a raise.

Another improper evaluation come about when my Division Engineer decided not to give the loaned-out engineer a merit raise because he didn't expect that person ever to be back. Since the Division is allotted a percentage of the payroll for raises in my company, he would rather give more money to those still under his jurisdiction and not waste it on a matrixed person, especially if he feels that that person would never return.

Still another situation occurred when an engineer was on loan for more than a year and suddenly his home department had a reorganization. Now this engineer didn't even know the new supervisor to whom he was assigned. The new supervisor never talked to him even though he basically still reported to him.

When I talked to this man he was confused and was considering leaving the company. He said, "I feel like a man without a home." When upper-management sets up a matrix organization one of the last things they seem to consider is the individual's position being matrixed.

A matrix system may work if the above problems are taken into consideration. People must be considered. They are not just numbers to be played with like pawns in a chess game. If the matrixed person were actually transferred to the new organization and if his raises and work evaluation goes with him, it may at least solve that problem. When the program ends, the engineer is then transferred back to his home base, that is, if there is work for him.

Another disadvantage of Matrixing is that the Project Manager no longer has complete control over the matrixed person. I had an incidence where a matrixed person under my jurisdiction decided he did not like what he was doing so he asked his home manager if he could return. Because he was a qualified person the home manager gladly took him back. This left me short of a good man.

In addition, the Project Manager of the new program cannot control or reprimand the matrixed person because he really doesn't work for him or give him raises. The engineer knows this and may take advantage of this situation.

So far, the matrix structure has been pictured as a bad system, but in reality there are times when a matrix organization is a good system. It depends on how the company is structured. The greatest advantage is, of course, the Project Engineer of a new program has a pool of talented personnel to select from and when the program is near it's completion he can send the engineers back to their home base, thus avoiding direct lay-offs.

The military uses the matrix system in their structure. Their approach influences many companies that have military contracts, to follow the leader and employ the same system as the military does. This was done when many companies followed the military's lead using the numbering system to rate employees (G11). So goes the military so goes industry.

Matrixing may be great for the military because they are not going to lay-off anyone and they would just transfer the surplus person to another organization. It works for the military because they are not committed to a design and to produce a product. They don't have to answer to their stockholders on poor profits.

Transferring people back and forth doesn't work too well in industry. When industry runs out of work they have to lay off people or lose profits. Like the military, however, matrixing can and does work in some companies. But I'm still not a great advocate of matrixing for the reasons stated above. In some cases, I do favor matrixing when a soft ware or system engineer is needed on a particular program.

1b. Projectize

A Projectize structure allows people that are assigned to work on a new program to be permanently transferred to that division. The division level would be used here as a starting point just to illustrate this structure. The Program Project Manager of a new program could be as high as a Vice President depending on the size of the program and the structure of the company. Usually, the Program Project Manager is at the division or department level in engineering.

The Projectize structure should only be used on very large programs. To Projectize a program of say 1 million dollars or less over a six month period, doesn't make sense. By the time a division is set up with the required personnel several months may have gone by.

It is very difficult to put a dollar value on when a projectize structure should be used or not, especially in these days of inflation. It would be safe to say that a program of 50 million dollars or more and a development period of two or more years is a good candidate to be Projectized. Maybe 40 million is high enough, it all depends on circumstances. A good example when Projectize could be used is when the company wins a new airplane contract and all departments would be under one organization.

When one starts a Projectized program, the Engineering Project Manager (EPM) and the Program Manager (PM), have to get together to develop a Program Plan. This is a very important. The PM may generate this plan by himself. Too often, the Program Plan comes after the program begins. This is not the best way. When the Program Plan is completed then the Project Manager has to prepare his own Engineering Project Plan. In some military programs these two plans are required as part of the data submittal for customer's approval.

The Engineering Project Manager (EPM), or simply Project Manager, now begins to accumulate several engineering departments (instead of individuals as in matrix) who would work on the new program.

Each Department Manager has to generate schedules, manloading, manpower spread curves and budget allocations within each department. The PM, the EPM, the Administrator, and the Finance department have to work out the budget allocation for each department, each engineering support organizations, for procurement, for production, etc. The full details and duties of an EPM and PM are covered later on in this Chapter.

1c. Service Type

The Service Type structure is when all of the design work is done under one Electronic Design Manager (EDM). These includes the Electronic Engineers, the Mechanical Engineers, the Printed Wire Engineers, the Software Engineers, the Technicians, the Draftsman and Computer Aided Design personnel.

The Systems Engineer doesn't necessarily have to be part of the design department, but it's probably a good idea that he is included or at least matrixed in this structure. This is the structure that I prefer and I used it to illustrate examples in Chapters I & II.

The advantage of this structure is that the design organization has full control of the design and development of a product. It can quickly respond to an emergency or to any modifications. It can handle small jobs as well as large jobs. There are no external empires to weed through to get things done. A Department Manager may be the highest level needed. A front line supervisor (EDM) could easily handle most of these jobs. This particular structure is what I am personally used to. In the 40 years of experience in design, I have worked at least 30 of those years in a Service Type structure.

The way this structure works is, when a new job comes into the company the PM sets up the Program Plan and the EPM prepares a Engineering Project Plan as stated before. The EPM then gets together with the Electronic Design Manager to map out the organizational strategy and who the players are according to the Statement Of Work (SOW) that was provided by the customer.

The SOW is a gross requirements document and the PM, EPM and the EDM should review it first. They then should renegotiate its contents with the customer to reduce cost and errors.

When the customer prepared the SOW in the first place, they usually use the "boiler plate" approach from other programs, and there may be some requirements that do not apply to this new program. These unrelated requirements should be "weeded out" before the design begins.

After this is done, the System Engineer (SE) prepares an Equipment Specification (ES) which he has to negotiate with the EDM and the EPM. The ES now becomes the equipment bible for that unit and, sometimes, it has to be submitted to the customer for approval before any designing begins. It depends on the contract. The ES is a very important document because all of the supporting organizations, including production, use the ES as their official reference document. All of this information was presented before in Chapter 1 and is only repeated here for reference.

In the original Quotation, all of the departments had submitted their costs and schedules. Now that the contract has been won, all of these organizations are required to resubmit them again with schedules to the EPM, so that the whole structure can be assembled and the proper budgets can be allocated to each department.

Years may have passed since it was originally quoted before the contract was won, so the old quote may not apply any longer. Inflation may have hurt the overall cost and new people are now assigned to the job. In any event, it should be requoted to assure the old quote is not too far out of line.

The Design Department may not have to wait for the ES to be completed before they can start the design. The EDM, in a Service Organization, can use the SOW as a temporary guide because he probably quoted the job in the first place from the Request For Proposal (RFP), so he has a good idea what's required. He doesn't have to depend on other departments to start.

The only time an ES has to be completed ahead of time is when the work is being sub-contracted out. Then it becomes necessary to provide an ES document for Vendors to bid on. This is one advantage of doing the design in-house under a Service Organization.

The ES, now becomes the technical legal document that a sub-contractor would design the hardware. The Prime contractor and the sub-contractor would have many meetings to negotiate the ES requirements because any bad agreements in this document could make or break a sub-contractor. It's the same situation between the Prime contractor and the customer negotiating an SOW. There may be errors in the ES or requirement document that cannot be met.

The Program Plan and the Project Plan described above generally applies to any organizational structure used and is repeated often throughout this book. The details of these two plans will be covered later in this Chapter.

The main advantage of the Service Type structure over Projectizing, is that the service type can handle smaller jobs of less than 50M dollars. Again, the dollar figure crossover depends on the company's structure and inflation over when to Projectize.

There is one other advantage that the Service Structure has over Projectizing, and that is work load. A pool of engineers in the Service Type would always be available in one organization because they are working on several programs. When a Projectized program ends, what does an EPM do with all the design engineers? Does he lay them off or transfer them? If he transfers them to their original departments or if he lays them off how does he get them back again if there are design problems or ECP changes in the program?

In addition, it is sometimes hard to keep the engineers busy throughout the program. There is usually a lot of idle time in a Projectized program because one has to wait for another department to complete their work before he can begin his. In the Service Type Structure, there is usually plenty of work because they are providing design services for the entire company on different programs. The design engineers do not have to work in series or wait for another department to complete their work first. They would be working on several programs at the same time thus having no idle time.

I have, for example, as a design engineer, worked on several electronic programs for three different airplanes at the same time, and doing a little research job on the side, whenever I had time. When an engineer works this way he is motivated, his mind is fresh and occupied, and he is constantly upgrading himself with new technology. He is certainly not bored, and because of this he is more efficient.

This overloading, however, can sometimes work against the engineer, because some individuals may be spread too thin and then nothing gets done on time. A good manager would recognize this situation and transfer the work load to others. Some engineers thrive on overwork while others become frustrated. A manager has to know the capabilities of each individual whose working for him and what the priorities of the program are, so that his engineers would be assigned to the highest priority jobs first, and not be overloaded.

1d. Independent Department

The Independent Department structure is very similar to the Service Type except that each of the design groups would have their own departments. The Electronic Engineers would have their own department, the Mechanical Engineers would have their own department, etc. They still provide a service as do the Service Type and they would be working on many programs at the same time, which is good.

I personally prefer the Service Type or the Projectize structure if it can be arranged. In this way all of the departments are under one control. When any other structure is used, such as the Independent groups, the Project Manager has little control over the other departments.

To illustrate this point, when I worked for an electronic company in an Independent Structure, I had to make a minor mechanical modification in an electronics black box. Whenever there is a problem or a modification to be made in an Independent Structure it is usually up to the Project Manager (EPM), as the interface engineer, to get things done. In this particular incident, I went to the mechanical engineering supervisor in another department to have a change made. I was told that his MEs were too busy and he doesn't know when he can get to it. Even when it was stressed that this job was extremely important and had to be done as soon as possible, the ME Supervisor didn't budge. Going to the ME's upper-manager resulted in the same story.

I decided to inform my own upper-manager and follow the chain of command, which should have been the proper order. That was a terrible mistake. Before my upper-manager would discuss the subject, he wanted to know all the details with charts showing schedules, cost, manpower and reasons for the change. He also wanted a memorandum prepared on all the details and addressed it to the mechanical division. All of these charts and memos took four days to prepare.

After the memo was sent, it took another three days for a response. They decided to have a meeting to discuss the modification. The Mechanical department finally agreed to support the change in a few days. By this time 12 working days were lost on the program. But the fun had just begun.

The production machine shop still had to make the part. The

machine shop supervisor told me to stack the drawings on top of a heap of drawings marked high priority drawings. Needless to say, it took another three weeks before the machine shop got started on my project. So much for priority drawings.

Altogether, it took a month and a half to do an eight hour job. This problem did not end here, however, it still had to be assembled by production, be tested by a separate test organization, and had to be inspected. In an Independent Structure the life of a Engineering Project Manager is frustrating to say the least.

The real problem with the Independent Structure is the interface between the designer and final product. The only interface between organizations is through the EPM. Another major problem is that the electronic design engineer doesn't even know if his design is working properly because he doesn't debug and test the first prototype. This job is done by the Test Organization.

The designer doesn't even interface with the Mechanical Engineer to assure that there is no thermal problems, and to determine if the Printed Wire Boards are partitioned in the box properly to minimize crosstalk and EMI.

The designer does not interface with the PWB designer to provide thermal information, high frequency sensitive traces, proper decoupling and proper partitioning of components and circuits. He doesn't know who makes the decision on partitioning of the electronics on an over crowded circuit card that would have to be split into two separate boards, for example. He doesn't even interface with the software engineer. The only person responsible to insure that all of these design requirements are correct is the EPM.

Each organization, in an Independent Structure, is self-governed who simply do their own designs. When something goes wrong they would hide behind their most famous explanation, "It's not my job." Live with that for a while as an EPM and one would be promoted to a Section 8 (loony bin) in no time.

When I asked "who's job is it?" they would use their second most famous expression, "Your job." The whole structure deteriorates into a company of people that don't want to take on any more responsibilities than they have to. They did what was asked of them and no more. It now becomes somebody else's responsibility. There is no central control unless it goes to the Vice president, which is too high.

Another weakness of the Independent Structure is cost. When there are so many separate departments involved on one program there is bound to be a lot of duplication of effort, especially when Quoting a job. An EPM has to accumulate all the quotations from all of the different engineering departments and he has to weed out the duplications. No matter how hard he tries, duplication of efforts would always exist. Each department would fight to maintain their own budget, even if it means duplication of effort.

1e. Structure Combinations

Structure Combinations is just combining some of the above different structures together. In some cases, this is the best compromise when putting a design organization together. Some companies have been using the Independent approach combined with Matrixing for many years and have been very successful at it. A combination of Projectize and Matrix is used a lot, where a few special people would be matrixed for a short duration. A Matrixed organization borrowing from an Independent department may even work.

Combinations with the Service Organization may be difficult. The object of the Service Organization is to have all of the design personnel under one manager. On one occasion, when I managed a Service Type organization, my design department did all the hardware design but the other departments did the software and systems design. There was constant bickering. My design personnel claimed they followed the system requirements and that the software people were constantly making changes to the requirements. This resulted in redesigning the hardware over and over again. The cost went out of sight. It was really no one's fault, it was just a bad structure by not having the design controlled by one manager. If the software engineer was matrixed to the design department it would have helped.

2. Organization Interface Chart

The recommended Organization Chart shown in Figure 8 on page 194, is an example of an interface chart. It has nothing to do with a company organizational charts. It is simply to show how all of the departments becomes involved in a program. The chart forms the same basic arrangement of any organization no matter what type of overall structure is being used, whether it be a Matrix or a Project type. These basic functions always apply.

The right side of the chart consists of non-technical support. These subjects are not completely covered in this book, even though they are important functions. It would take another book to cover them properly.

Remember one thing, there is always an equivalent person in the military for each of these functions, and they would only talk to their counterpart within the company. For example, the Master Scheduler in my company, wanted to discuss financial problems with the Naval Finance department while he was in Washington. The Naval Finance man refused to discuss finance with him so he had to call back to his company's finance department. The information was the same, but the Navy had to hear it directly from the Finance department. That's the way it works in military programs.

2a. Vice Presidents

There is usually two Vice Presidents in a company that are

involved in one major program, one for engineering and the other for Administrations, and both report to the President. There could be more than one program and each requiring a Vice-President.

In a Projectized program, having two Vice Presidents may cause a lot of friction. It is better to have one Vice President in charge of everything because he has better control of the program and many a dispute can be resolved in his office. The explanations of the departments that follows assumes there is only one Vice President.

The Vice President (VP) is the overall program manager but he usually does not get involved in the engineering details. He is, of course, responsible for the program and receives the status of the program from the Program Manager's Office and the Engineering Director. The VP may have several other programs under his command as well, which he also has to monitor. I will not spend too much time on this subject because I was never a VP or closely associated with one.

2b. Program Manager's Office

In general, the Program Office is responsible for many programs and one Program Manager would be assigned to a particular program. The assignment usually comes after a new program has been awarded to the company. A different PM may have been involved in the initial quotation but may not be the same person to carry out the final program. There could be several men working for the PM, depending on the size of the program.

2c. Program Manager

The Program Manager's (PM) duties is discussed in more detail later on in this chapter. As one can see from this chart, the PM coordinates with the Project Manager (EPM) and all the non-technical functions of the program. At no time, should he interface directly with engineers. Unfortunately, this happens too often. In the company that I retired from there was no PM assigned to a particular program and the task was given to a Master Scheduling individual, as a coordinator, who was making engineering decisions. That turned out to be a very **bad situation.**

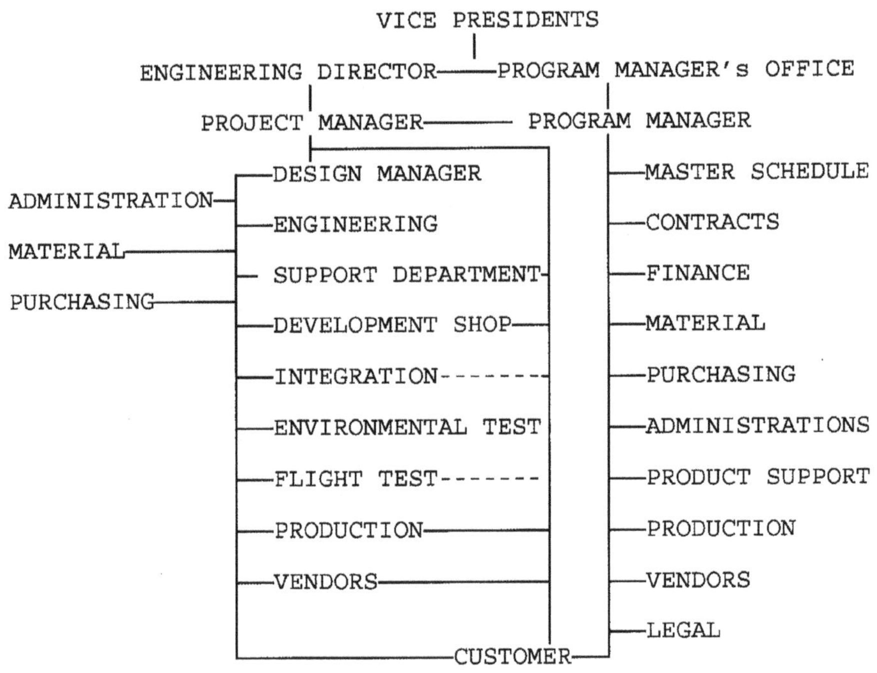

FIGURE 8. ORGANIZATION INTERFACE CHART

2d. Master Scheduler

The Master Scheduler prepares the overall delivery schedule for the program. He has to know the internal workings of the company, who the players are, and when things are supposed to happen. This person should be heavily involved in the program and should attend all design reviews to be able to update his schedules immediately if there are slippages. In this way there are no major surprises in the overall delivery schedule. If he knows ahead of time that there would be a schedule slippage in design he has to inform the various departments down the line so that they can "gear up" at the proper time with a new schedule.

There are several other maneuvers or options that he can take to maintain the end item schedule in spite of the in-fighting. Master Scheduling provides a very important function in the electronic field. He is the coordinator of the program but only over schedules.

2e. Contracts

The word "contracts" is self explanatory and will not be expanded on in this chapter. The Electronic Design Manager does not become involved with contracts on programs from a customer (military). Only the PM interfaces with the company contractors and with the customer. An EDM may hear the repercussion from the contractors office when he misses a schedule or over-runs the program.

The Contract office would interface with the EDM directly when a design item has to be sub-contracted out or when the Design Manager is working a job for the military. CRADs, for example, are research programs that are contracted directly with the customer and, in this case, the EDM would require assistance from Contracts.

2f. Finance

Finance is also self explanatory and will not be covered extensively in this chapter. The PM interfaces with Finance when generating the quotation and during the contracted programs. Finance provides overhead costs and converts the manhours to dollars when the quote is submitted. This conversion may change when the contract is won due to inflation or other overhead cost changes. Finance monitors the cost of the program and alerts the PM of any over expenditures.

2g. Material

The Material organization, on the right side of the chart of Figure 8, is only concerned with material to be purchased for production. A list of material is provided by Engineering so that the Material department may do a search and determine if the items are in the company's stock. A Material person has to determine if the paper work is proper, if the drawings are released and if the parts are Non-standard. He has to know if a Specification Control Drawing

(SCD) had been prepared in order to control and make a part.

The Material department then would assign a purchase order number to the part and submit it to the Purchasing department. Material is also required to sign off all release drawings. Production Inspection would not buy off a production unit without all the proper signatures. Some companies don't have a separate Material department and use the Purchasing department to cover this function. Also, some companies may provide a separate Material group to service Engineering development only, as shown on the left of the chart. He basically has the same responsibilities as the production Material person except on prototype work he doesn't have to be concerned with SCD's or release drawings. Material in some companies, are responsible for incoming deliveries of parts and equipment and would record the cost and dated. Government Furnished Equipment (GFE) must be accounted for and are stored by Material.

2h. Purchasing

The Purchasing department on the right side of the Interface chart, is for production. Advanced Materials Order's (AMO) are provided to Purchasing from engineering so that they may order long lead items. Connectors are notorious long lead items, sometimes it takes up to eight months for a delivery.

The one problem with military parts today is that most supply houses do not stock parts like they used to. In the 50's and 60's, big warehouses were used to stock military parts because the military was the principal customer in the electronic world. Today, with the advent of computers, TV's, and VCR's; the commercial industry now dominates the warehouse market.

If a military connector is ordered one may have to pay a production tooling charge because the supplier has stopped producing that part and would have to reopen up his production line to make that part again. As a matter of fact, the supplier may agree to supplying the part to a buyer and then delay the production start-up until he gets more orders from different companies for that same part. The supplier is not going to tie up his production line on minimum orders. This is where it takes an experienced buyer who knows when he is being purposely delayed. With the Russians reforming more and more and with the US military spending less, the warehouse situation is expected to get worse.

The biggest problem now is obsolete parts. A military aircraft's life expectancy is supposed to be ten years. That no longer applies, because some of my designs are still flying 30 years later with no end in sight.

The question now is, how does a supplier provide spare parts to the military when the parts are obsolete and won't be made any more? Well, some smart entrepreneurs took advantage of this market and they started Hybrid companies. They would produce almost any chip that was manufactured before, except now the price has increased tremendously, maybe up to ten or twenty times the original cost, and the quality may not be the same. So purchasing parts for the military can be frustrating. Designers and managers must be aware of this when they are selecting parts.

For production parts, the first thing a buyer has to do is go out for bids. Military contracts require that all sub-contract work, including parts for production, must be competitively bid to spread the wealth around and not have one large company monopolize the distribution industry. It is supposed to help small business.

As a matter of fact, a large Aerospace company cannot even buy parts from a large distributor without showing just cause. The bidding alone could take up to six months, so this is why an AMO is required, to get a jump on any long lead items. Limiting the suppliers, however, has the effect of increasing the costs of many parts.

As was mentioned in Chapter 1 and should always be a consideration, is that the cost of certain parts is relatively minor in comparison to the cost of labor. This is why it would be better to place orders for parts early, even though they may be dropped from the design by the engineer. If one considers lost time due to delays, or waiting till the design was firm, many manhours would have been wasted.

Parts are usually the cheapest item in a military program especially in the early stages of a program. I have never been reprimanded for over-spending for parts, but over-running manhours would immediately alert upper-management.

As was stated under Material above, production parts must be fully approved per contract and re-screened to be accepted for military programs. Re-screened parts means that each component has to be individually tested under extreme temperatures before it is delivered. Generally, a qualified outside house is used to re-screen parts.

Some companies also use separate buyers to handle prototypes and development parts as shown on the left side of the Interface chart of Figure 8. Prototype parts don't require re-screening and may even be of commercial grade. If the prototypes are refurbished for production then the parts must be replaced with re-screened parts.

Another thing to consider for prototype development is, when an engineer is looking for a part he would call certain suppliers to see if it is available. Now this is where the fun begins. An inexperienced engineer would take the supplier's word that the item he wants is available and would be in production. The engineer would incorporate this part in his design and find out much later, possibly in the production stages, that those parts are not going to be manufactured. The supplier misrepresented the information to the engineer.

What happens is, the supplier has a certain number of these parts in stock on a shelf somewhere in the country or in Canada. They would deliver X number of pieces to satisfy the prototype needs to make a sale and knowingly they got rid of dead obsolete parts.

Engineering purchasing can sometimes be a dirty business. This can even happen to experienced engineers. Don't trust anyone! An engineer should first do a little investigation on his own. Make them send a written statement that the item would be manufactured. Maybe go through a buyer. The supplier usually would not tell a buyer that a part would be in production if it is not, because he may be libel in court, but he would lie to engineers. When an engineer calls a supplier the first thing the supplier would ask is, "who are you?" When he says he is an engineer the supplier knows he has a free ride.

2i. Administration

The business administrator, in the electronics world, provides a very important function. On the right side of the Interface chart the Program Administrator prepares the financial tools to monitor the entire program. This Administrator receives all of the quotations from each department during the quotation process, including the quote from the engineering Project Administrator shown on the left side of the chart. He assembles all the data and helps prepare the final quotation. He may have to interface with each department to discuss their quotations separately. He does not, at any time, try to reduce or influence any quotes from these organizations. This is the job for the EPM or PM.

After all the numbers are in and the totals are accumulated, he goes over it once again with the PM and the EPM. The quotation then is distributed to upper-management and the Finance department for review. Upper-management may reject the quote because it is too high so the quote may bounce back into the PM and EPM's hands and the cycle begins all over again.

When the job is finally awarded to the company, the PM and EPM would repeat the quoting procedure and again would ask all departments to redo their quotes to insure every department is covered and that there are no major cost deficiencies. The Program Administrator would again accumulate all of these quotes and present them to the Program Office. The PM and the EPM would probably have to go back to each department to trim down their quotes again. It's a hassle but that's the way it works. If the program is awarded the same year there is no need to requote.

The Program Administrator prepares finance schedules for each department, showing budget spreads and accumulated manhours to be used. The Program Administrator works closely with the Program Manager. He has to set up technical and financial meetings. Some meetings may be with the customer.

This Administrator would prepare the agenda for these meetings and may take the minutes. In the PM's meetings with the different departments, the PM is more concerned with the schedule and the budget. What sometimes follows, is usually a shouting match between the PM, the EPM and the different departments because of schedule slippage or over-runs. There would, of course, be some under-runs and the administrator may recommend a shifting of budget to get the program back on course.

The Project Administrator on the left side of the Interface Chart of Figure 8, is more involved with the individual departments. For engineering he would only be concerned with the EDM's internal schedules and budgets. There is no set rule on how many Project Administrators are required. Depending on the size of the program, each department may have their own Administrator or one Administrator may handle several departments.

These Project Administrators should also work very closely with the Program Administrator. In Chapter I, there were several charts that the Project Administrator and the Electronic Design Manager had to prepare together for engineering. These charts are the tools that the Project Administrators uses to monitor internal schedules and budgets.

Many Engineering Managers feel they do not need administrators because they feel that they can do it themselves. In most cases engineers make very poor administrator. They would have enough trouble with the technical aspect of the program without worrying about doing the accounting work.

One reason some Engineering Managers prefer to do this function by themselves is so that they can manipulate the budget and not have to account for it. Check and balance is again the name of the game. Someone has to keep these managers on track.

I once worked for an upper-manager who was against the idea of working with administrators on preparing schedules and budget spreads, because he couldn't hide his over-runs and shifting budgets between different programs. I forced my upper-manager to be ethical whether he liked it or not. It almost cost me my job. He did re-assign me, however, to get rid of me.

2j. Product Support

During the quotation period Product Support should become heavily involved. Product Support consists of Manuals, Maintenance, Logistics, Field service, Ground Support Equipment, and Automatic Test Equipment. This is a very large portion of the proposal. My definition of a good design has always been, "The equipment is only as good as its repairability." If the equipment fails in the field and is difficult to maintain and repair, it is a poor design.

As was mentioned many time before, `Mean Time Between Failure' is the key requirement for the military. When the military is in a combat situation, it must have reliable equipment and they must be made easy to repaired. Logistically, spare components and modules have to be provided with the proper manuals and controls to keep the equipment going.

The PM interfaces with Product Support organization during the quoting stages and after the program is awarded. Since this is a substantial portion of the program it is imperative that the PM, Master Scheduling and Finance monitor their activity closely. The EDM may have to interface with Product Support by providing drawings and test instructions.

Automatic Test Equipment (ATE) is a separate department in Product Support because it requires designing Interface Devices (ID) to interface with the Unit Under Test (UUT) and the ATE. It also requires special software to program the ATE to automatically test the UUT. There are three ways to handle ATE:

1. The ID could be designed by the Design Manager's department on the left side of the Interface Chart and the Software be done by the ATE department.
2. Both the ID and the Software could be handled by the Design department.
3. Both the ID and Software can be designed by the ATE department.

I favor the number 1 approach even though there may be some bickering between departments, as was the case I mentioned above, but this can be resolved. In number 2 there is no check and balance. Number 3 is alright except the work may not be steady enough to keep a design organization in operation. Also, there is no check and balance.

To keep a design organization intact a lot of ATE work would have to be required. In my last company many job-shoppers were hired to do the software. The hardware was designed by a separate department, but was taken over by ATE using job-shoppers. The problem is, when the job is over what happens to the help? When a job is ending, the first people that are released are the job-shoppers. Then, how does one follow up the design? The ATE and the ID designs would not be completely proven out until production units are in the field. There would be changes made, so if the job-shoppers are gone who would make the changes?

The reason I prefer number 1 is by having a separate design organization they would always be available for support. They would always have plenty of work in the company to maintain a decent man level of full time employment. These employees are always available to follow up the design and make modifications.

This same ATE organization is responsible for Ground Support Equipment. In this case I prefer number 2 because the initial UUT was made by the design department and they know how to test it. This also applies to outside suppliers of a UUT.

2k. Legal

The legal department works with the PM and with contracts to insure the legality of a contract. They also become involved with sub-contracts with other companies.

One time, my department was awarded an electronic black box design for our aircraft. The job was bid satisfactorily and I was told that I would get the assignment. For some reason, the Purchasing Department assumed that the Project would go to the same Vendor who did a similar job on another aircraft, and they started proceedings with that company. Purchasing should not have done this because it's against company policy. In-house work should not compete with outside work.

Legal got involved and decided the contract had better go out or the company may be libel. I don't know how other companies handle this type of a problem, but in my company, the Purchasing department, not engineering, actually run the company. They seem to get away with anything they want to do. It cost me a million dollar job.

As an engineer, I had a constant battle with Purchasing because they are only concerned with their empire. The more sub-contract work the company does the bigger their empire.

In the "make or buy committee" in my company, my Engineering Design Department was not represented, so many `buy' decisions were made that could have be done with company personnel. The loss of production income is never considered because Purchasing has the most influence on the `buy' decision. They are only concerned with more work for their own empire.

So, when a sub-contract design goes out for bid's, Legal plays an important function and is very busy protecting the company. In another situation, my upper-manager bought computer equipment with research money. That was illegal. The computers should have been purchased with capital equipment funds, which is part of overhead. We could have been sued by the government but the Legal department stepped-in in time to resolved the problem. Check and balance again. Unfortunately, I was heavily involved in that mess because my upper-manager somehow was on vacation. I had to take the flack from Legal and my Director of engineering while my boss enjoyed the sunshine in Hawaii. So goes the life of an EDM.

2l. Production

During the quotation period, Production is asked by the PM to provide the manufacturing cost of X number of units as called out in the RFP. In some cases there may be small quantities ordered, which may not be cost effective to tool up in a production line. In this case it may be more cost effective if engineering handled the job.

When there is a large order then Production would be asked to quote. This does not mean that when the job comes in, production would automatically receive the job. Other factors may come into play and the PM may decide to go else-where for his production. It has happened. However, the PM does need the original production quote to price the job.

Sometimes, in the RFP, it would be requested that Production's quote be separated from the engineering quote, or that the Production quote would be submitted at a later time. The military may only want to buy a few items and then decide not to spend the money for production. They need the production cost for their records and possibly, at a later time, they would have the funds to go into production. There is a lot of `ifs' in this business. This is one reason why the RFP has to be read by the PM, EPM, Contracts, Finance, Master Scheduling, and the EDM to insure that the company is answering the RFP properly.

When the job is finally awarded, the PM and the EPM would ask Production to re-submit their final quote. Master Scheduling would then set up a scheduling program for production. Unfortunately, production always has to wait for Engineering to complete their tasks before they can begin work. Slippage in engineering schedules causes a greater slippage in production, because of the lead time needed to receive production parts and in ordering of any special test equipment. This is one reason why a Production representative should be present at all design reviews.

One of the last job I was on, the PM did not want a production representative in the design reviews because he was thinking of going to another facility in another state for production. This was understandable, but it sure made a mess of the production schedule when a new production house was brought in later on in the program. The PM thought he was going to save costs but in reality it cost more and resulted in late deliveries. He didn't understand that the Design Department already had a working relationship with the first Production facility and to change horses in the middle of the stream would introduce numerous new problems.

Production interface with the left side of the chart is mostly with the EPM and Engineering. The Liaison Engineer (LE) is the coordinator. Interfacing is usually in the form of drawings, parts list, AMOs, and Functional Test Requirements.

The First Article Test (FAT) would be made by production to prove to the customer that the unit would meet production testing using production equipment, and it would also pass Integration Laboratory Testing. This gives the customer a comfortable feeling to know that all the follow on Production devices that are tested with production test equipment would work in the field.

2m. Vendors

On the left side of the Interface chart the EPM, EDM and Purchasing would interface with all Vendor work required for development. A Specification Control Drawing (SCD) would have to be produced by the Systems engineer for the Vendor. The size of the job would determine how it is to be handled. Vendor type work now being discussed, has to do with the support of an in-house design and is handled a little different than the Sub-contract work of an electronic unit that is to be developed externally. Sub-Contract work requires an ES instead of an SCD.

Engineering plays an important roll in the selection of this Vendor. For the prototype work the engineer has the option of going to any Vendor he chooses, but when observing the right side of the Interface Chart, Production Purchasing doesn't have this option. He has to put the item up for bids to meet Military contract requirements. Printed Wire Boards are a good example of items that would go out for Bids. The design engineer may have his prototypes made by one Vendor but then another PC house may out bid the first Vendor for the production contract.

My company has a list of qualified suppliers for many parts. In the case of PWBs, these houses were initially inspected by qualified people and had to be approved by them. They also have to be certified by the military to manufacture these items. My company had about ten PWB houses that were approved, and it's a cost advantage for the Design Engineer to use one of those Vendors for his prototype work. Tooling costs doesn't have to be paid twice if the Vendor is selected ahead of time by engineering and is the same for Production.

My PWB engineer had a case where he thought he was using an approved PWB house for our prototype work because we used them before, but he found out later that they were already disqualified for sloppy manufacturing. This is why interface between organizations is important, to catch these situations at the beginning in the design reviews and save problems later.

The biggest problem now in military electronics is obsolete parts and companies going belly-up. As discussed before, for economic reasons, a Vendor may decide not to manufacture certain parts any longer. He may have to go out of business. I have had situations where I had to redesign the electronics of several PWBs because a PWB manufacturer actually went belly-up. The original designs were already twenty years old. Vendor selection can be frustrating sometime and very risky.

2n. Customer

Customer interface is usually through the Program Manager. Technical information may be discussed by engineering but he must be careful because anything that he agrees on with the customer or, for that matter, with a supplier, becomes binding in a court of law. Any transmission of documents should go through the PM. This is especially true when dealing with a Sub-Contractor. Never agree to provide specific goods or services, unless all parties are in agreement over costs or the feasibility of such items.

One doesn't assume anything until it is thoroughly checked out. As was mentioned before, it is not easy for a PM to represent engineering to make technical decisions, and conversely it is not easy for a designer engineer to make contractual decisions. Dealing with the customer can be very complicated. This subject is covered in more detail when the Program Managers duties are discussed later in this Chapter.

The left side of the Interface Chart deals also with the EPM, the EDM and his design organization. These will also be discuss later on in this chapter when the EPM's duties are covered in more detail.

2o. Organization Wheel

Another way to illustrate The Service Organizational Structure is with a Wheel as shown in Figure 9 on page 209. As one can see, the Design Department is in the hub of the wheel and interfaces with all of the other departments through the spokes of the wheel. The Project Manager also interfaces with these organizations from the top of the rim. The Program Manager and the Customer are located outside of the wheel because they do not interface directly with organizations that are located inside the wheel but they do interface with Production from an administrative point of view. The PM is not shown. The user is actually the military in the field. The customer is located in Washington, DC, who provides the contract.

All of the spokes are pointing to the design department in the hub because the technical interfacing is done through the EDM. These other organizations in the spokes were also shown in Figure 8 as Support Organizations. Figure 9 is self explanatory, but it is important to understand that everything revolves around the Design Department in a Service Type Organization.

It essential that the design organization be of the highest quality. The whole program depends on the communication between the design department and the different support organizations. The Project Manager or EPM interfaces with these other organizations but mainly with regards to financing and schedules. The technical interface is always done through the Electronic Design Manager. The EPM would be able to converse with the support engineers on technical matters at the design reviews.

The PM would also be present at these design reviews and may provide some indirect technical support. An experienced PM can usually contribute in these reviews. One problem that has to be guarded against, is that some PMs tend to want to control the meetings and even make technical decisions. Even though he is in charge of the program he should not be dictating the design. This has happened too often. The meeting should be controlled by the EPM. The EPM and EDM have to make sure that this does not happen by using their best diplomacy. They should, of course, listen to any suggestions made by the PM but the final implementation should be made by the EDM. The meetings should not become a major power struggle.

The EDM may interface with the support engineers at any time during the program, but not all of them should be invited to the reviews at the same time. It would take too long to cover all issues if all were present.

Each week only certain subjects are put on the agenda as they become appropriate and timely in the schedule. Part of the design review schedule should show when each support department's subject would be discussed. Certain support departments should be present at all design reviews such as; Reliability, Quality Assurance, Testability, and Production. The Support Organization's responsibilities were covered in more detail in Chapter II.

3. PROGRAM MANAGER

The Program Manager is the administrative force of a program. As was shown in Figure 8, he interfaces directly with all departments on the right side of the chart and indirectly with the left side of the chart through the EPM. The PM's position is considered the most complex position in an engineering program. He coordinates directly with the customer and with upper-management, including the President. He is indirectly in charge of the program. He should never, however, make engineering decisions unless he has consulted with the EPM and the EDM first.

Many a PM has agreed to provide additional electronics to the customer not knowing the repercussions in the design. As an EDM, I had bid on a fixed price program for an electronic test box and later the PM told the customer, without my knowledge, that for the same price that I quoted, additional capabilities would be provided in the test box. With this extra capability they had little trouble selling the contract. He was proud of himself for the way he won the job.

When I was informed of the extra capabilities in the box, I was furious. He simply walked away and said, "It's too late now, it has to done." Not only was the budget now too low but, because of this additional work, the enclosure increased in size by 1/3 and the weight exceeded the Mil Spec carrying limits. I lived with that bad decision for 18 years until I retired. I cannot stress enough on how important it is to work together. When one party goes it alone, it usually becomes a catastrophe.

On one of my last major jobs, before I retired, the ATE Department sold a 50 million dollar program to the Navy. It was quite a feat. They sold the program on their own, except that I bid the hardware portion of the design. The problem was, when the program started there was no PM assigned, there was no organizational structure, and there was no Program Plan.

After one year the Navy began asking about the Program Plan. On the technical level these ATE engineers were very good and have plenty of experience in the ATE field, but they had no experience in running a complete program under Mil Specs. They did not create an Organizational Structure which resulted in overruns and eventually caused late deliveries.

As was mentioned before, the Program Manager interfaces with the EPM and with all of the departments on the right side of Figure 8. The left side of the chart is mostly engineering development activities. The PM would be indirectly involved with all of the departments on the left side only during the design review meetings. He would conduct his own weekly meetings with the EPM and most of the departments on the right side of the chart. Quality Assurance may also be included in the PM's meeting. It all depends on how the PM wants to handle his program.

In the Aerospace company that I had worked for 30 years, Electronic boxes were treated in two ways; as an in-house design as part of the Aircraft avionics, or under a separate contract directly for the customer. If the box is part of the aircraft electronics, then the Design Department is somewhat treated as though they were another supplier. The PM is not involved in in-house designs that are part of the aircraft.

The Program Office is normally in charge of the whole aircraft, and several PMs may be assigned to different sub-systems. One PM may be put in charge of many sub-systems or only over one. When an in-house design is to be made directly for the customer only one PM would be assigned to administer that program.

The functions outlined in the Interface chart of Figure 8, still apply, but in the above ATE case, Master Scheduling carried two hats, Master Scheduler and acting PM. He just used the name Coordinator instead of PM. The EPM also carried two hats; EPM and Systems Engineer.

This was not the normal way that my company does business. It does have an Engineering Procedures Manual that describes the proper way of running a program. It was totally ignored by the ATE upper-manager.

A PM and a System Engineer should have been assigned to this program. By given Master Scheduling this extra authority many items fell through the crack. Even though he was the coordinator, his main concern was to meet the schedule, regardless of whether the equipment worked to its most reliable requirements. That was his weakness.

The same is true of an EPM doing the system's design. A System Engineer, in my estimation, is the "cream of the crop" in engineering, and most EPM's are not qualified to wear that hat. When an electronic box is required, the designer has to rely on the System Engineer's analysis to provide the best Equipment Specification.

Many times an ES was written by an EPM during the design or after the design was completed. This, obviously, would cause unnecessary changes to the design. In several other cases engineering designers wrote their own ES. This approach is completely against all policies because the ES could be doctored to suit the design. The ES should be written by the System Engineer to provide proper `check and balance'.

On small jobs, having no PM or EPM may be fine. There are a lot of small Panels and Junction boxes that are used on an aircraft that do not require a full blown organization. An EPM, however, could be assigned to several of these smaller boxes. When the box is complicated and expensive the procedure outlined in the Interface Chart of Figure 8, should be followed. There are no short cuts in this business. PMs do not get involved with Panels and Junction boxes.

In general, each company has their own internal way of doing business and as long as their policies are followed the job usually runs pretty smoothly. There is the right way, the wrong way, and the company's way of doing business. The company's way is the only way because they have been in the business for a long time, and the policies they have outlined in their Procedures Manuals were created from past experiences and have worked for them before whether one agrees with them or not.

The second method discussed above is dealing directly with the customer. This is when the military would put out a Request For Proposal (RFP) to design and develop an Electronic box. In these cases, a PM and a EPM should be assigned to the proposal and the Interface Chart procedure outlined in Figure 8 should be followed very closely. The following paragraphs provides a breakdown of the PMs duties and his responsibilities:

3a. The Program Manager's duties

The main function of a PM is to be the administrative manager of a program. He is in charge of the program but should rely on the EPM for technical advice. Some PMs are technically capable because they may have came up through the ranks in engineering and they may have a tendency of making engineering decisions. It is hard not to interfere in engineering decisions when a PM was once there. A PM must avoid making engineering decisions and should have a good relationship with the EPM. By using his best diplomacy he can make some technical suggestions and any good EPM would acknowledge them and may even implement them through the EDM. The PM's duties start with the Quotation. There are many ways of soliciting a new job. Request For Proposals (RFP) are published by the military in the Commerce Business Daily and other publications. The RFP may come directly from the military if they are interested in the company's product line. Other companies may put out an RFP for subcontract work.

It is not the responsibility of the PM to go out and solicit new business but he may if he wishes to. Winning a job is every employee's responsibility. Remember, quoting a new job has the highest priority of any company. It sometimes means temporarily pulling people off of an existing job to participate in a new quote.

When the RFP does arrive the Program Office has to determine if the job is in the company's line of business. Does the company have the expertise to accomplish this task and would it be profitable? Once a decision is made to bid for the job, then the PM's work really begins. (For the sake of discussions, the Service Type Structure will be followed).

A PM would be assigned to this task by the Program Office who then has to notify Engineering so that an EPM can also be assigned. A copy of the RFP should be given to the EPM, the EDM, and all the departments that are involved in the program. The new PM now has to prepare a Preliminary Program Plan that would be submitted as part of the proposal. This PM may not be the same person who is finally assigned to the job.

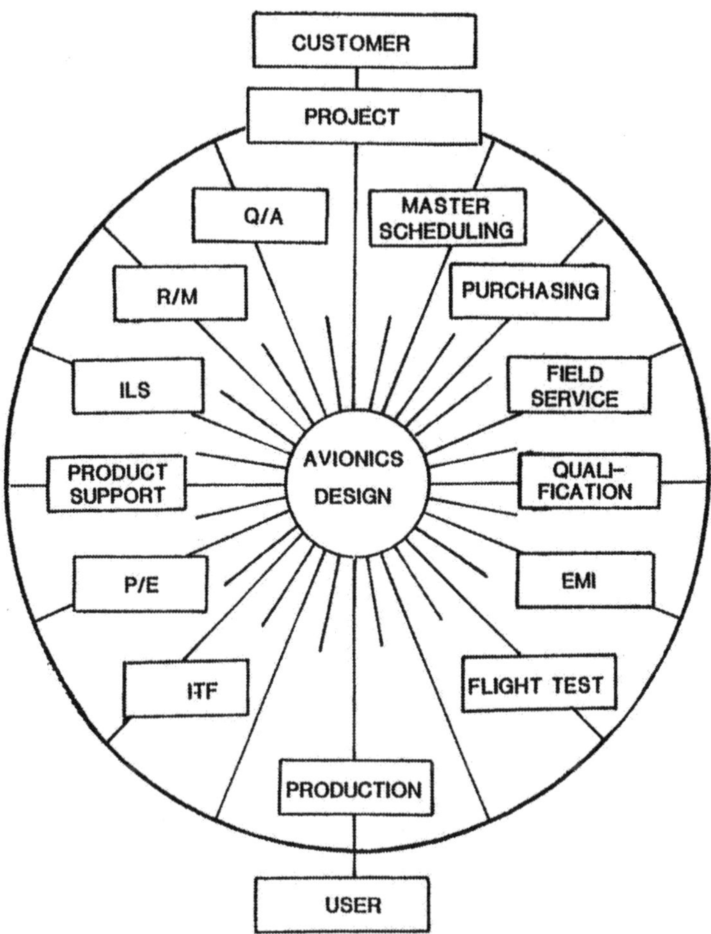

Figure 9. Organizational Wheel

3b. Program Plan

The preliminary Program Plan is normally the PM's responsibility, but he should go over it with the EPM and the EDM for technical clarity before he publishes it. This Plan establishes the requirements and sets guidelines for consistency of quality, content, and format for the program. The Preliminary Program Plan is first produced to answer the RFP quickly so that the customer can evaluate the company's method of doing business. When a contract is received this plan would have to be finalized and re-submitted as a Program Plan which then becomes the controlling document that:

1. Provides background data, and indicates the purpose and objectives of a program.
2. Consolidate applicable contract and program directive, provisions, or references and requirements of the customer, company procedures, and documented management decisions.
3. Defines responsibilities, identifies major tasks, and gives specific management direction in design, test, and certification of a product.
4. Assure compatibility with other objectives, actions and instructions, considering interface and flow of work between company organizations.
5. Establishes charted schedules showing timetable and significant milestones for all organizational activities.

The following items should also be included in the Program Plan:

1. Description of the program
2. Organizational Chart
3. Overall level 1 & 2 schedules.
4. Engineering responsibilities
5. Budget Charts
6. Department assignments
7. Budget Allocations
8. Tasks to be accomplished
9. Qualification
10. Integration Tasks
11. Production procedures
12. Functional Test Assignment
13. Preproduction Performance Test
14. Design reviews with customer
15. Data management
16. Drawing control and release
17. Non-standard parts control
18. Data submittals
19. Equipment Specification
20. Purchasing
21. First Article Test
22. Deliveries
23. Product Support
24. Logistics
25. Manuals
26. Training
27. CDRL schedules
28. Spares

In the Program Plan the PM should prepare a short statement on each of these topics, not necessarily in the order shown. These statements are basically descriptions of what the responsibilities are of each organization and what hardware and documentation would be provided to the customer. The Program Plan is the governing document that is to be used by all of the organizations in the company that are involved in the program for bidding purposes. The EPM would use this document to produce his own Project Plan.

Some companies may not go through all of the above exercises because it doesn't have the time. I have quoted many large jobs in three days. Talk about WAG's. Unfortunately, the original RFP sometimes gets delayed in upper-management, hassling over whether to bid the Proposal or not and doesn't get to the bidders till just a few days before submittal time.

A good PM would have a Program Management's meeting with all of the departments to discuss the tasks in the Program Plan and answer any questions. After all departments have examined the RFP and the Program Plan they would all submit their quotations. The EPM would go over the Engineering Quotes separately and massage them before submitting them to the PM. All other organizations would submit their Quotes directly to the PM. All Quotes would consist of cost, schedule and manloading at a minimum.

The Quote would go back through upper-management for approval and finally to Finance. The departments that submit quotes directly to the PM are those on the right side of the Interface Chart of Figure 8. However, depending on the company's structure; Master Scheduling, Contracts, Finance, Material, Purchasing and Legal may be part of overhead, and do not charge directly to the program. In that case they do not have to submit any quotes, but would be included in the program by finance as overhead which is usually a percentage of the overall quote.

3c. Contract Awarded

After the contract is awarded to the company, the PM would have to go back over the RFP and the latest SOW and, if necessary, revise his Program Plan. After all, the contract may be awarded many years after the RFP was originally submitted and a new PM may be assigned to the program. The first Plan was also a preliminary one so all of the organization now have to go over the RFP and SOW and re-submit their proposals again.

At one time I won a job five years after my Quote was submitted, and the people that worked on the original quote had retired or were transferred. The re-quote can sometimes be shocking, especially if it was a Ruff Order of Magnitude (ROM). The problem is, the ROM is supposed to be an estimate but it never turns out that way. It usually ends up being the final quote to the customer and engineering is stuck with it.

When one wins a contract it's pretty difficult to renegotiate the costs later with the military, because the company probably won the contract on the merits of its low bid in the first place, especially if it's a fixed price contract. When an EDM is over budget, for example, and tries to explain that the Quote was a ROM and he needs more budget, it would go by deaf ears. There is no sympathy in this field.

After the contract is awarded, the PM's work really begins. He now has to go over all of the new quotes again, and massage them to get the price down close to its original cost. Invariably, everyone's quote would be extremely higher the second time around. This increase happens because they now have more time to examine the new Statement Of Work that was finally put out by the customer.

The SOW may not have been available before and is usually an expansion of the RFP with a lot more detail. For the same reason, the customer had more time to produce the SOW than he did for the RFP, and try to slip in a few more requirements that the company was not aware of.

The first thing the PM, the EPM and the EDM have to do is examine the SOW to see how much more information has been added. A comparison is then made with the RFP. The SOW is then red marked as to what additions were made and what items should be removed or renegotiated with the customer.

A meeting with the customer is then set up so that the PM can present his views on the subject using view graphs or slides if necessary. It is imperative that the PM be well prepared because this may be his last shot at reducing the requirements or increasing the cost.

What usually happened is, when the customer was preparing the SOW he would refer to another similar program and produce, what I mentioned before, a "boiler plate" SOW. The similar program may not apply to the new program so the PM has to re-negotiate with the customer and correct any anomalies or the program may be in trouble before it begins.

For example, in the ATE program that I referred to before, the ATE department bid the job themselves not using a EPM or a PM. They negotiated with the customer directly and won a large $50 million dollar contract. That was great feat but, they did not follow any particular structure, nor did they include any of the support organizations in their quote, such as Reliability, Maintainability, and Thermal.

They did not understand that the Electronic Design Manager requires these organizations to support them and to monitor the design and provide check and balance. The Design department does not release drawings, either. They do not submit non-standard parts.

As one reads through this book he will be coming across those three important words "Check And Balance" over and over again. All of industry evolves around those three words. Even with the U.S. government there are three check points; the President, Congress and the Supreme Court.

The ATE Manager said to me that he thought the support organizations were part of my quote. I told him that I also work for the same company as he did and I have to follow company procedures. Those department don't work for me. When this job finally came in, I was not the EDM and the new man was stuck trying to meet the SOW.

In the electronic business, check and balance is furnished by the support organizations and the customer. As was mentioned many times, the military has the equivalent support department in Washington who would only converse with their own piers in the contracted company. These military departments would be conversing with their equivalents periodically to insure that the design is on the right track.

Because the ATE department did not follow the above structure, they did not know what items in the SOW could have been negotiated out. As a consequence, the new EDM assigned was constantly asking the Navy for waivers in the SOW in order to meet their design requirements.

The end of the program resulted in overruns and major schedule slippages because the SOW was not renegotiated before the program started and all the players were not included in the quote. My company had to eat the overruns and lose some of their profits.

After the SOW is renegotiated, engineering and the other organizations still have to go back and determine if they can still do the job for the quoted price. If the price is still too high then the PM and the EPM have to go back to these organizations and try to reduce their costs or the PM may have to make a decision of cutting out some of the less important support functions. Engineering may have to look at other ways to reduce the design costs. One can only reduce the cost just so much before quality is impaired.

There is a saying used by many subcontractors, "Bid low and pray for Engineering Change Proposals (ECPs)." A lot of companies went "belly-up" using that philosophy. It's not a good idea to reduce the cost drastically just to win a contract and hope there would be ECPs to recoup one's loses. It's a risky way of doing business.

Renegotiate with the customer. Be honest and explain the errors in the SOW at the beginning. Maybe, a suitable cost compromise can be reached. One of the misconceptions that many companies have is that the customer is hard-nosed and may cancel the contract if there is an admission of under-bidding or technical errors.

The military is well aware of many of the supplier problems. They have been through it many times. All one has to do is talk with them. It is surprising how they will cooperate. Remember one thing, they want and need that design and they know which companies are capable of producing it. They want the program to be successful within a reasonable cost.

Now that the price has finally been agreed upon the EPM has to prepare his final Project Plan and the

Systems Engineer has to prepare an Equipment Specification (ES). These two subjects will be covered later. Once this is done, the design begins. The PM normally would remain with the program to the end, unless he is given another assignment. He should attend all the design reviews and may even attend the EPM's upper-manager's reviews.

He would chair his own meetings with the EPM, Production, Master Scheduling, Purchasing, Test, Q/A, Administrations, and who ever else he needs to provide him with the information he needs to make his own presentation to his upper-management. With this information he may have to make a presentation to the President depending the size of the company.

In the early part of the program the PM's meeting should be held weekly, and then possibly less often as the program progresses. The President's meeting could be monthly. The PM would also have to chair meetings with the customer quarterly or sooner, at the customers facility or at the company's facility.

All communication between the company and the customer is through the PM. This was mentioned many times before, and is very important because a commitment from any employee to the customer is binding in a court of law. Lets face it, the PM is responsible for the controlling the contract. The president or his upper-management would come down on him if the program has contract problems. Any program may have problems and most of the time it can be resolved if they are mentioned early enough. No surprises!

To be a PM, in this business, is very difficult. Some of them are promoted up the ranks directly from engineering. Some came from Administrations. Some came from the military with similar experiences. Like all professions there are good PMs and there are poor PMs. The PM's personality is especially important. He should be sophisticated, extremely intelligent and exceedingly patient. He has to deal directly with the customer, so he must know the military business and how to converse with them. He has to have the deepest respect of all those below him and those above him. He is the `supper star' of the program and everything depends on his relationship with the customer. Everyone that deals with him in the company should treat him as a Very Important Person (VIP).

4. Engineering Project Manager

The Engineering Project Manager (EPM) is another difficult position. He is considered the technical manager of a program. All of the engineering problems go through him. He has to interface with the PM, with Production, with all of the support organization, with sub-contractors, with purchasing, with engineering and with the customer.

Even though he is in charge of the technical end of the program, he does not dictate the design. One problem with this position is that no one works directly under him. He doesn't have authority over any particular department or individual. To get things accomplished he has to fen for himself. On top of all this, he has to produce many charts, attend many meetings and make presentations to his upper-manager on the status of his program. Some of these charts and data submittals he has to prepare for his company are as follows:

o Man-hour Tab runs; week and month

o Performance Charts (Earned Value)

o Configuration Allocation Reports

o Company Forecasts (twice a year)

o Level 1 through level 4 schedules

o Weekly Status Reports

o Weekly Look Ahead Charts

o Weekly Milestones Charts

o Engineering Cost/Schedule

o Performance Measurement System

o Man Power Spread

o Engineering Definition

o EPM 6th week review

o Project Plan

o Program Office Memo's

o Action Items - uppermanagement

o Major Technical Problem Reports

o EPM Design Reviews

o PM design Reviews

o Customer Design Reviews

o Performance Report

o Program Directives

o P.P.I's, TA's/TVNA's, TRACS

o Trip reports

o Director Variance Analysis Report

o Program Tracking

o Cost Estimates

o PPI Man/Material/ODC Projections

o Individual Work Assignments

o Drawings, ECO's and LEO's

o Rework Instructions

o DRCN's

o Forecast Variance Sheets

o Matrix Charts

o Drawing Trees

o Sunrise/Sunset meetings

o Master Scheduling meetings

o Specification Compliance Analysis

o Specification Control Drawings

o Engineer Project Variance Analysis

o Cost Performance Index

I have purposely provided this list to demonstrate the amount of work it takes to be an EPM in some companies. My argument against most of the above tasks is, "How many charts and reports does an EPM have to prepare to tell his upper-management that his program is in trouble or on track."

These graphs and Tab sheets are supposed to aide the EPM, but instead it just gives him an enormous amount of work. It may take days to produce just one of them. These tasks are just too much for one man. Yet, upper-management seems to think they are needed to define and track a program.

I will not interpret these charts because each company is different and some may require even more tasks. Most of the tasks are self explanatory. These assignments are supposed to be tools to help the EPM. They were created by professors in colleges as the theoretical classical approach to provide procedures for monitoring and controlling major programs. It had to be invented by non-engineering personnel because many assignment are just redundant exercises. I, obviously, do not approve of most of these tasks and, as a result, I came up with my own tracking system described in Chapter I. To be truthful, I don't know what half of these charts mean.

The above tasks are just many more reports and charts that an EPM has to produce for upper-management. On top of it all he still has to do his own engineering work. He still has to get involved in the program and be the Project Manager. His main duties consist of preparing proposals, being involve in the parts buying, drawing release, filing clerk, sub-contractor evaluator, Fact Finder, Shipping, Trouble Shooter, and Task Assigner. In addition, the EPM has to be somewhat of an expert in Systems Engineering, Publications, Field Services, Testing, Drawings, Graphics, Electronic Engineering, Mechanical Engineering, Software, Thermal, EMI, Maintenance, Reliability, Packaging, Finance, Technician, Technical Advisor, Lawyer, Contracts and Logistics to name a few.

As one can see, it takes many years of experience to become a good EPM and understand all of his responsibilities. The EPM is a person who usually comes up the hard way. There isn't any short cuts to learn this position. Unfortunately, most companies do not realize how complicated this position is, and don't select the proper person or provide him with enough help. Upper-management's weakness on this subject will be discussed later.

I was an EPM for several years and to be honest, an Electronic Design Manager's position was a lot easier for me. Maybe it was because I was in the design field longer and was more comfortable there. One thing I found out as an EPM's position was he can never casually walk to a meeting or to another department because he doesn't have the time. There are just too many duties for him to cover. He has to attend sunrise and sunset meetings and design reviews every day. It is nothing for a EPM to put in 12 or more hours a day and work six days a week with no extra pay.

The EPM is next on the firing line after the EDM. The whole engineering structure evolves around him. One of the most famous sayings that was mention before on taken extra responsibility was, "It's not my job; it's the EPM's job." When I was an EPM and I would ask someone to take on a particular task, I would always be given back that famous response. People would not take on a task that they are not directly responsible for because if it gets `screwed up' they don't have to take the blame.

There was a case, mentioned in Chapter I, of the man who refused to ask questions from the production manager that was located just next door to him. It's not his job, he said. This type of behavior was due to upper-management's scare tactics. In this one particular company they used many scare tactics such as, "You will be voluntarily dismissed," syndrome. Too many Hitlers would cause employees to do minimum work and not to go beyond their duties, even if it's for the benefit of their company because being wrong or making a mistake is worse than trying to improve the product and one could lose his job.

4a. The EPM's duties

The main function of an EPM is to be the Engineering Manager of a program to control all of the engineering departments including the EDM. He doesn't do the actual design but can contribute through the EDM. An EPM has to accept the position that the PM is really in charge of the program because he has the money and answers to the president. There are many conflicts that may occur between the EPM and PM over whose in charge of engineering and the EPM has to assert himself that it is his function.

A good PM would work through the EPM on engineering issues. When a PM by-passes the EPM and goes directly to engineering, conflicts would be generated. Conversely, an EPM should never by-pass the EDM and give orders directly to the designers.

This protocol should be followed throughout the program and between departments. If it isn't, the working individual doesn't know whose orders to follow. Chain of command may not be perfect but it is still the best way of doing business. I said to my engineers previously, if they do receive a direct order from an EPM or a PM, do as he asks but report it to me. I will resolve or challenge the orders later, quietly in my office with the EPM or PM.

With reference to the Interface Chart in Figure 8, the Project Manager or EPM interfaces with the PM and with the departments on the left side of the chart as shown. He would interface with these departments directly or indirectly through the EDM. This chart illustrates the basic interfacing chain of command and should be followed to maintain a smooth running program.

When the RFP arrives the PM has to get together with the EPM to discuss strategies. The EPM should also help the PM prepare the Program Plan. Before responding to the RFP, the EPM provides a copy to all engineering departments so that they all can submit their quotations.

The EPM would go over their quotes to see if they are overlapping on certain functions and to determine if their prices are not out of line. Chances are both of these problems would exists. It would be up to the EPM to go back and negotiate with the different organizations to insure that each engineering quote is within reason and that they are not duplicating a task or that they didn't leave some tasks out of the quote.

When I first became an EPM I tried to reduce some department's quotes. I discovered it was not an easy chore. Each department insisted that they need all of the budget they had submitted and cannot reduce it at all. This is when the haggling gets rough. It's the EPM's job is to make sure that engineer's quotes do not cause the company to lose the program.

Being a former engineer, an EPM has a pretty good idea on the amount of manhours each department needs to do their job. Some departments would simply take a percentage of the Electronic Design Engineer's quote. The EPM may have to rely on past quotations or actuals from other similar programs to scope the program. The administrator should be able to get that information for him. There may have to be a lot of unpopular meetings to resolve this problem.

Some department managers had refused to talk to me about their quotes because they felt it was already at rock bottom. After some overlapping was pointed out, some managers did finally reduce their quotes, but it still wasn't easy. Under no circumstances should the EPM modify any of the other department's quotes without their approval.

As Figure 8 shows, the EPM interfaces directly with all the support departments, production, Vendors, and with the EDM. For quoting purposes and budget allocations the EPM also directly interfaces with Integration, Environmental Test, and Flight Test if it's for an aircraft. All other technical interfacing with these departments are done indirectly through the EDM.

The EPM interfaces with the Customer only as the PM's assistant. He is the PM's technical advisor in all reviews with the customer. If he can't answer a technical question, he should immediately call the EDM for answers, even if he is out of town.

4b. Support Departments

As was mentioned in Chapter II, in the military there are several departments that monitor the electronic design and control the maintenance records of the units in the field. These military departments prefer to deal with their counter part in industry. Many companies follow the ways of the military and are structured the same way. These departments provide the most important function in the design and contribute to "check and balance."

All the Departments that provide Quotations to the EPM are as follows:

o Design Manager
o Human Factors
o Safety
o Components Technology
o Testability
o Manufacturing Liaison
o Quality/Assurance
o Reliability
o Maintainability
o Weights
o Thermal
o Stress
o Electrical
o Producibility
o EMI
o Development Shops
o Technicians

- o Environmental Labs.
- o Integration Lab.
- o Drafting
- o Drawing Check Department
- o Flight Test
- o Drawing Release
- o Inspection
- o Functional Test

Depending on the company's structure, there may be even more departments that would be required to provide quotations to the EPM. Organizations such as Product Support, Manuals, Automatic Test Equipment, Training, Field Service, and Logistics have to submit their quotes directly to the EPM instead of the PM. All of these quotes would include manhours, material, schedules, and manloading as a minimum.

Once the quotations are accumulated they should go to the PM who would probably be unhappy with the final cost. So, the cycle begins again, and the EPM has to go back to engineering to reduce their quotes further. Believe me, it's no fun. The major problem with this procedure is time. As was mentioned previously, by the time the RFP gets to the bidding departments after many delays in upper-management, there usually isn't enough time to do a good quote. This has happen to me time and time again.

In some cases, to meet a deadline of a quotation submittal, the company president or upper-manager may decide to reduce the quote by simply taken an arbitrary percentage off of each department's quote of what he thinks it should be. The EPM then has no say so and he's stuck with those numbers.

By upper-management quoting this way it can cause quite a problem for all departments. This has happened to me as a EDM on several occasions. On one job, my quote was reduced by an unbelievable 50%. Unfortunately, we won the job. When I was finally turned on to do the design, I went as far as I could with the budget they gave me. I then informed the PM that all work would stop at a specific date unless I get more budget. I was crying "wolf" all the way through the program but they just ignored me. They claimed engineering has the reputation of always crying for more budget when they really don't need it. So I let it happen and stopped all work. Some how they got the message and produced more budget. Let's face it, without the Design Department there was no program. Management certainly should not quote a program from the top of their heads like they did. But, it happens. My company, of course, had to eat the overruns.

It's not very clear, but for some reason people seem to think that hardware designing and development should be the cheapest part of a program. After the design of a piece of equipment is supposedly complete, the design department is expected to just go away. They don't understand that there is follow on work to be done. For some reason, no one seems to complain about the cost of software. I think it's because many people do not understand software. It's a mystery to them.

Electronics is a strange business. There is always lost time. What does the EDM do with his engineers while they are waiting for parts or during the fabrication of prototypes. An engineer cannot be put on a shelf and be removed as needed. There would always be some unaccountable lost time in a program, and there would always be rework after Integration Testing or Flight Test. Rework is that part of the quote that is usually eliminated by the PM, but it would always have to be done.

It was once remarked by a PM that, "The EDM is just fattening the quote up with unforeseeable problems." Non-Engineers cannot understand why the design cannot be done right the first time. I discussed this subject extensively in Chapters I & II. They don't understand that an Electronic design is just one big variable because of so many unknowns.

Military electronic design is usually the result of new inventions or innovations. The design may have never been done before but it is expected to be of the highest quality and most reliable.

It's not like commercial design where many companies make the same product such as a TV or computers over and over again. They just update their product year after year. When an advanced fighter is awarded to an aerospace company much of the electronics that goes into that aircraft would have to be designed from scratch. It's not very often that they can use shelf item equipment.

4c. Contract Awarded

After the contract is awarded, the first thing the EPM does is assist the PM to update the Program Plan and then prepares his own Engineering Project Plan Summary. This Summary is basically, the instructions to all engineering as to what their responsibilities would be and provides an outline of the tasks that each department has to accomplish.

Using the updated Program Plan and the Summary, the EPM has to go back to all of the above departments to update their quotes. The chances are a new PM and a new EPM would have been assigned to take over this program. Also, the engineers who originally quoted the job may also have been replaced. This usually happens when a job is awarded many years later.

When all the new quotes are accumulated the hassling begins again. If the new total costs are out of reason the PM and the EPM may have to drop some of the support organizations from the team as mentioned before, and double up on their duties by other departments. This is a bad way to do business because there would now be an inexperienced person doing another man's job. The tasks are still required and the military equivalent would not be supporting the job. Sometimes, though, it has to be done.

After all the quotes have been renegotiated and accepted, the EPM and the PM would now have to allocate the budgets to the various organizations. These budgets would have to be updated periodically because things have a habit of changing.

Some departments would find they don't need as much budget as they originally quoted for some unforeseen reason. Some departments may ask for more. It becomes a balancing act for the EPM. Some tasks may have to be dropped altogether and some new tasks would have to be added to the program. Nothing comes easy and it may result in overruns. A good PM and EPM would foresee these problems and possibly save the budget from being overrun or causing a late delivery schedule.

4d. Project Plan

The Engineering Project Plan (EPP) is generated by the EPM which contains the complete detail plans and schedules for all of engineering design tasks and development activities including support functions by other company organizations. This plan is basically, a more detailed Program Plan and an extension of the Project Summary, which originally defined the engineering activities. The SOW is used to determine most of these activities. An EPP may be required by the customer as a part of the data submittal. This Plan has to be completed before, but not always, any prototype work begins and should be included in the schedule as part of the Program Plan.

One of the first charts the EPM has to generate are schedules and manloading. He would have to go back to all of the departments to get their detailed schedules and manloading. These organizations should include names of individuals and their tasks. From all of this data the EPM can now generate Level III and Level IV schedules. Level III schedules show the overall engineering milestones when each major item or the department's activities are concluded.

Level IV schedules are more detailed, covering each sub-task such as a drawing, a design task, a printed wire board layout, a software program, etc. The Level IV schedule should be very flexible in the schedule. It is really a guide line for the design department to know when an item has to start and when it is has to be completed.

These schedules were generated by the EDM and would be updated by the week because they may have to be changed during the design and development stage. There may be problems with one design, or the designer may be waiting for parts. He may elect to work on a different design that was scheduled later in the program. If anything, the Level IV schedule provides a list of all of the tasks that have to be performed and hopefully on schedule, but should never be considered concrete.

Level IV schedules cannot be fixed because there are just too many unknowns in this business. I "blew my stack" as an EDM once, in a meeting when my manager reprimanded me in front of his upper-management and the vice president because I changed a drawing's completion schedule. I tried to explain that Level IV schedules are only a guideline to help the EDM monitor the design.

My boss was an Aeronautical Engineer and, obviously, he didn't know much about electronics and how unpredictable it can be. In the Aerospace business upper-management is strictly political especially when a Mechanical Engineer or System Engineer is in charge of Electronic Design Engineers. They say it doesn't matter. I say it does.

Not having Level IV schedules can also be detrimental to the program. On the last EPM job I had, a Design Manager wouldn't produce his Level IV schedules because he felt it was for his internal use only and not for publication. He didn't want his dirty laundry exposed to world and wanted to hide any delays in his department. The only way I could track his work was to make my own Level IV schedules.

Needless to say, in the Division Program's meeting, this EDM was made a fool of when my schedules were presented. The EDM looked foolish in front of upper-management when he tried to explain why certain items were overlooked and were not listed on any schedule. When he was asked when a particular drawing was due he didn't even know the task existed. It was a little underhanded of me but I had to do something to get the EDM's attention.

The EDM was reprimanded by the Vice-president and was instructed to have all of his Level IV schedules completed by the end of the week. He never forgave me for this, and he made it very difficult for me throughout the rest of the program. He went out of his way not to cooperate. He had twenty years with this company and I had only two, so I didn't have a chance. If I had complained to his boss first, maybe it wouldn't have been so damaging to him. My maneuver sort of backfired, so I just received a lesson in poor diplomacy.

The Project Plan should specify what tasks are to be performed by each department showing budget and manloading charts. It should cover the following:

o mechanical drawing

o drawing release with signatures

o Contract Deliverable Requirements List (CDRL).

The System Engineer would use the Project Plan and the SOW to produce the Equipment Specification. This is why I recommend that the EPP be done when the Prototype effort begins to give the EPM enough time to produce all of the above tasks and charts first.

4e. Meetings

Even before the Project Plan is completed the EPM should schedule weekly design reviews with all departments. This includes designers, support departments, production, master scheduling, etc. All of the design review meetings are chaired by the EPM, but sometimes they may be chaired by the EDM. Some EPMs prefer to have separate engineering meetings with the EDM to resolve internal problems first and then have the design review with all of the other departments present. If it works that way, fine.

The first design review is the `kick off' meeting. The EPM would get together with the EDM to prepare for this meeting, because a good portion of the presentation would be technical. This is a very important meeting because it brings the whole team together at the beginning so that all departments know what tasks would be in the program and how they interface with each other.

After the `kick off' meeting only those pertinent departments would be requested to attend. Who is invited would be determined by the EPM. Certain individuals, however, should attend all of these meetings. These are the EDM, Master Scheduling, the PM, Engineering Designers (as required), Reliability, Maintainability, Thermal, Production, Engineering Liaison, Quality Assurance, Integration, Product Support, and Administrations. These meetings are very important. Even though the meetings are costly in manhours they do solve many problems up front, which would save many hours in the end.

When I was in a Staff position acting as liaison for Product Support, I attended some of these meetings where only three people showed up; the EPM, the EDM, and myself. I was acting more like a PM. Now how can one conduct such a meeting properly when all of the action items have to be dealt with second handed, and many important issues have to be resolved by "I will look into it."

Obviously, many items were overlooked and had to be resolved later with rework. The EPM claimed he wanted to save manhours by not having all of the departments attend and because they were busy on other programs. This design review, with only three men, turned out to be disastrous and in the end the program was very costly.

The very last job I worked on before I retired, was conducted in the proper manner. I was the EDM on that program and the EPM was a recently retired Naval Commander with little experience in that position. The ex-commander knew his limitations and relied on my experience on company procedures on how to conduct these design reviews. As the program progressed the ex-commander began to realize that an EPM's task is very complicated. By the time the program ended, however, he became a well educated EPM. This program was not smooth by any means. It had many design problems, but by having the required personnel present at different meetings, all of the departments were made aware of any predicaments. All the problems were eventually resolved early in the program.

The detailed Level IV schedules of this program had to be altered many times, but the final delivery was on time and within budget. Had I been the type of EDM who was afraid to admit there were design weaknesses during the program, there would have been many major problems later. And, because the meetings were conducted properly with the right departments present, the whole program was a success. The final product was reliable, easy to maintain, and the Navy was very pleased because the new design saved them millions of dollars.

To conduct a design review, an EPM should prepare an agenda which is to be sent to those who would be invited in advance so that they would be better prepared to discuss the items on the agenda. If a department engineer cannot make the meeting he should send an alternate. To start the meeting an attendance paper should be passed around so that each one present provides his name, his department, and his telephone number.

The first item presented in each meeting is the minutes of the previous meeting and if they have been settled. Then next, all previous Action Items should be discussed and either resolved or carried over. The EPM would then ask the EDM to cover the status of the design phase of the program. It is preferred, if there are any questions they should be brought up immediately and not wait till the end. In other major meetings with the customer, for example, where there are so many individuals present, it is better to hold their questions till after the presentations are over or the meeting would never be completed.

The EPM meeting is within the company and if there is yelling and screaming during the meeting, so be it. If there is a problem it should be resolved immediately, but the EPM has to control the meeting.

When the EDM has finished his presentation in the design review, the EPM should go through the rest of the agenda. Then he would go around the room asking each individual for their status and if anything disturbs them. It's surprising how many new issues come up that weren't even considered by the EPM or the EDM before.

There is always one minor problem whenever there is an open table discussion. There would always be an individual who wants to dominate the meeting with nonsense trying to impress someone. It is up to the EPM to recognize this situation and either put a stop to it or never invite that person again.

While all the discussions are going on, the Administrator or a delegated person should be recording the minutes and new Action Items that are generated. Each Action Item should also contain a start date, the name of the responsible person, and an expected completion date. Several days after the meeting is over a copy of the minutes is sent to all those that were present and to their managers, if so requested.

My philosophy in running a program has been, my door is always open to all and I would gladly explain any issues or even the design to anyone associated with the program. I always tried to be truthful and let everyone know when there was a problem. I don't believe in holding issues back or concealing problems because, in the end, it will bite me back. I have always believed in this philosophy, and it has never hurt me once.

As a matter of fact, when one keeps the other departments and the customer informed, they feel they are a part of the program and would try to help rather than hinder the program. I was well respected for this, especially by military personnel, because I was always willing to discuss any issue with them, even on the phone. If things looked bad in the schedule or with the budget the military would always cooperate with me and help resolves those issues. No surprises!

When I was an EPM on another program, my manager said to me not to discuss certain issues with the Navy. "We only tell them what we want them to know," was his statement. What a bunch a bull. As far as I'm concerned, the Navy is paying for that program and they have the right to know all aspects of the job, good or bad. Keeping them in the dark doesn't resolve a dammed thing. It's always more cost effective to resolve a problem at the beginning and not later in production, even if it means involving the customer.

Many managers and companies have released a job knowingly there were deficiencies in the design, and would try later to get more budget from the customer as Engineering Change Proposals (ECP) modifications. The same old story "Bid low and pray for ECPs." A lot of deficiencies can be hidden and corrected through ECPs. Some deficiencies are justified because they were not caught during the design phase, but to deliberately hide them and wait for ECPs is a Cardinal Sin, in my estimation. The Government should make a Contractor pay a fine for pulling that stunt.

4f.Tracking

One important function of an EPM is to monitor and track the engineering schedules and budgets. Each week the EPM should examine the schedules and budgets of each engineering branch. The status of these two items should be brought up in the design review each week. In this way, as a problem arises it can be resolved immediately before it gets out of hand.

The level III schedules are usually sufficient for the EPM to track most programs. This would alert him of any milestone slippages, and he can address it at the meeting or directly with the department that is in trouble. The EPM must have control of the technical portion of the program. The level IV schedule is, generally, the responsibility of the EDM in the design activity, but the EPM should also be aware of his schedules and work with the EDM to insure they don't go astray.

Many companies use a single charge number for all departments to record the hours they have spent on a program. This may be alright for small jobs because the budget may be tracked by each department number. Generally, this charge number is sufficient for the EPM to track the budget on small programs. However, on large programs, each department may be composed of many sub-groups so it becomes difficult to determine how much each sub-group has charged and for what function. In Chapter II, I discusses my unique tracking procedure which covers all of the sub-functional tasks of a program to help the EPM and the EDM track the budget over a major program right down to each individual. I suggested, on large programs, that this tracking outline be used or an equivalent. It not only would help track the present program but, in addition, would provide the EPM with a documented record for bidding new programs.

There are two other meetings that the EPM would have to attend and be prepared to defend. These are the PM's meetings and meetings with the EPM's upper-managers. If the PM has been attending the design reviews then his meeting should run smoothly because he was made aware of the problems as they arise. But there are times that the PM is out of town or just could not attend the design reviews. This is when his meeting could get nasty, because problems have occurred that he was not aware of.

The EPM's upper-managers meeting is when he has to make a presentation using all of those wonderful charts and tab sheets previously listed. I still maintain that most of those charts are not really necessary to track a program and I believe that the tracking charts, manloading spread charts and monthly budget charts are all that is necessary to support a program. The only indicators that upper-management are really concerned with, anyway, are schedule and budgets. The rest is just academic.

As an EPM, I had to make monthly presentations to upper-management on my program. My first experience with this meeting was frustrating. It took me over a week to prepare all the charts that was required, and I still had to do my engineering work. It took me awhile to learn how to by-pass some of these requirements. I learned that if I did not present an inconsequential chart no one would ask for it. So I left several charts out. It did back fire on me once and I promised to have it ready at the next meeting, which I didn't. I told them I forgot. Sometimes they forgot. The underhanded things I had to do just to get my job done without wasting my time producing unnecessary charts.

5. Upper-Management

What is upper-management? This depends on the size of the company. In a small company the president may be considered the only upper-manager. In a large company, such as aerospace, upper-management could be from a Department's engineer's position and up to the president.

Upper-managers can make or break a small company, and could cause a large company to lose a major contract. Why then do so many companies have some poor upper-managers? At least that's my consensus after being associated with so many upper-managers in different companies for the last 40 years.

This then brings up the question, what is the best way to select a manager? Isn't there a classical method of selecting managers? I believe there is. I feel that the classical model of a company that has a fool proof way of selecting manages was Bell Telephone Laboratory (BTL) now known as Lucent. I worked for BTL during my first 10 years as an engineer. BTL was not only a major technical company, but it was one of the best run companies, I thought. Many of the technical and organizational procedures in this book, have been extracted from my experience at Bell Telephone.

5a. Selecting Poor Managers

Many upper-managers at my aerospace company received their position through the `buddy-buddy' system or through hiring friends from other companies. Many managers got to their position because they happen to be at the right place at the right time. I do have to say that many upper-managers did earned and deserve their position and went on to become fine managers. Unfortunately, they were in the minority.

I have witnessed, for example, a manager was demoted because he `screwed up' a job. Later he was transferred to another program only to be promoted back up again. Politics, is generally the name of the game in upper-management.

As I mentioned before, in my last company when a new program began the upper-managers of that program put out a memorandum to all departments that they need qualified engineering personnel. They would request that each department release only talented individuals. But these departments, however, would generally transfer their weakest engineers. On one program that I was familiar with, some mediocre engineers were eventually promoted to supervision and others became Department Managers, simply because they happen to be at the right place at the right time.

When I was selected as a Supervisor, for example, I was simply asked if I wanted the job. In three weeks I was an EDM. Does this make any sense? There was no hearing and no investigation. The EDM before me was selected the same way but he was misplaced. This other man was a good engineer but he had no experience in a Design Department. I'm happy to say I thought I became a pretty good design manager.

Even Lee Iacocca noticed, when he took over the Chrysler Corporation, that there were too many empires in the company with a lot of redundancy. Many of Chrysler's Managers were more concerned with maintaining their empires than doing what was best for Chrysler. This problem still exists with many companies, especially in aerospace.

Some managers are selected because of their educational titles. Professors from Universities are especially in demand for upper-management positions, simply because of their title. Some of these professors are good, but most have little or no practical experience in the electronic or aerospace business.

It's common practice to use their title on a Quotation because it sounds very professional. Many quotations have gone out to the customer with PHDs and MAs listed as participants in the program, even with their pictures. Actually, they were just figure heads and their resume's were only used to impress the customer.

Another source of potential upper-managers are retired military personnel. Some of these men make good managers but some don't. Yet, the incompetent ones somehow obtain key upper-management positions. Most military retires, however, know their limitations and would adjust accordingly and become fine managers.

I had given an example of a recently retired Commander who became a good manager when he relied on my inputs on how to conduct an engineering meeting. He knew his limitations at the time but was willing to learn. This man, eventually, became a very capable Division Engineer.

What was wrong in my company was many poor managers survived using politics and friendship. For an upper-manager to play this political game properly, he has to learn a few basic rules. First, he has to be a yes man. Never express an opinion unless it is specifically ask. Go along. Don't make waves. Agree with his boss. Be prepared to provide all kinds of statistical data on a program. It's ironic, but these types of managers seem to hang on for years before they are discovered, and in most cases they are never discovered. I had witnessed this phenomenon occurring too often.

Most good technical engineers do not `cut the mustard' in the political wars of upper-management, so they avoid that game. As was mentioned before, my upper-manager was an Aeronautical engineer and his boss was a Structures engineer. Now how in the world can these individuals understand electronic problems. "Just keep me out of trouble," was the remark my upper-manager made to me. What if I did have a problem? Can I depend on him?

One basic problem with upper-management is their desire for power. In most of these positions these individuals have full authority. No one can challenges them. Again, there is little "check and balance."

Each Department Manager has a boss that is supposed to provide the check and balance but in the political arena the rule of thumb is, don't make waves. Maybe a private conversation may occur, but nothing ever changes. A Division Engineer had said to me once, "I wish I could control my managers better. They are sometimes so unreasonable." Who's running the company, anyway? A bad manager can go on, unchecked, doing it his own way and not his company's way for many years. It happens.

5b. Example of Poor Upper-Management

When I worked for an electronic firm as a EPM, my upper-manager was a newly hired ex-System's Engineer who didn't have the vagus idea of what the EPM's responsibilities were. Yet, he imposed countless tasks of paper work and charts on me instead of supporting me on the required work of producing electronics. In this particular company upper-management would have their conferences, make major decisions affecting the EPM's work, but they would never consult me. As I said before, they don't know what goes on in the trenches.

One major bad decision these upper-managers made was the elimination of our development shop which I mentioned before. "All development assemblies would now be done in the production shop." The reasons they gave this order was, they felt there was a duplication of effort of having two development shops. It would be more cost effective to have only one shop because production personnel salaries were lower. It would also be easier to control the drawing changes in production and there would be production commonality between Printed Wire Boards.

In the upper-managers point of view, these reasons seem logical. This suggestion was passed on to the Vice President of Engineering and he approved it immediately. He didn't ask engineering what we thought of it. Now I had to live with that bad decision for years. Had they even discussed it with me, I would have torn that whole concept apart in the following manner:

5b1. Drawings

Every time there is a minor drawing change in a production development shop; the drawing has to go through the Check Department, Quality Assurance, Parts Control, Reliability, Maintainability, and countless other departments for signature approval.

When a design engineer is in the development phase, he may be making changes by the day. After the prototype is assembled and tested, there would be more changes. When the unit is tested in the Integration Lab, there would be more changes. When the unit goes through environmental testing, there would be many more changes. When the unit is placed on the aircraft and Flight Tested, there would be more changes. It goes on and on.

I admit that drawing control was always a problem, even with the old system, because a change may be red lined onto the drawing and later somehow falls through the crack and never gets incorporated in production. This is a problem, but the solution is not to abandoned the development drawing procedure but to solve it.

If a company has a professional Computer Aided Drawing (CADD) system, the drawing control is relatively simple, because the development drawings would be locked in ROM and can only be copied, thus a changed drawing would be updated with the letter "a" for instance. This would, in addition, provide engineering with a complete history of the drawing changes which may be required by the customer. Drawing Control Procedures were discussed extensively in Chapters I & II.

5b2. Duplication of Effort

Upper-management sees two organizations having the same capability, production development shops and engineering development shops. They look at the amount of savings there would be if the two organizations were combined, except they do not provide the same functions. Production development shop's basic charter is to develop test equipment and electronic aids for the manufacture of hardware. Engineering development shops are to assist engineering in the design, build breadboards, assemble prototypes, and make quick changes to meet the development schedule. There is no comparison in the type of work required by these two different development shops.

The major problem in using production development shops is that the assembly line girls are trained to assemble and wire using production line techniques and procedures, with wire tables using removal and replacement documentation. This is the normal and accepted procedure especially when a delivered unit comes back for rework.

In order for a production line girl to make changes she has to have updated released drawings with all the proper signatures. She has to have the drawing changes incorporated into her wire table, and has to have the unit inspected. The unit has to be electronically tested in front of a test inspector. It then has to go back to the shop for conformal coating of the PWBs, and re-inspected and tested again. All new parts have to be pre-screened and inspected before they may be installed in the unit. This is the normal procedure in production and has to be maintained.

As mentioned in Chapter I, I covered the story of how it took two months to make one wire change on a PWB using the production shops. In an engineering development shop it would have only taken 1 hours to do the same work. Engineering cannot use production facilities for developing their prototypes. Prototypes, should never be treated like a production unit. That's why they are called prototypes or by the three `Fs'; form, fit, and function.

5b3. Commonalities of PWB

Every company can see the advantages of commonality of production items to be used on different programs. There is a lot of profit to be made, the item is cheaper, and it can be used on many other aircraft. Sounds good.

The problem is, making a change on a new design is now governed by an old design and visa versa. If one needs to add an extra chip on a board of an old design, it has to be done in such a manner that it does not alter the electronics on other applications. This can have a domino effect, because this same PWB may be used on 10 different aircraft or programs.

An investigation has to be made first to insure that the other programs are not affected. As an EPM, I was constantly required to review and sign drawings that were being changed from other programs to see if it would affect my program. Needless to say, because time was so precious, I lightly scanned those changes and later discovered that one change did indeed affect my program. An EPM, under these conditions, has to take the time to go over each drawing change carefully before he signs it off. It was very frustrating for me.

Commonality is fine when it's for the same program because there is some control. But even then, when the changes get out of hand one may have to abandon the commonality because of the domino impact. Management shouldn't tie the engineers hands up when he is trying to get a design out. The intentions of saving money using commonality may be good, but don't straddle the designer and the EPM with unrelated work.

Some engineers have spent countless hours trying to maintain commonalities on a PWB. It was upper-management's decision to use commonality, and so it shall be. Engineering, in the above case, was not allowed to divert from this decision and in one case only half of a PWB was utilized for my program. Because of this, the complete board still had to be tested by my technician no matter whose program it was used on.

The large printed circuit board shown on Figure 29 on page 305 is an example of commonality. On fourth of this board is for the Display Control Panel. The rest is for some other purposes. The problem is the entire board has to pass environmental testing and production FTP. A smaller DCP board could have been used requiring a small power supply. If a change is made on the DCP portion, would it affect the other circuits?

5b4. Production salaries

Production salaries are lower than those technicians used in an engineering prototype shop. This is true. When upper-management plays the numbers game it becomes obvious to them that a lot of money could be saved using production personnel. And, again, because upper-management doesn't know what goes on in the trenches, they do make quick decisions such as, "Here after, Prototypes would be made in the production development shop and that's an order!"

They would never ask the troops for help because they are upper-management and they know all things. I have been working with prototype development shops for 40 years and now, all of a sudden, they are too costly. Obviously, for the reasons that were pointed out above, the cost of using a production development shop is much greater than it would be by using a separate engineering development shop because time is money.

I know that all this sounds like I'm very bitter with upper-management, and many would assume it's all hogwash, but it happens. By the same token I have worked with many good managers who were not only respectable but technically competent. One may be lucky and not have to experience the same squawks that I have had.

5c. Proper Selection of Managers

As I mention above, I feel that BTL's method is a good model for a company to select its managers. It is a simple approach. When there is an opening for a first line supervisor, for example, several Department Managers would submit names to a special promotion committee. At least 3 names, but no more than 10 shall be submitted. The committee would judge these people on four merits:

o Technical Ability 1/5
o Education 1/5
o Experience/age 1/5
o Personality 2/5

Notice that Personality carries the greatest weight. On the other hand, his technical ability, his Education and Experience should be directly related with the managers job offered. If there is a tie between two individuals then the `Buddy Buddy' System may be used to influence the final decision. By this time, both men are equally qualified. All supervisor promotion must be approved by the upper-managers and the Director of engineering.

New hires should rarely be made new supervisors. They should be required to accumulate a few years within the company first before they may even be considered. There is no way to truly judge a new hire on the above merits from a resume or by recommendations.

Upper-Managers are selected about the same way supervisors are, except they are judged even more critically. At no time shall an upper-manager be promoted to a position when he is two or more steps lower. No skipping allowed. Each individual must have held the lower level management's position for at least 2 years before he shall even be considered moving up. Promotions to upper-managers must be approved by at least two levels above but not lower than the Director.

These rules are not hard fast and, of course, there would be times when they cannot be followed. In those few cases, the promotion must be approved by both the Vice President and the President.

The main concern with promotions is to eliminate the `Buddy Buddy' system to and reduce the possibilities of having poor managers. There is no perfect method of selecting managers but the above approach does provide some "Check and Balance" which is one of the main goals of this book. Check and balance has to be used in all aspects in a company as well as in the design.

5d. Make or Buy

In the Aerospace industry most companies have the capability of producing many of their own aircraft components. Fortunately, the government is one step ahead of this and can foresee that there could be a monopoly in their contracts, so they require at least 40% of the expenditures of a major program are to be spent with sub-contractors. This gives smaller companies a chance to bid for sub-contract work and spreads the economy around throughout the country. This is an acceptable practice, but there are times that it is best to build some items in-house for internal profits and to maintain capability. There were many times in my company when 60% of the contract were given to suppliers. That is, of course, too much.

Unfortunately, some aerospace companies rather buy than make electronic items. I was stuck in this situation for many years and never got out of it. In my company's directive it was the policy to make an item rather than buy, if it was within the company's line of business. Yet, upper and middle-managers ignored this edict and preferred to buy any required electronics by dominating the "Make or Buy" committee excluding the design department. It was a closed meeting.

The type of in-house electronics I am now alluding to is Interface Electronics and not the standard aircraft avionics such as Radio's, Transmitters, Navigation, Computers, etc. These equipment are the standard products of many companies and they are experts in their fields.

Interface Electronics is a field by itself, because one has to know the inter-relationship of an aircraft. In a new aircraft, there would be many units that are Government Furnished Equipment (GFE) and sub-contracted equipment. Most of these so called black boxes, however, cannot communicate with one another because they were selected from standard off-the-shelf items and were specifically designed for other programs. It is the Interface Equipment that provides the communication between boxes.

"We are in the Aircraft business and not in the Electronics business!" This statement sounds sensible, but the problem is how does one interface the electronics. Someone has to know something about Interface. To start out with, in an aircraft program, the System's Engineers had to come up with the overall configuration in the original proposal. They had to select the types of electronics to satisfy the customer that would provide the proper payload for that aircraft.

After the aircraft contract is awarded, these System Engineers have to generate a systems block diagram connecting all of the black boxes. When they investigate further they discover, for example, that the Navigation System that was selected by the Government (GFE) would not communicate with the Displays or with the Computer (GFE).

They cannot ask the producers of the equipment to modify their designs because they are shelf items and are being used on many other programs all over the world. These suppliers would have to charge astronomical prices for any modifications of their boxes because they were delivered years ago and are still working in the field. So, an Interface Device (ID) has to be designed and developed to provide the communications between boxes.

If the decision is to buy the ID from an outside supplier, an Equipment Specification would have to be generated first which then has to be submitted as part of an RFP to all sub-contractors to bid on. The ES takes about three to four months to prepare before in can be released. Then all of the bidders would have to examine the RFP and submit their own proposals. The company then has to examine these proposals for technical competence and cost. These possible sub-contractors are then invited to make oral presentations of their proposals, etc. Months later a vendor is then selected from this list of suppliers. Altogether, this whole operation takes about eight months to a year to accomplish. Can a year's delay to start the design, and maybe another year or two for development be tolerated in new program? Usually not.

Unfortunately, with this arrangement, the lowest bidder may be awarded this job, and generally doesn't know all the interface aircraft requirements. Secondly, interface electronics is the last element that have to be developed. Suppliers have to rely completely on the Equipment Specification, which could have mistakes or not be complete.

The main point I'm making is, sub-contracting may take a year just to get started and because they are many amateurs in the field of Interface Electronics, the IDs may have to be re-designed or modified several times because the ES is not complete or the unit doesn't meet the ES. In that case the supplier would ask for more funds.

As a EDM in the Interface Department, my organization has designed Interface Electronic equipment for 35 years. In this organization my engineers were experts in the Interface Field and know the inter-relationship of Aircraft Electronics. It always comes back again to the same old subject, upper-management doesn't know what goes on in the trenches. The point of all this is, that one should be made aware of all of the maneuvers that goes on in upper-management. One should keep an open mind and also be aware of all the political tactics in this field, because it may affect his job.

5e. Possible Solution

Unless a similar method of selecting managers as recommended above is used there may be no solution to the `buddy buddy' system because upper-management would not change their methods of selecting and promoting managers, and because technical engineers don't necessarily want the job. Most engineers are not politically motivated to get into that arena. How then does upper-management receive the proper information to make a proper decision on who should be a manager?

They can discuss such items with lower-managers who are involved in the in-fighting, or have a qualified Technical Staff Engineer who has many years of experience in the design world, examine any of the company's selections or any major decisions that would affect the Design Organization.

Using a Staff Engineer, however, could back fire. My upper-manager, on another occasion, was not a technical type even though he was a graduate EE. He made the mistake of relying on one Staff Engineer who had never designed a thing in his life. This man was a politician and had everyone convinced that he was the authority on electronic design. I argued with this man constantly about the proper procedures in design but was unsuccessful. He was too good of a con artist. I finally had to resort to underhanded maneuvers and bypass him to get the job done.

In another situation, a Director of Engineering was transferred from a production division from another state to the my facility. Not knowing what my organization was involved in, the new Director decided that his organization would no longer do hardware work and would only do systems and software work. With one pass of a wand, thirty five years of expertise went down the drain.

Years later, he found out that my hardware organization was responsible for many electronic black boxes that were still in existence, still flying, and had to be upgraded and maintained. But, it was too late by then, because there was a mass exitus of qualified engineers who were responsible for those early designs. The design organization and its capabilities were lost, and the company now had to rely completely on out-side suppliers to modify these equipment using ECPs.

The final solution that they thought would solve the problem was to tell the Navy that the equipment was obsolete and couldn't be upgraded, so if the Navy still wanted that capability they would have to pay for all redesigned boxes. Talk about running the cost up when all they had to do was modify the old equipment, that is, if they had kept the original personnel. The cost of redesigning would be in the order of millions of dollars. In the end the Navy said, "No way! My company had to eat those redesigns in order to keep the original contract alive and the planes flying.

As far as Make or Buy is concerned, a good company should have a electronic design organization that is controlled and supported by upper-management. In my organization we had no home and were just tolerated because we did provide a service and sometimes we had to save a program. We were like firemen and were only called to put out a fire.

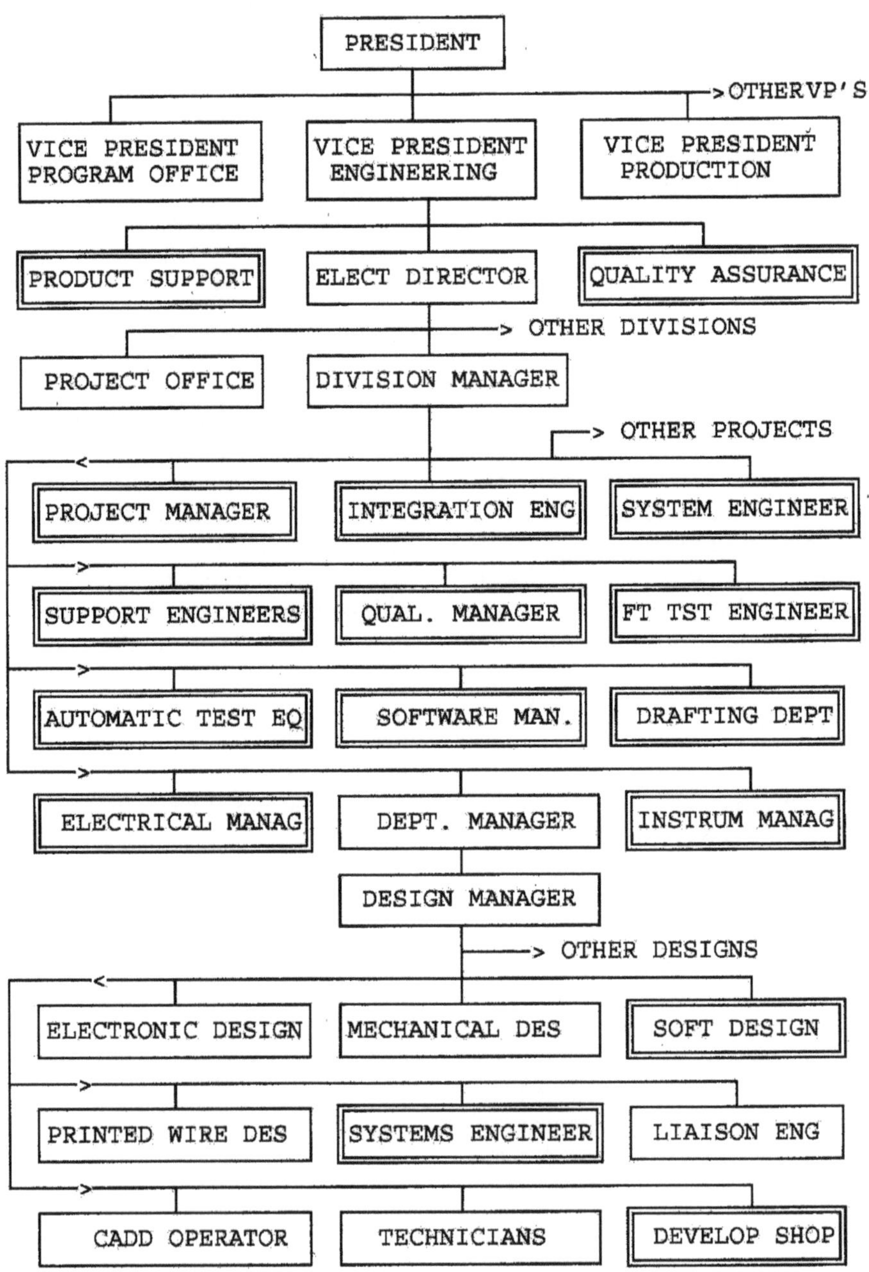

FIGURE 10. ELECTRONIC ORGANIZATIONAL CHART

6. ORGANIZATIONAL CHART

The Electronic Organizational Chart ("OR" Chart) of Figure 10 page 236, is what I consider to be a reasonable Departmental breakdown for a large company. This is just a suggestion to show how most of the company players are within the structure. Some departments were omitted, such as the Check Department which may come under the Drafting Manager.

This chart could be used by any size company. A small Electronic Company, for example, may select just some of the categories for their "OR" Chart. Obviously, a small company may not require a Flight Test Division, etc.

The President of this "OR" Chart has one or many VP's. In a large aerospace company, for example, VP's may consist of Structures, Shops, Aerodynamics, Hydraulics, and others. Since this book stresses mainly electronics, I will only concentrate on the design and development of electronic equipment.

The chart shows a Electronic Director heading up all of the electronic departments, however, this may have to be broken down further, depending on the number of programs that are in-house. If there are several major programs going on simultaneously, the Director may assign a different Division Manager for each program. The Division Manager may also have several projects under his responsibility which could be going on simultaneously, including Research. Research is also not shown on this "OR" Chart.

The Project Office and the Support Division all report directly to the Director because they would have several managers working on other programs. In this way the Director can matrix managers, remove and re-assign managers as necessary. Also, one program may not be able to support the different Support Engineers. A Thermal Engineer, for example, could be working on several programs. The lack of budget is the main reason for recommending Matrixing in this particular Chart. If the program were very large, then it could be Projectized with no Matrixing. Boxes shown in Figure 10 that have double lines signify a possible Matrix position.

The Matrixed Departments would report to the Division Manager, and would be assigned to him from other Divisions or Directors. This would give him more control of the program, and any problems that arise during Integration, for example, can be resolved at the Division level without going up through the Director or as high as the Vice President. However, these matrixed Engineer would always have the option of going through their own upper-management and by-passing the Division Manager, if he feels their is no cooperation. This does provide a check and balance when an impasse or disagreement occurs.

The only departments that report directly to the Electronic Division

Manager are Department Managers who may have several Design Managers reporting to him. All other departments are matrixed in this chart.

The Design Manager and his design team, as shown, were discussed in detail in Chapter I.

7. COMPANY BUDGET

The Company Budget is one subject that I am not an expert on, but there are a few important points that one should considered.

When a large company is setting up the budget for the year there are several items that have to be accounted for that do not appear on any Quotation. Some of these funds are Bid and Proposal (B&P), Research, Training (Classes), Traveling, Publications, Advertising, Pool, Employee Benefits, Medical, Insurance, Employee Savings, Retirement, Facilities, Reproduction, Payroll, Company Computers Services, Company Maintenance, Catalogs, and a host of others. Most of these are covered under the company's overhead.

The problem is, when one is quoting a job the company would tack on an overhead cost by at least 8.5 times and add it to the quote. In 1965, the total labor manhours used for quoting purposes, including overhead, was 14 dollars an hour. In 1990 it was over 120 dollars an hour.

One cannot completely blame inflation because the average pay rate for all employees in a company in 1990 was approximately $18 an hour as compared to $5 an hour in 1965. That is not 8.5 times. Overhead is now a killer and would be getting worse.

What happened? How can a company compete for a contract with this kind of overhead? Is it mainly due to the evolution method of quoting Military Programs? I cannot answer that questions, but the influence of the military not to include service charges as part of the quote had some effect. This meant that more and more services had to be charged to overhead.

To reduce overhead my company moved their production facilities to another state because they expected cheaper labor costs. But that didn't happen. Overhead does not change much from state to state. They discovered later that the cost of overhead became even higher after they moved.

The main purpose of this discussion is to provide as much background as possible in the quoting arena. Even though overhead is not an engineering function, it does affect one's capability to compete. The main overhead funds that concern engineering are B&P, Research, Pool, Training, and Engineering Computer Services, and one must be aware of all of these factors when bidding a new job.

7a. Bid and Proposal

Each year the company has to set aside a certain amount of B&P money to go after new business. The budget can be broken down into two categories, company funded or customer funded. If the customer funds the B&P effort they may provide the funds directly or the company may supply the funds up front and they would be reimbursed later from the customer. In some cases the company would obtain a loan for this task. B&P is essential if the company is to compete in their primary line of business. Without it the company would eventually be out of business.

If the company is funding the B&P costs, then it comes out of overhead. Non-solicited proposals, for example, are costs that also would be company funded. When budgets are being formulated each division would submit a B&P budget to finance, with charts and statements explaining the reasons B&P is required. All of these budgets are accumulated by finance and presented to the President. These would be massaged and eventually a B&P budget would be established and allocated accordingly to each Division. Many B&P proposals that engineering submitted may be excluded. One has to accept what they are given.

This does not mean that a particular division cannot receive more B&P budget. If it can be shown that a new program is worth bidding on, the company may provide more budget to that division. The company is flexible that way and would always hold back a reserve of B&P money for this reason.

In either case, whether the B&P funds are company funded or from the customer there is one problem that has to be guarded against, and that is over spending. There is always a tendency to load up a B&P charge number of many managers, especially if the B&P is customer funded. Many managers use this as an opportunity to save their regular budget and overload a new quotation effort with people. This is when the PM and the EPM have to intervene to insure that this does not happen. Charging to B&P and working on another program is against any company's policies, and may result in dismissal. And it is illegal if the government is funding it. One has to be careful over this miss-charging.

It is also illegal for an engineer to charge to one program while working on another. Check and balance is the only solution and the Division Engineer's administrator has to monitor this cross-charging to insure it is not being violated. This abuse could result in losing a major contract and being fined by the government. It is very serious!

In many case Finance would allocate a fixed B&P budget to a Department for a specific proposal and it is up to the manager to stay within that budget. If more budget is needed the manager would have to submit reasons for the increase. Finance would revisit the B&P budgets periodically to determine if it is being used up too fast or over expended. Some divisions may have to update and resubmit their B&P proposals every three or six months.

Usually the budget is allocated on a man level for a particular proposal. It is very difficult to break the cost down for B&P because of so many unknowns. A manager may assign two, three or even ten men to work on a proposal, because these engineers would have to create or actually invent the design, and they still have to provide schedules and manloading charts.

7b. Independent Research And Development

Research budget (IRAD) is handled through the Chief Scientist or equivalent. Each Division would submit research proposals to the Chief Scientist using a special format. The main issue here is that the proposal has to do with the company's line of business. It also has to have some value to it. Too often, a research proposal is submitted that is not in the company's line of business or comes under the heading of "it's nice to know." Research is not just for education, it is supposed to create new thoughts and bring in new work. Research could be theoretical in nature or it may be actually designing and developing hardware.

The budget would be allocated by the Chief Scientist for each Division. Research requires quarterly and final reports and possibly an oral presentation. The most important aspect of Independent Research And Development (IRAD) is that the Government may refund up to 80% of the cost depending how well the project scores in the report. The government could very well provide no refunds at all if it is a poor research program or badly run. This is why the Chief Scientist has to evaluate each program to insure it is worth funding.

Too many departments think of research as a method of getting more budget. It's nice to have a research program going on to support some engineers, especially when another program had just ended.

I have used research programs many times in that fashion but I had to be careful not use it as a "Ping Pong ball." That is, shuffling people in and out of research jobs as necessary to save employees. This creates a poor program because there is no continuity on the job using so many different people working on the same IRAD at different times.

When I was an EDM, my Division Manager considered research as his primary programs. He would `Ping Pong' the engineers who were working on a contract back and forth to research. This disrupted the real contract work which led to a poor design.

The question is, which is more important Contract work or IRADs? Obviously, contract work is, because it is the company's main line of business. The company is in the business to make money. There is no profit in research. This Division Engineer would pull my men off a contract job in the middle of the design to support his pet research projects and have someone else take over. It became a mess for me. Duplications of drawings, over budget, late schedules, and continually reworking the hardware had resulted. It was one of the few times I lost control.

Another type of a research program is Contract Research and Development (CRAD). These programs require that some sought of hardware be produced to demonstrate the possibility of new technology or to prove out an IRAD. The equipment doesn't have to be qualified to meet Mil-Standards. The above discussions on IRADs still apply to CRADs.

7c. Pool

Pool is a charge number that was created for upper-management. Because management is involved in so many different programs it is very difficult for them to charge to each program directly. A manager may spend 15 minutes on one program and 20 minutes on another, etc. His time card would look like a major tab sheet if he charged directly to each program. To simplify this recording of budget hours for management, the Government have agreed to allow a pool charge for each division.

The way it works is, when one bids a job a percentage of cost would be added to the quote to cover Pool charges. If there are several programs going on in the division, each of these programs would contribute hours to the Pool charges. Pool is a direct charge to the contract and not part of overhead.

The way expenditures are covered is, as engineering uses up budget on a particular program a percentage of Pool is charged to that program. In other words, for a 40 hour week, a manager's distribution of his Pool hours are weighed on how his engineers are using the budget for that week on each program.

If there are 10 programs going on in the division and engineering is using 20% on one of those programs, then for that week the manager would charge 20% of his Pool charge to that program. The manager would use only one pool charge number on his time card and the calculations are done automatically by finance. A tabulation of hours charged to each program is provided so that an EDM can monitor the pool hours being used on each of his programs. It just makes it a lot easier in time keeping for managers.

The only problem with this arrangement is, if a manager has already exceeded his allotment in Pool for a program, that manager would probably continue to charge to that program. There could be up to 10 managers and secretaries using the same Pool charge and it doesn't take long before the main program is over-budget.

My first encounter with this Pool arrangement was when I first became a EDM. I knew about Pool charges but I never worried about it before as a Design Engineer. I also didn't know how it worked. I assumed the Administrator would monitor it.

On one particular program, I was monitoring my own budget very carefully so that my department stayed within a tight budget. After eight months in the program the Administrator told me that I was already over budget even though I had another four months to go on the program. I was stunned because my records showed that my department had more than enough budget to complete the program.

What happened was, there were too many managers charging to my Pool charge. It ate up my budget. What a shock. I thought I was fat, dumb, and happy with plenty of budget left until that happened. I just received a major education from this experience, but fast. This is the kinds of things that happens when a man is promoted to management without being properly trained first. That's one of the reasons I wrote this book. Budget tracking is one of the most difficult tasks for a design manager and if one doesn't know all the manipulations that go on in finance he can get hurt.

Another problem that one has to guard against is overcharging by another department who are using the same charge number. My ME, for example, had a job done in the shop and told them to use his charge number. That was fine, but the shop mechanic sometimes didn't stop charging even after he was through. The old shop story was, "if you have a charge number write it on the shop wall." This way the whole shop can use it. This was not a joke, either. It happened.

In one case I had a shop supervisor suspended for allowing this mischarging to take place. He should have been fired. I finally had a meeting with the new shop manager and we agreed that they should do a job only with written authorization by me or my boss and within specified hours. I know that if the shop says they can do a job for 30 hours one can bet they only need half that time. But, at least, now there was some control. More hours can always be renegotiated if necessary.

As an EDM, I had spent countless hours tracking down people from other departments that are charging to my budget. They act as though it was blank check. When these individuals were challenged they would simply say, "Sorry, I'll get right off that charge number." But, by now, how does one determine how many hours were legally charged and how many were incorrectly charged.

On another research job, I had to close the charge number down and open up a new one just to get the leaches off of my budget.

For contract work, the PM and Finance provides authorization documents on charge numbers to each department. For research work, however, there is no PM or EPM assigned, so the EDM has to monitor all the charges by himself.

Pool is a good tool for Managers to use when they have to charge to many programs, but the Administrator and the EDM have to monitor their own programs weekly. I finally told my Administrator to give me a print out of all Pool charges on all my programs each week. That was the only way I could control my budget.

7d. Training

Training and education of employees is an overhead charge. The company cannot justify using a direct charge to the customer for training and upgrading their personnel. Training for the maintenance of a black box, for example, is a direct charge because it has to do with a deliverable product. The training now being discussed, is for educating company personnel with new skills to improve their capabilities. Graduate courses for higher education should be encouraged by the company, but it is still charged to overhead.

Training to improve skills is especially important in engineering. As a EDM, I required that all of my employees be trained to operate our company's CADD system. Over a 5 year period I had sent over 50 employees to take the CADD course. CADD is essential in today's design world. This is one reason why overhead may be the same no matter what state the company moves to.

There are many different internal training classes given by the company that employees may take to improve their skills. As a matter of fact, at BTL, 1/3 of the evaluation for raises was based on continued education. When I worked for BTL I found out that I was getting poor raises because I wasn't furthering my education. It was back to school for me and in a hurry after I discovered the lack of continuing my education was the reason for my poor raises. Any engineering course was sufficient for credit whether it be at the University or within my company's educational system.

7e. Engineering Computer Services

The CADD systems mention above, have to be maintained and upgraded with new software. The evaluation in CADD is similar to a home PC where new and better software is put on the market. Improved software, higher speeds, and better equipment are always being upgraded and installed. More and more departments require work stations where only draftsmen or artists were used.

I, at one time, had over 20 terminals under my jurisdiction. To ensure each terminal was not being neglected I invited other departments to use my terminals. The machines were used on a first come first serve basis so if a station is open and not in use I told them to just sit down and start working.

At one time my people were so busy I had three shifts going and I still had to send some men to another building to do CADD work. My objective was to keep the terminals operating to justify their existences. As the saying goes, "If one doesn't use it he will lose it."

In my company, Computer Services provides the upgraded software and maintains the equipment. This is considered overhead. The question is, should computer time be charged directly to a program, or to overhead? My aerospace company's management bounced back and forth on this issue. First it was charged directly to the contract, later to overhead, and then back again. It's my opinion that there are already too many charges on overhead causing it to be so high.

Why should a program that doesn't use CADD, have to be charged at the same overhead rate as someone else's computer time?

Computer time should be charged directly to the job, in my opinion. The trouble is how does one monitor and bid this unknown computer time. One way is to use previous data on a similar program to estimate the computer time for the costs of any new programs.

Computer services also provide programming for technical development and research. Again, these services should also be charged directly to the job to reduce overhead. There are many other services that should be charged direct to a contract instead of using overhead, such as reproduction.

The more services that can be charged directly to a program, the lower the overhead charges would be. In this way, engineering must account for all services when bidding for a new job and would not have to pay for services they are not using. In any event, whatever the company's policy is, that's the method one should follow whether he agrees with it or not. Don't try to reinvent the wheel.

8. HIRING AND FIRING

8a. Hiring

Most companies have a Personnel Department that advertize or solicit new employees. If a manager requires more engineers, the EDM would probably have to get permission from upper-management to hire more people. The Department Manager or Division Engineer would notify the departments Administrator in writing or verbally, the type of engineers that are needed and their classification; EE, ME, Software, etc.

My company used the numbering grade system for classifying all engineering that covered salary grades from 1 through 10. Above 10 was for management with the President being at 20.

A young undergraduate would be hired as an Associate Engineer with a salary grade of 1. A graduate student at 2. Depending on experience and the previous job held a new hire could be hired up to a 10. Most companies do not hire people directly as a 10. They prefer to promote from within. However, I was hired by an electronic company as an 11 as a EPM, so it doesn't always apply. It all depends on what talents the company is looking for.

I personally like a mix in talent with an organization similar to a professional ball team. A good sound baseball team, in my estimation, is composed of %20 older an seasoned players, %60 experienced players and %20 rookies. In this way the older players can stabilize the younger players and provide them with professional instructions but the majority of the strength comes from the %60 middle baseball aged group.

I feel the same arrangement applies to an engineering team. If there are too many apprentices the group may turn out poor designs. If the group consist mostly of old engineers the group may be designing equipment with archaic ideas. The group needs middle aged engineers who have the expertise and are willing to invent new ways using the latest technology in their designs, and still stay within the RFP requirements. The young graduates would learn from these engineers especially the older men about reliability and quality. Each young graduate should be assigned to assist an established engineer and should not be given major responsibilities.

Why then should a company hire young graduates? It's very simple, the older men would eventually retire and some of the middle men would be getting older. It takes about five years out of school to become an experienced designer, sometimes more. So a salary engineer grade 1, in five years, would at least have been promoted to a salary grade 4 by then. Similar to the baseball team example, don't just hire for the short run, hire people that would be with the company for many years.

Then, of course, it's the economy of supply and demand. If the company receives a major new contract it would need more help. But the company doesn't want to overload itself during dry spells. It becomes a juggling act. This is where "check and balance" comes into the picture. An EDM would want to increase his members to do the job, but he has to convince his upper-managers to warrant this increase in manpower and still be careful of not over hiring.

If the department has a steady supply of research work that can be depended on for many years, it becomes a little easier to move men around between contract work and research but neither program should be jeopardized over the other.

If the design organization totally depends on contract work then it would have to be careful not to over hire.

8b. Interviews

Generally, the personnel department would screen out new hires before the department is asked to interview them. On new graduates, our personnel would only recruit engineers from qualified schools and only accept applications from the top twenty percent in their class. This does not always guarantee the best applicant. I mentioned before, that many of the top students would be board as design engineers because most of the work is tedious.

When I recruit young graduates, I'm looking for men and women that are mechanically minded. Are they willing to do hand soldering? Do they work on their own cars? Are they good at drafting and know anything about CADD machines? Are they only interested in doing just research work? It really depends on the position that is open. I hired a girl EE who preferred hardware. She turned out to be a very capable engineer working on Automatic Test Equipment, programming the test station to test black boxes.

I interviewed a "B" student onetime and asked him many questions. He was a technician in the service for two years and he enjoyed working with his hands. He also said that he was on the college golf team. I said, "Oh, in that case your hired."

"Just because I play golf?" he asked.

I was kidding him of course. He had a +2 handicap. I guess it is obvious that I play at golf, but I'm not a golfer.

So, the EDM has to weigh the manpower situation and sometimes it takes a gut feeling on who to hire and how many. If they have been screened by personnel the chances are they would make good employees. So it's a matter of personality and intuition on the managers part.

Pre-screening by personnel doesn't always work, however. I was interviewing an engineer for a salary grade 8 position onetime who had tremendous talent according to his resume. He was very impressive in his interview and I almost hired him. My Division engineer was very impressed and left it up to me to make the final decision.

I had a gut feeling that this man was giving me a Con job so I began calling all of his previous employers. These calls proved me right so I turned him down. He became very belligerent about it because my Division Engineer practically told him he was hired. My boss shouldn't have done that. I finally told him another person was selected for the job and if it comes up again I would consider him. He didn't believe me and began yelling at me on the phone. Boy, was I glad I didn't hire him.

What made me suspicious of him was his jumping around from company to company. His average stay per company was one and a half years over a fifteen year period. He was pre-screened by personnel but, obvious, they didn't check his past employment. I called personnel on this and they said they had warned my Division Engineer that he seemed to have a personality problem. He ignored them and did not mentioned it to me. I guess that's why he left it up to me. If we hired him and he turned out bad my manager could always blame me. Silly, isn't it. Engineers sometime act like children.

8c. Firing

The other thing a manager has to consider when hiring an individual is, if he turns out to be a lemon how does one get rid of him? My company had a six months probationary period when hiring, where a manager could release any engineer for any reason, whatsoever. After that it takes an act of God to get rid of him. So, it is very important that a manager hires a good candidate. Especially look at his personality. Nothing worse than having an engineer that can disrupt an organization.

I had a young graduate that was hired directly by my Division Engineer before I was transferred into his division. After he was assigned to me two years later, I discovered that he was very incapable and lazy. He would report late often. Now the question arises, how do I get rid of him. When I wrote him up, my Division Engineer reprimanded me. He said that I was not a good manager and I didn't train him properly. After that scolding I then gave him only menial tasks to do to keep him out of trouble. I finally was able to transfer him out of our department. A year later his new boss yelled at me on the phone for not telling me how bad he was. I didn't lie to him. I just didn't give him all the details. It was just a little white lie, I thought.

This is not the recommended way of getting rid of a poor engineer. Had my Division Engineer respected my evaluation of this man, we could have fired him the proper way. The procedure is to write him up every time he commits a foul up, such as being late to work too often or not doing his assignment. He must be confronted about his faults and explained the consequences. My company has a special form to be filled out for this which must be shown to the individual under question. He must sign the form and agree to amend his ways. If he violates his faults over three times it can be submitted to the personnel department who would then start the proceedings to fire him.

It doesn't end there, however. We laid off a Mexican-American once and he took our company to court claiming bias. My company settled by paying him two years back salary and had to rehire him again. He refused to work for my boss, however, because he claimed he was a racist. This was not true because we had five mexican descendants working for us who agreed with me that we were correct when we let him go. They knew he was no good.

8d. Layoffs

Unfortunately, every company has to, at some time or another, layoff people. It was one of the most difficult times for me. There are several reason why there are layoffs. One reason is the anticipation of winning a major contract and losing it. I was almost laid off when my company lost the Lunar Landing bid to Grumman. We over-hired for that contract and transferred too many people to prepare the proposal by predicting a win. When we lost many people went out the door. I was lucky because I was picked up by another design division.

I lasted 30 years after that with that same company because the electronic division was well run and had new work coming in from all the different aircraft my company produced. We also had plenty of IRADs and CRADs to keep us busy. I was lucky.

Canceling a contract for one reason or another is another cause for layoffs. This could happen when a company has won a major military contract and after a year or two, congress or the military decides to cancel or reduce the production quantity.

Some contracts were lost because of poor results or over priced for the production aircraft. I was again on the layoff list when this happened to our company. A helicopter program was canceled because it became too heavy and too expensive. The cancellation was not justified by the Army because they continued to add armament to the payload which also increased the cost. It happens.

My company also had a commercial airline that was losing money every year. They finally closed the program down and sure enough many layoffs occurred.

Every company handles the layoff procedure differently. The way my company handled it was by using a stacking arrangement. For each classification, each man received stacking points. The way he received points was, he would get one point for every month he was employed and received a set of merit points for ability up to 160. He was stacked according to how good he was from 60 to 160 points in 20 point increments.

What save me was I was always stacked at 160 or at the top of the list. The layoffs would be selected from the bottom of the merit list first. If a poor employee, for example, had ten years with the company he could receive 120 points for employment and the minimum of 60 for merit for a total of 180 points. He would be difficult to layoff. The procedure was that he would be compared with the next man in line of the merit list. If the next man was stacked at 80 merit points and had only five years with the company, his total points would be only 140. Out the door he would go.

The way the poor engineer accumulated ten years was by being laid off and rehired again. He would be laid off and go to work for another company and would later be recalled because the recall was in the same order he was laid off at. It took three years to be away from my company before one would lose his recall rights.

Telling employees that they were being laid off was my toughest job. Sometimes it was easy but sometimes it could be rough. I had an employee argue with me because he thought I was given preferential treatment to a certain other individual. I told him he is being laid off according to the stacking. His stacking wasn't high enough. He never bought my reasoning and was definitely mad. A year later he was rehired but wasn't so mad at me anymore. He turned out to be better than average engineer at his salary grade and his stacking did go up. He finally understood the stacking system.

Stacking was generally done by the Department Manager or Division Engineer. It was done twice a year and, believe me, it could be brutal. The yelling and screaming was sometimes a little out of hand. Each department wants to hold on to their best men and wants them stacked merit wise as high as possible. Young graduates were all stacked the same because there is no history on their ability. If a man is a salary grade 3 or higher he is stacked in his own salary grade according to his ability.

I had a situation where my salary grade 4 engineer was stacked at 120. Another department wanted to stack a girl with 3 years experience, higher than him. My man had 30 years experience. They tried to convince me that I didn't have to worry about my man being laid off because his total was very high. That is not what stacking is all about. It's supposed to be strictly on merit. I settled with them by having both of them stacked at 120. I certainly wasn't going to lower my man.

Bumping out engineers is also done according to the stacking position. A man can bump another man in another department if his stacking number is higher and has the same classification. However, the one that bumps another man must be able to do the other man's job or he could be fired. He does have the option of accepting a layoff if he thinks he can't do the job.

The problem is, in my company, many engineers are place into the same classification with different talents. A structures ME, for example, couldn't do the job of a black box packaging engineer, yet they could both be in the same classification. In addition, each engineer regarded stacking as a report card. It wasn't meant that way but if one is stacked low it does mean he is poorly rated and may reflect in his raises. Every year engineers would argue over their stacking because it was so important to them. Some would go into arbitration over it with the union.

CHAPTER IV
ELECTRONIC DESIGN

This Chapter is devoted to actual designs. They are included in this book to assist the prospective manager on the things he has to anticipate when consideration new designs. Hopefully, these examples will be helpful to an engineer as he advances and becomes a Design Manager.

1. Doing the Design

Doing the Design is not an upper-management's responsibility but more of a Design Manager's duty. Upper-management should be aware of some of the issues in the design, especially on schedules and overruns. This chapter is strictly a supplement of Chapter I, and is primarily illustrated here to help understand some of the pitfalls that can happen when designing electronic boxes. This chapter covers mainly hardware design and provides some theoretical analyses.

Chapter I, covered all of the many tasks that a Design Manager has to know to run a department.

Chapter II, covered the entire sequence of a program, from when a Request For a Proposal (RFP) is received and progresses through the phases of Quotation, Design, Integration, First Article Test, Production, and finally to the delivery of a product.

Chapter III, consisted mostly of Company Management philosophies on how to run an engineering organization.

The objective of Chapter IV is to familiarize the reader with what it takes to do the actual design and to help him tackle many of the tasks required to produce a product. There is actually more work in producing a product than just the design. This chapter was not intended to be very technical, but on some occasions I have introduced some complicated mathematical formulas mainly to prove a theorem or an equation and to emphasize a design criteria.

When I became aware of all the different support departments that were involved with a design, their functions were not very clear to me. For the first 10 years as a designer I resented these organizations. I felt they were infringing on my design, trying to influence my creation. Check and Balance was the furthest thing from my mind because I knew that I was a good designer. "Why do we need these people, anyway?" was my presumption.

I later transferred to a secret organization as a Flight Test Engineer. My new job was to determine if all new purchased EW equipment performed properly in flight. Since the plane was a single seater I had to rely on the mission recorder to be able analyze the operation of their equipment. Of course, the suppliers resented me and avoided me like I was the plague.

That was my first encounter of why Check and Balance was necessary.

After four years at this position, I transferred to another EW organization as a Systems Engineer but on a very large reconnaissance airplane. My job was to prepare the Equipment Specification for a very complicated EW system which included antennas, communications, Radar, computers, and displays. This position alone was very educational for me and taught me what a Systems job entails. I found out the hard way that a SE should not agree with a Vendor's terms without consulting with the Program Manager. I also discovered that System Engineering was a very difficult position and was not my bag.

I was later offered the job as the Design Manager to run my old department. That's when I learned that their was no management course or training program for new managers. I stayed in management for ten years but the last four were brutal. My organization was transferred to another department that worked strictly on research programs.

That's when I found out that research engineers haven't a clue of what it takes to design and produce military equipment. They never heard of such things as ARINC case sizes, CDRL, Preferred Parts List, Non-Standard Parts, Sem-E size Printed Circuit Boards, Support Organizations, Built-In-Test, Reliability, Drawing submittal in Micro film, etc. I was made acting Department Manager over twenty Research programs and still had my own military contracts worth millions of dollars to monitor. My hands were tied by my Division Manager who made my life difficult.

I finally had it with this situation and left the company altogether and became an EPM for another company. Up until then, I always thought an EPM's job was a piece of cake. That's when I discovered all of those crazy charts that had to be submitted.

After one year in this rat race I went back to my old company, but as a Staff Engineer in the Automatic Test Equipment Department. These people also never heard of Non-Standard Parts, Drawing check, etc. After all of these experiences I am now convinced that "Check and Balance" is very important.

1a. The Design

This chapter is devoted primarily to my engineering experiences in electronic design and some of my previous designs are covered. I am presenting it in this fashion to provide examples of what it takes to be a designer by showing the reader some of the obstacles that I have had to endure. One may also find himself in the same predicament some day and some of my experiences may help solve a possible dilemma.

1b. History

Back in the twenties, when I was born, there was very little electronic industry. Everything was mechanical; iceboxes, victrola, telephone, etc. Even lighting a room was done by utilizing gas pipes that were fed throughout the house. There were some bad fires as a result of this method. Only the rich could afford a radio, electric light or a refrigerator.

In the early thirties my Dad bought our first tiny table top radio. The man that installed it ran an antenna wire from the radio in the living room through the entire length of the house and out the back yard for another 25 feet to the top of our garage. He called this wire an aerial. Antennas were only used by insects then.

I was still hand winding my Victrola, playing old Caruso opera's because that was all we had in the house. I never did own a phonograph machine until after World War II. Don't feel sorry for me because I had a great young life. We had no drug or alcoholic problem and we just enjoyed life. Then, World War II changed the whole electronic world.

An Electronic Engineer in the forties and fifties had to be more of an all around person. He had to know all aspects of the field. He had to know audio, radio, TV, communications, and even RADAR. In today's world everyone now is a specialist. It's like the medical profession. The general practitioner can no longer master all of the medical fields anymore.

The invention of the transistor in 1948, caused the greatest impact over the electronic world that we have ever seen. The first transistor invented was a Point Contact. This type was not a great invention because it wasn't going anywhere. It didn't work too well as an amplifier. Its positive gain characteristics, though, made it an excellent oscillator.

After I went to work for Bell Telephone Laboratories in 1952, I was given a dozen Germanium Junction Transistors that were hand grown by a Physicist. He grew them himself in the laboratory but didn't know if they had any practical application. He first gave them to my boss who intern gave them to me to see if I could make them work as an amplifier. This was my first exposure to these devices and I didn't know exactly where to begin.

I tried vacuum theory first on these devices but that didn't apply at all. I talked with some other engineers who had designed circuits using Point Contact Transistors and they gave me a pretty good education. Using trial an error I was able to design, I thought, the first Junction Transistor amplifier. It was a poor amplifier because it took 10 transistors to acquire a gain of 20 db. It certainly didn't look practical at the time, so I didn't think the Vacuum Tube had any competition. One Vacuum Tube could easily provide 40 db of gain. They were eventually improved with higher Beta characteristics which proved to be better than vacuum tubes.

Electronic switching was when the transistor really took off. Vacuum Tube switching wasn't that practical because the switching levels would be on the order of 180 volts for a high and 50 volts or so as a low. It would take a large room full of electronics with refrigeration cooling to accomplish what a small personal computer can do today and it would cost millions of dollars.

I remember hearing the old timers at BTL saying that the transistor is not here to stay. Many wouldn't even try it. I hope the young engineer will never allow themselves to fall into that pit-fall of not keeping an open mind on any new innovations in the design world. Electronic technology is forever changing and one must keep up or fall to the waste side.

I almost fell behind myself several times when I went from vacuum tubes to transistors, to Integrated circuits, to digital, RTL, DTL, TTL, C-Mos, Schottky chips, Fast Schottky chips, Computers, Boolean Algebra, micro-processors, memory devices, software, AutoCADD, Surface Mount ICs, electronic work stations, fiber optics and finally to my own PC. Every time a new device was invented I had to take special classes or simply teach myself how to use them. It wasn't easy, but I was proud of myself for keeping up. Many engineers didn't and discovered they couldn't compete in the design field any longer.

1c. Specialization

It is almost impossible to be an all around engineer in today's world of electronics. I was lucky, in that respect, because I slowly evolved from vacuum tubes to transistors, etc. So, which specialty in engineering does one want to follow? In Chapter I, I discussed several possibilities such as Systems Engineer, Project Engineer, Support Engineer and Software Engineer, but I didn't brake down design engineering into different specialties. The following are some of the fields one can specialize in as an electronic designer:

Specialization

ANALOG	DIGITAL
Audio	Professional Computers
Phonograph	Solid Displays
VCR	DVD Recorder
Telephone	Telephone Switching
Cassettes	CD Cassettes
Sub-audio	Automatic Test Equipment
SONAR	Digital Navigation
RADAR	Digital Voice
Television	DVD Television
Radio	Personal Computers
Transmitters	Machine Shop Controls
Antenna	EW Analysis
Micro wave	All Types of Controls
Comm.	Comm. Controls
Video	VCR'S
Tube Display	A/D Converters
Inter-Com	AutoCADD
Navigation	Micro-Processors
Acoustics	ASW
Tube Display	DVD Radio
Motor Con.	Calculators
Tube Displays	
Voice Conversions	
Lamps and Relay Controls	
Motor Control	
Machine Shop Controls	
Power Supplies	
Direct Current	

These are a few of the options that one may choose to specialize in. Almost every analog item on the left could be replaced with a Digital item on the right. This is the way of the future such as DVD video recorders and DVD television. Unfortunately, one may not have any opportunity to select the specialty that suits him. When a student graduates from college he may not know what field he wants to follow. This often happens. When he lands a job he just can't tell the new boss he doesn't like the kind of work he was assigned to. At least not right away.

In my first job at BTL, I was doing statistical analysis, using a lot of math with an old fashion mechanical calculator. I am not a desk man. I'm a Laboratory man. I like to work with my hands. So after a year at this, I asked to be transferred into a design organization. They reluctantly agreed and that's when my design career really started. I was lucky to be working in a company where there was so much research and development that I couldn't help but learn.

One maneuver that BTL did for young graduates was to move them around to different departments every few months and after a couple of years they knew exactly which field they wanted to follow. This satisfied the individual and he made a better employee for the company. One may not get this opportunity. He may have to actually leave the company to pursue a vocation. After all, he would spend most of his life in this profession so he should really take the time to find a niche. If he likes what he's doing he will be a happier person. When I started out I always considered myself an all around engineer. I was willing to tackle almost any design job. I don't think that option is available today.

One subject an engineer must be reasonably good at is writing. English was my worst subject. During my young life in high school I avoided the subject. I did just enough to get by. Rather than working at it and improving myself, I hid behind my ethnic background. We only spoke Italian in my house and I would use it as an excuse. Yet, my older brother, Tommy, excelled in English and he became an outstanding Systems Engineer in computers.

At least one must be able to do is to communicate with others in the field and be able to make technical presentations. At first I was poor at making presentations because I tried to imitate the better speakers in my class. I discovered later there was no need for me to be a great speaker in engineering. I learned that if I would just use my own New York accent, speak in a slow but normal fashion and not try to be someone else, I usually got by pretty well.

Writing and speaking is usually the weakest trait of most Design Engineers. They prefer to work in the Lab to tinker with their designs and be left out of the politics. If one is of Latin, European, Asian, Arab or of Oriental background he had better learn the English language as best as he can. English is the language of the United States and it may hold one back if he can't communicate. One doesn't have to be a professional public speaker. Most people would accept an accent, as long as he is technically correct and knows his business.

1d. Starting The design

How does one start a design? How does one create something from nothing? How large should the enclosure be? These are some of the questions that plague all young engineers. Where does one begin? I have been kicking these questions around in my mind for this chapter and I am not sure I can provide a direct answer to these questions, but I can provide some simple tools to help a young engineer become a good designer and where to start.

If I had loaded this chapter with a lot of boring formula's and calculations I don't think I would have delivered the message that I'm trying to pass on. I think the best way is to present an example design approach. In other words, I want the reader to understand how I think and how I would tackle a new design. If one can think as I do he would probably do alright. It's not that I am perfect, it's just that I have 40 years experience under my belt. I have already learned by my mistakes.

In order for me to accomplish this communication I have to start at the beginning by presenting design problems as I encountered them over the years. This may give the reader some ideas on how to tackle a design.

Many times I was stumped on a design and just by talking about it to an associate it triggered a thought in my mind and I immediately knew the answer. Help from ones peers is not a sign of weakness. It's just smart business. It's called, team work.

Electronic design is very exciting and rewording but it can also destroy an individual. I am not going to cover every design that I was involved in, because that would take another book. I'm only going to cover those designs that I feel will help the reader think and improve his own capability.

As I mentioned, in March of 1952 I went to work for Bell Telephone Laboratories (BTL) in New Jersey and left after 10 years to live in a milder climate. The education and experience that I acquired at BTL was far superior than anything that I could have receive anywhere else. The "on the job" training that I received provided me with a foundation that stayed with me the rest of my career.

The most important design practice that I learned at BTL was design for reliability. When one is designing electronic hardware his mind should always be aware of the little things that would make the design more reliable. The engineer would automatically allow for high temp, power dissipation, testability, overloading, maintainability, etc.

As I mentioned in Chapter III, a designer is like a ball player that has to play often to be sharp because he would then do things by reflect actions. An experienced engineer would have done so many different designs so often he won't even have to think about reliability. It's imbedded in his subconscious mind and he would automatically allow for it in his calculations.

One other thing that I learned at BTL, was that engineering calculations should be very close to the measured results. Every design should require a final report that epitomizes this comparison. This is important, especially in Military contracts where the customer wants a demonstration on how one had arrived at the designs.

1d1. Example of a simple circuit

With reference to the simple relay driver and lamp driver circuits on Figures 11 & 12 on page 259, let us examine some obvious design faults that a young engineer may overlook. The first thing that should come to his mind when designing even the most simplest circuits as these is reliability over extreme temperatures. Let us assume the design has to meet extreme temperatures between +71 to -55 degrees celsius per Mil-Std 5400. The selection of a 2N2219 transistor as shown seems like a good prospect because the relay draws only 200 ma under steady state conditions. This transistor can handle 800 ma max and is full Mil approved. Some obvious faults with this design are:

Fault 1; In Figure 11 all relays are considered to be an inductive load. This means that the relay momentary turn-on current would be as high as 5 x the rated value of 200 ma or 1 amp, higher than the maximum drive current of a 2219.

Correction; A stronger power transistor should be used to drive this relay. On the other hand, the lamp driver of Figure 12, using the same 2219, requires only 40 ma steady state. In this case, the turn-on current of a lamp happens to be 10 x the rated load or 400 ma. So the 2219 transistor used in this application worked fine.

Note that the 400 ma is only a momentary surge and would not be calculated in the overall power dissipation summary under continuous operation.

Fault 2; One thing wrong with the lamp driver circuit of Figure 12 is a poor first stage driver circuit of Q2 and R3. At room temperatures the circuit would work fine but at -55C the DC beta of the 2219 for Q3 may go as low as 20. The base drive current of Q3 requires 20 ma. The 10k ohm resistor R3 that is connected from the collector of Q2 to 28v can only provide 2.8 ma base current for Q3 far below its drive current. In my designs I would always assume an even worst case condition and presume the Beta of Q2 to be as low as 10.

Correction; R3 should be reduced to 1k ohms to allow this circuit to work at low temperatures. This would at least provide 28 ma to the base of Q3 at minus 55 degrees celsius to maintain a lighted lamp.

The question always arises, why does one have to keep a lamp lit at -55 C? Nobody would be in the room to look at that lamp at that temperature. First, it's required in the Spec. Secondly, suppose a pilot had to make a quick flight out of Alaska in the winter. He doesn't want to sit there on the runway for an hour waiting for readout circuits to warm up in his cockpit before he can take off. It can happen, so design for it.

Fault 3; The next thing wrong with Figure 12 may be due to a transistor characteristic called ICo which is the internal leakage current of Q2 and Q3. At high temperatures, ICo increases considerably so much so that the saturation voltage of Q2 would increase (open up) and may cause Q3 to turn on inadvertently.

Correction; To avoid this, a diode may be added to the emitter of Q3 or add a series resistor in the base of Q3.

Fault 4; The third thing wrong with the lamp circuit of Figure 12 is connecting a three input NAND gate to drive Q2. Most NAND gates are not designed as drivers.

Correction; A proper buffer driver stage should be used after the NAND gate or use an open collector NAND gate with a pull up resistor.

Fault 5; Notice that the input of Q2 of Figure 12 is connected to the NAND gate through the connector J2. This is bad practice. One should never have an input stage floating on a board which depends on its bias from another stage located on another PC board. If the board is to be tested separately a bias has to be provided. Another thing that can happen is if the connector pins of J2 becomes dirty Q2 would be in a state of limbo. It might even burn itself out.

Correction; A base resistor should be added to Q2 to pull the base up or down as necessary as it was done using R2 of Figure 11.

Fault 6; The resistor combination R1 and R2 of Figure 11 may work at room temperatures but Q1 would quit at -55 C. The input TTL Inverter SN5404 would only provide about 3 volts at the connector pin J1. As one can see, this voltage is padded down to 1 volt at the base of Q1 by R1 and R2. The relay driver transistor would be base current starved at low temperatures and 1 volt may not be high enough to trigger Q1 on.

Correction; The best approach is to use an open collector Inverter Integrated circuit such as the 5405 but reduce R1 to 2k ohms and pulled it up directly to 28v as the collector resistor.

There are many engineers that wish to specialize in a particular field. In the next series of articles, I provide the reader with examples of my designs to help the reader select a field.

Some of these are Amplifier designs, Infrared Systems, Oscillator designs, RF Receiver, Outer Space Communications, ASW, EW, Fault Warning Systems, Communications Controls, Displays, Navigation, and Packaging.

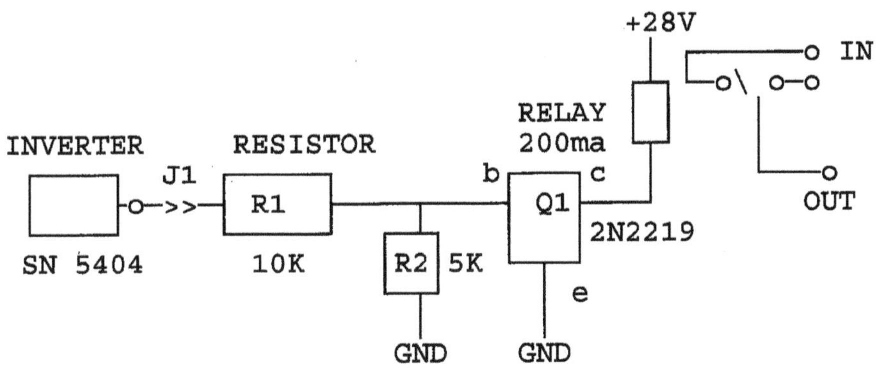

FIGURE 11
RELAY DRIVER

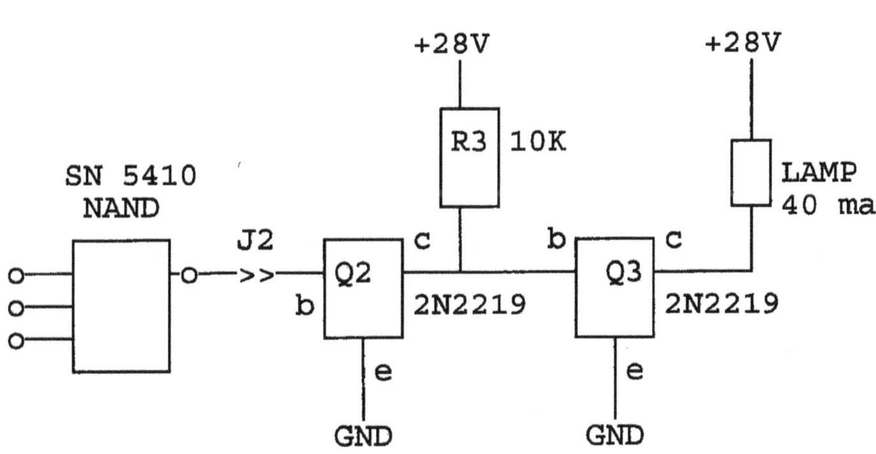

FIGURE 12
LAMP DRIVER

I am using these two circuits as simple examples because they where originally used by one of my young engineers who did not allow for maximum temperatures. I didn't discover this weakness until qualification testing took place and the circuits failed at low temperatures. What I'm trying to get across now is, one must consider MTBF and environmental conditions if the circuit is expected to work at extreme temperatures.

Maintainability is an other item that must be considered in the design. One has to say to himself, "I'm the guy that's going to repair this unit, I had better make it easy on myself."

This statement can be compared to an automobile design. If the engineer that designed the auto had to do the repair himself, the average consumer wouldn't have to practically take the engine out to replace a fan belt, for example. And so it is with an electronic design. One has to consider "easy access" for repairability and testability.

The next major item I learned over the years is simplicity. There are always many different ways to make a design. An engineer's first approach may not be the simplest, but then when does one shoot the engineer and stop him from making changes. This is sometimes a difficult choice to make.

I was involved in a design where there were 50 chips on my PC board. Two months later, my boss looked at my schematic and insisted that I use another approach just to save 4 ICs. His approach did reduce the number of chips, I have to admit, but to re-layout a board just to save 4 chips is foolish and costly. Hindsight can always be used to reduce the chip count.

I can always look at anyone's design, after the fact, and see a better way to save components. When one is heavily involved in a design he generally never has the luxury to go back and reduce components. This is why he should spend the time to insure the first approach is a good and simple. If it is simple the chances are that it would also be reliable.

I once assigned one of my engineers a simple design task. He then created a monster out of it making it very complicated. I asked his assistant, why he went to such extremes when a simple approach I had outlined would do the job. His answer was, "he would be bored doing it your way." Unfortunately, he was too far down stream to stop him to do a redesign so I let it go, that time. It didn't happen again, I can assure the reader.

2. Vacuum Tube Feedback Amplifier

I know that vacuum tubes are outdated but this next design does prove a point. My first design assignment at BTL in 1953 was to assist in the design of a Vacuum Tube Audio Amplifier with a gain of 100 db for a Military program. My second assignment was to design a 100 khz, 60 db amplifier. The first two things that comes to one's mind, for both of these applications, are stability and Noise Figure (*3). Stability is obtained using multi-feedback technology. Noise Figure is determined by how much noise is being generated by active circuits at the front end.

The first thing one has to do is select a special low noise Vacuum Tube at the front end. But it must be understood Vacuum Tubes do become Micro-phonic with age and generate more noise. These low noise tubes were replaced years later with low noise Germanium Transistors and later by Field Effect Transistors. Feedback theory applies to all amplifiers whether the circuits are vacuum tubes, Op-Amps or with transistors.

Feedback Amplifiers, in today's world, are no big thing. One simply purchases an Integrated Circuit (IC) that is appropriate for the job. The message here is, does one know enough about the theory of feedback amplifiers to buy the right IC that provides the right amount of feedback? Does one know what Noise Figure is and how to handle it? Does one know what constitutes a feedback amplifier? Does one know what a Bode Plot or Bode Step is (*1)? Does one know what the Nyquist Criteria is (*2)? When an engineer designs an amplifier from scratch he has to be familiar with all of these terms and how they apply.

A simple purchased order of a chip today has replaced a three month design task. It sure is a lot easier today. The point is, does one understand the principles enough to make the proper selection? When one looks at the availability of IC amplifiers today there is over 100 to choose from. Which one fits the application?

Feedback amplifiers can be made to be stable if the proper precautions are taken. The main object is to provide enough signal feedback into the circuitry to stabilize the amplification so that it may be operated over a wide temperature range using the proper components. Each component, including transistors and resistors, have difference characteristics, tolerances, and temperature coefficients. When one puts all of these together in a feedback amplifier, and don't allow for all of their variations, he may wind up with a beautiful oscillator.

How much feedback is required or what type should be used depends on the application. Too many engineers would go to the extremes in their design. They would use local feedback with two or three multi-feedback paths. Remember, try thinking simplicity.

In the 100 kHz, 60 db amplifier mentioned before, another engineer used a multi-feedback design with 60 db of gain and 60 db of feedback. When this engineer was transferred to another program I was asked to complete the amplifier design.

After I assembled a bread board of his amplifier I noticed it was very unstable. It could be made to oscillate simply by passing a hand over the board. To stabilize it I had to reduced the multi-path design to only one single feedback path, and reduced the amount of feedback to 40 db. I had to add extra stages of amplification, which did require more real estate, but then it resulted in a stable circuit that would operate under any component variable conditions.

I first went through a calculated redesign using feedback technology of opening up the feedback loop and checking the phase shift at the high and low ends of the band, inserting extreme tolerance values of components and even adding more feedback than required. How much feedback is required? Fifty percent seems to be the maximum number in most cases. A trial an error approach on a bread board may be the final determination.

Simplicity, again, may cost more in the end but, as I mentioned before, in a military program reliability or MTBF is the name of the game. I think it was a challenge for the other engineer to attempt the design of a feedback amplifier using multi-feedback paths with minimum transistors, but then he created an unstable circuit.

3. Feedback Amplifier Design

Let us examine the transistor schematic shown in Figure 13 on page 264. This is an example of a three stage single loop feedback amplifier. I am not going to get into all the details of this circuit but just enough to illustrate some of the fundamentals of feedback amplification. There are two kinds of feedback used here. Local feedback and single loop feedback. Local feedback is accomplished for each stage by the emitter resistors R8, R10 and R12. The object of any feedback circuit is to remove or reduce the effects of variables in the circuit.

Loop feedback accomplish stability when an increase or decrease in amplification occurs in a transistor due to variations of its beta effect between different transistors and over temperatures. If the beta increases due to higher temperatures there would be more amplification. This increases the negative feedback back to the front end through C6 which subsequently reduces the gain of Q1. The overall band pass gain would now remain fixed. In other words the feedback loop would soak up any gain variations due to beta, component tolerances, or changes due to temperature. In this way the overall gain of the amplifier remains constant.

Multi-loop feedback is when the feedback takes several paths, such as feeding back to the emitters of Q1, Q2, and Q3 from different collectors. I will not discuss the design of a multi-feedback circuit because I feel it is too complicated and not worth the effort. So one might save a transistor stage or two using multi-feedback, so what. I still say keep it simple and reliable.

3a. Mathematical Approach

By using a reasonable size resistor in the emitters of Figure 13, the estimated gain of each stage is RL/RE, which is the resistive load divided by the emitter resistance. In the second stage, for example, the gain of Q2 would be:

$$G2 = \frac{R4 \times R5 \times R11 \times B(R12)}{R10\ [R44 + R5 + R11 + B(R12)]}$$

or the parallel combination of R4, R5, R11, and Beta x R12 divided by R10. For the moment, let us ignore the shunt network of C5. The variables in this equation are the Beta's and the internal impedances of Q2 and Q3. Transistor Betas vary from 100 to 1,000 between selecting the same type of transistors and can go down to 20 at low temperatures. So if R12 is big enough, say 100 ohms, the lowest the input impedance of Q3 could ever be is 2,000 ohms (B x R12).

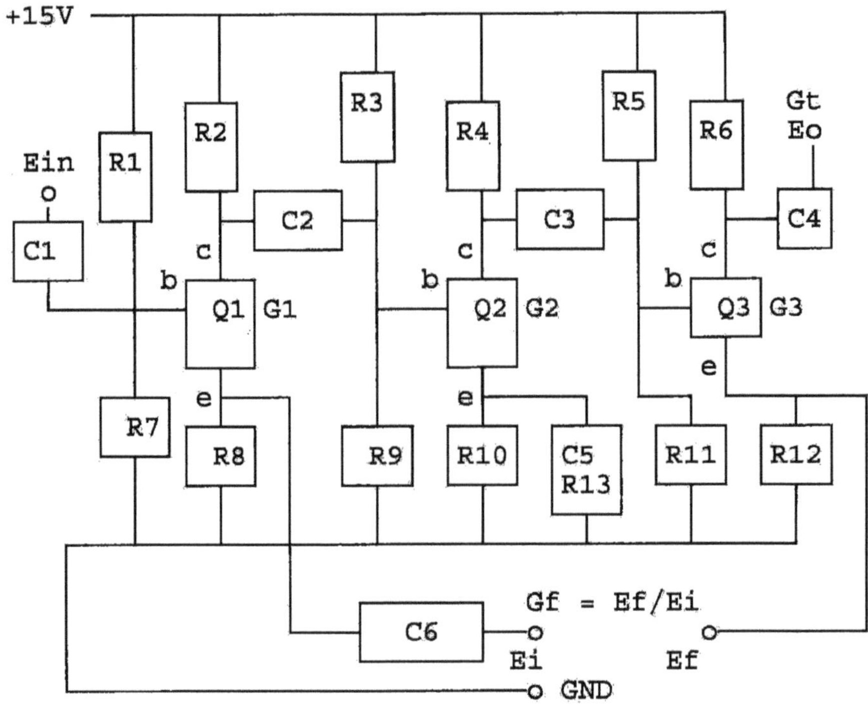

FIGURE 13 - FEEDBACK AMPLIFIER

*1 H. W. Bode, Relationship between Attenuation and Phase in Feedback Amplifier Design, Bell System Tech. Jour.,Vol. 19, p. 421, July 1940.

*2 Nyquist, Regeneration Theory, Bell System Jour., Vol. 11, p 126, Jan. 1932.

*3 H. T. Friis, Noise Figures of Radio Receivers, Proceedings of IRE, Vol. 32, P.419, July, 1944.

*4 Reference Data For Radio Engineers, by ITT, page 766, 1956.

This impedance may be too low when one is trying to get some gain out of this stage. The total parallel impedance may be as low as 500 ohms. If one needs 60 db of gain and 30 db of feedback or a total of 90 db, then each stage has to able to contribute 30 db of gain. Local feedback by itself is not enough to stabilize this circuit.

This was the basic circuit that was given to me in the above example where the engineer used Multiple feedback paths for stabilization but using only two stages. It was obvious to me that more stages were needed in this diagram to achieve the gain and to stabilize the design by using only a single loop feedback as shown.

As a three stage amplifier the feedback signal Ef must be 180 degrees out of phase with the emitter signal of Q1 at the center of the overall band pass to reduce or increase its gain.

Now, what causes the amplifier to go into oscillations? It could be at the low end or the high end of the frequency band. The networks of C2, C3, C5, and the associated transistor internal capacitances would each have frequency break points that could contribute to a phase shift at both ends of the frequency band where oscillations may occur. The principle factor of (*2) Dr. Nyquist theory is; as long as the open loop gain is less than "1" when the open loop phase goes through 360 or 0 degrees the circuit would not oscillate.

The way one calculates and measures feedback is to break the feedback path at a convenient point. If the point at Ef is chosen, a signal generator is imputed to the emitter of Q1 through C6 and the output is measured at the emitter of Q3. A gain response curve and a phase curve are then produced.

I would suggest that all gains be converted to decibels so that the total gain of each stage is additive rather than multiplied. If the open loop gain and phase curves show that the phase angle passes through the 360 degree point when the gain is -10 db at both ends of the curve, the amplifiers would be stable. This would be a pretty good design as it provides a cushion of 10 db in gain variations at both ends of the frequency spectrum.

The Dr. Bode step (*1) is a method of producing a wider frequency band by using special networks in the loop, thereby changing the phase and delaying the 360 degree crossover point at the high end of the frequency spectrum. This step is commonly used in video amplifiers where wide frequency bands are required.

I designed a video amplifier that was used at the front end of a special oscilloscope. The special scope was to be used as test equipment in tight quarters aboard a Submarine. The frequency band was from 5 Hz to 40 mHz. In those days that was a challenge. In today's world, one would simply select a wide band Operational Amplifier IC, but the above feedback theory still applies. A 60% feedback Op-Amp Integrated

Circuit may still sing at high or low temperatures if one is not careful, especially when he designs an amplifier that is into the mHz range.

3b. Graphical Approach

There are two ways to formulate a paper design and determine the gain phase characteristics of a feedback amplifier. It can be done by calculations or graphically. I prefer the graphic approach because it is faster and one does not have to go through reams of calculations every time a change is required in the phase shift to insure the amplifier does not sing. Both methods are basically the same, except the graphical method requires less math.

Step 1 - Calculate the pass band gain of each transistor stage. This gain is purely resistive since the reactive components do not get into the picture at the center of the band.

The open loop response using the graphic approach is as follows:

The gain of Q1 at room temperature is:

$$G1 = RL/RE = \frac{R2 \times R3 \times R9 \times 100R10}{(R2 + R3 + R9 + 100R10)\ R8}$$

$$\text{or } G1 = \text{Approx.} = \frac{R2 \times R3 \times R9}{(R2 + R3 + R9)\ R8}$$

The gains of Q2 & Q3 are derived in a similar manner.

Step 2 - The open loop gain is:
Go = Eo / Ein = G1 + G2 + G3

Step 3 - The Feedback gain is:
Gf = Ef / Ei = G1 + G2

Step 4 - The total closed loop gain is:
Gt = Go - Gf

Step 5 - Plot the low frequency break points (*2) for each stage. The low frequency break point of Q1 is composed of blocking capacitor C2 in parallel with R3 & R9. Stage Q2 break point is C3 in parallel with R5 & R11 and C5 in parallel with R10. Stage Q3 is basically out of the picture in the feedback loop response. Note: Beta x Re of Q1 & Q2 are assumed to be too large to affect the break points at room temperatures. These emitter resistors would come into play only at low temperatures.

Now all of the frequency break points in the feedback loop are plotted for each stage to show gain response and phase shift. Because we are computing gain in DB's the responses and phases are additive. The sum of the low frequency and phase are plotted on three cycle logarithmic graph paper.

Step 6 - Plot the high frequency break points for each stage. The high frequency break point of stage Q1 is the parallel combination of R2, R3, R4 and the internal capacitance of Q2. Stage Q2 is the parallel combination of R4, R5, R11 and the internal capacitance of Q3.

The Bode Step at the high frequency area is produced by inserting special networks such as C5 in series with R13. C5 shunts R10 to increase the gain of Q2 at a specific frequency point which has the affect of delaying the phase shift and expanding the frequency response. R13 would eventually take over and the high frequency step reduces. As in the low frequency break points, each gain response and phase shift is plotted.

Note: The frequency break point is defined as the frequency point when the response of an RC or RL network is 3 db down, then the phase angle is 45 degrees. The break frequency of each RC or RL network is found using Reactance Charts or:

Fb occurs when R = Xc = 1/wC
or = XL = wL

Note: 2. The plotting was done on 3 cycle logarithm graph paper so that all of the response curves for each of the stages become additive.

Plot all the response curves and the phase angle on the same graph paper. When one does it this way he would find it very easy to plot the final response and phase curves by selecting different frequencies and adding up all the levels and phase at that frequency.

Step 7 - Breadboard the design and measure the feedback open loop characteristics using an appropriate signal generator with a meter or scope. One may have to juggle a little to get the right amount of feedback an phase shift. When the designer is satisfied put the amplifier into an oven and check the gain and phase at both ends of the band at +71 degrees C and at -55 degrees C. Repeat these same tests with high and low beta transistors. I always kept a hand full of high an low beta transistors in my desk draw just for these tests.

In the calculations one should have accounted for the tolerances and temperature coefficients of components. It's desirable to try a different assortment of resistors and capacitors in the oven. Altitude, usually has no effect on the gain except that if one uses plastic transistors they may crack.

Step 7 is often by-passed by many designers who believe that bread boarding is a sign of weakness. They say, a good engineer should be able to calculate a reliable design without the need of expensive bread-boarding and laborious testing. Some joke. This is why I previously stated that if a design engineer thinks that designing is a lot of fun and is not willing to do all of the tedious tasks of bread boarding and testing he had better look for another vocation.

I have designed many amplifiers using this approach and never got into trouble. After I completed the electronic design for a box and don't hear any repercussions from the customer or from field services I know my design is reliable. No news is always good news, they say.

4. Infra Red Acquisition System

I now want to discuss another situation that is associated with feedback amplifiers and low noise figure. It was in a design that I was involved in when I first went to work for an Aerospace company. My first assignment was to assist an engineer in the design of an Infra Red (IR) Tracking System for the Air Force. See Figure 18 on page 283. The Infra Red cell input has a very high impedance of greater than one meg ohm. This impedance is the limiting factor for the Noise Figure for that cell. In other words, the maximum sensitivity of this IR system is limited by the noise that is generated by a high biased resistor used for each Cell. Each resistor would generate thermal noise when electricity is passed through it. The lower the resistance the lower the noise, but if the resistance is too low the cell would not be biased properly which results in poor range. The thermal noise of a carbon resistor at room temperature is:

2 -20
(*4) $E^2 = 1.6^{-10} \times 10\ R \times BW$

For a 1 meg ohm resistor and a BW of 1,000 Hz, the resistor noise is:
E = 4 micro-volts

The thermal noise for a wire wound resistor is approximately 1/10 of a carbon resistor or = 0.4 micro-volts.

The second consideration in this design is gain. The cell signal may be down into the micro-volts level, so 100 db of required gain is not uncommon for these devices. Here again, the engineer that I was assisting, was using multi-feedback paths and disregarded component variations. He did not go through the process of opening the loops and checking them. He ignored phase shifts problems. And, he ignored two major items that are most importance and that is crosstalk or pickup.

I got into some heated arguments with this engineer over his design and he pulled rank on me and had me assigned to another program. So, I let it happen. This IR System, as shown on Figure 18 page 283, was composed of 16 IR channels in a special circular array of eight positions to be able to track aircraft. It provided 4 quadrants, each composed of four cells. When this engineer fired up the system he had nothing but trouble. Some circuits began oscillating and those that did not oscillate had crosstalk between channels.

He worked with a technician for another month trying correct the problems and instead he turned in his resignation. He bailed out. The upper-managers were not aware that he was having problems and made him all kinds of offers to stay. He finally left and I was assigned to take over the design.

The first thing I did was to re-do all the amplifier designs and tested the feedback loops to insure there would be no oscillations. His design also had poor sensitivity so I had to re-design the front end with a special boot strap emitter follower to get the front end impedance high enough to match the cell bias requirements. These cells are very high impedance devices and the impedance of one emitter follower was not high enough which would load down the cells causing loss of sensitivity.

After I got rid of all the oscillation problems, I then had to work on eliminating crosstalk, but before I did that, I had to clean up the power supply and provide proper de-coupling to each amplifier DC power source. Interference was creeping in through the power lines. Crosstalk, one must understand, is more of physical problem. Each channel was mounted on very large PC boards and they were installed too close together with no shielding between them. So, I separated the boards in the chassis and inserted Mu-metal shielding between the boards. That's all it took to make the system work. Magnetic pickup can only be shielded with Mu-Metal. Aluminum is good enough for RF but not as good at low frequencies as a magnetic shield.

The IR system was later tested by the Air Force on a mountain top and it did tracked aircraft successfully. Crosstalk was also covered in detail in Chapter II under EMI and TEMPEST.

The pre-amplifiers used at the front end of the IR cells above, were special low noise Vacuum tube types. These became micro-phonic with time and had to be replaced periodically. The emitter follower that the other engineer used was loading down these vacuum tube cathode followers causing a high Noise Figure and sensitivity losses.

The boot strap emitter follower that I finally used was a double emitter follower, so that the input impedance was beta squared times the emitter impedance. I later designed a very low noise pre-amplifier to replace the vacuum tube version using Field Effect Transistors. With this device I was able to improve the sensitivity by 10 db over the vacuum tube version and eliminate the micro-phonic noise problem completely.

5. Master Sweep Oscillator

My next assignment at BTL was to design a stable high frequency Master Sweep Oscillator that varied from 1 mHz to 5 mHz. The frequency could not vary, at a particular setting, more than 10% over the full military temperature range. The key to this design is the "Q" of the tuned circuit. The higher the Q the more stable the frequency. To accomplish the highest Q possible the shunt and series resistances of a tuned circuit have to be virtually eliminated from the circuit. To increase the shunt I used the boot-strap emitter follower circuit to take advantage of its high input impedance of beta squared times the emitter impedance. This circuit eliminated most of the variations that would occur from the internal parameters of the transistor. Field Effect Transistors were not invented yet but they could also be used in a boot strap configuration.

The next item to be designed was the inductor of the oscillator circuit. This particular sweep oscillator used large motor driven plates to provide the variable capacitance. The inductor then had to be fixed and stable. The coil was mounted on a two inch fiber core with pre-formed grooves and tightly wound. The series coil Q is:

Qs = XLs / Rs

Where Rs is the internal series resistance of the coil wire. A high Q came about by using a large diameter wire as practical to reduce the internal resistance.

The parallel Q is equal to:

Qp = Rp / XLp

Where Rp is the parallel impedance across the coil. This is where the high shunt impedance of the boot strap emitter follower comes into play. This design was environmentally tested in 1956 over extreme temperature ranges using different components to insure its stability. It is still in operation today. The total Q then becomes:

Qt = Qs + Qp

An interesting trick I incorporated with this design was in the method I used to transfer this variable frequency over a long distance for superhet requirements. If a emitter follower were used to drive the high frequency signal, its source impedance had to be very low to match the 50 ohm coax cable impedance. This put a burden on the emitter follower output power transistor stage and loaded it down.

To overcome this problem I used a small 10:1 step down transformer to drive the 50 ohm coax line at a 1 volt level and then at the other end of the line I reversed the transformer and stepped the signal back up. The transformer I used was a miniature type so it could be mounted on a PC board. There is always a way to overcome a problem and in this case small transformers did the job.

6. Tracking Receiver Design

Another design I now want to introduce is an RF receiver shown on Figure 21 on page 314. There are plenty of text books on receiver designs but by going through some of the problems I encountered, I feel it would help the reader in many different ways. I am sure that other engineers would have taken a different approach.

At BTL I worked with a 30 mHz transistorized IF strip design for several years on a Military program. When I changed jobs in 1961 to an Aerospace Company, I was asked if I had worked with IF strips before. I said yes, so they gave me an assignment to design a complete transistorized dual channel 60 mHz receiver. I almost hit the floor. I never designed a complete receiver before.

This Aerospace Company had a major contract to design and develop a Polaris Missile Tracking and Data Processing System. The main object was to track the missile during re-entry blackout which occurred between 70,000 to 170,000 feet, and record the missile parameters when normal communications ceased. See also Figures 19, 20, and 37 P. 321.

When we analyzed the ionosphere we discovered a window

at 5 GHz. We used this frequency range to transmit data from the missile which was received by remote stations on mobile truck trailers on land and on ships in the Pacific. The 5 GHz signal was then reduced to 60 mHz in the remote station using Parametric Amplifiers (Par-Amps), and a Klystron Oscillator. It was then reduced by a Superheterodyne receiver to a 30mHz IF strip.

The first version of this receiver was a purchased vacuum tube type and was delivered to the customer. Later the customer wanted to increase the number of stations for greater coverage. A second receiver was in the process of being assembled when our Project Manager discovered that the company that supplied the vacuum tube 60 mHz receiver went out of business. We had to deliver a system in a couple of months.

So, I was selected to save the program and design a transistorized version. I had a month and a half to design and build a dual channel transistorized 60 mHz receiver using a 30 mHz IF strip.

I decided to modularize this receiver by putting the RF sections in brass cans and mounting them on plug in boards. All of the RF sections were connected via coax lines. Dual channels meant each had to have a wide band 60 mHz amplifier, a switchable local oscillator, two IF strips, two AM detectors and two separate audio and digital differential amplifiers.

In those days there weren't too many silicon RF transistors to select from. Germanium was the only way that I could get the frequency high enough. Germanium, however, is limited at high temperatures. I asked an got a waiver on the temperature because the receiver would be in an air conditioned environment of a trailer.

6a. RF Amplifier Design

The front end frequency response of the RF amplifier had to be wide enough to accept frequencies of 59 mHz to 64 mHz. The gain of this amplifier had to be 60 db, because the Par-Amp output signal level was low and there was additional losses in the coax cables. The Noise Figure also had to be low at the front end not to degrade its sensitivity.

The approach that I took was to use fixed capacitors and Inductive slugs for the tuned circuit and I staggered tuned each stage in a Chebishev filter arrangement. The disadvantage of this filter approach was that the ripples on the top of the pass band cannot exceed 3 db, which meant more stages were needed to smooth out the ripples. If the ripple is too high there would be more than 3 db of sensitivity between the high and low peaks when the operator selected different frequency bands.

The stray capacitance for those transistors were fairly high so I couldn't use reasonably large fixed capacitances. The output stage had to provide at least 1 volt rms to drive a Superheterodyne bridge detector.

6b. RF Package Design

These transistors were packaged in a metal can as shown on Figure 21, and the collector was not physically connected to the case as the more conventional transistors are. This provided an advantage of mounting the transistors into a metal clip upside down like dead bugs and attaching them to the case of the metal container. This arrangement cut down the effects of the uncontrollable stray capacitance and kept the transistors quiet (no oscillations).

I honestly believe this was the first time a transistorized receiver was ever packaged in plug in modules and, for that matter, it may have been the first transistorized RF receiver ever built in the United States in 1961.

This type of packaging is best suited for RF because all the components were soldered directly to the brass case providing a good ground plane. Since all of the metallic modules were mounted on a plug-in board the dc power for each channel had to be decoupled. This was done by using feed-through capacitors on each metal can. However, after I fired up the receiver I had nothing but trouble due to crosstalk. I lost a week trying to solve this problem and it turned out to be I didn't have enough decoupling on the power lines between IF strips.

The feed-through capacitors I used did not provide enough decoupling, so I added small inductors (the size of a 1/2W resistor) in series with the dc lines. The crosstalk problem immediately went away. This problem drove me crazy because there was no indication that the signal from one channel was interfering with the other channel. I did not have the sophisticated test equipment in those days to help locate the problem. I discovered it purely by accident.

When I am doing a design and have problems I will try anything, no matter how stupid a thought it is because I may get lucky. Time was against me and I was in a panic mode. I lucked out. I always remember one thing about this business, electronics design is really simple and, if I kept my cool, I would eventually solve the problem. I have never found a problem I couldn't solve, providing good design practices were used in the first place.

I also want to inject another thought at this time, and that is packaging is part of the electronic design especially in the RF world. This is why I do not favor the Independent Department Structure, mentioned before, because the EE should be involved in the design of the package. If one leaves it up to an ME packaging engineer there would probably be all kinds of pick-up and crosstalk problems.

I have met some good ME packaging engineers who have had previous experience in electronics and understand the electronic requirements, but I still think it has to be a team effort when it comes to packaging. The team may be composed of EDM, EE, ME, and the PC engineer. As a manager, I would always present the package configuration to all of the engineers involved and to the customer.

6c. Gain Control

Feedback Amplification theory that I have expressed before does not apply in RF technology. Tuned circuits is the name of the game. Because there is no feedback used, stability in RF circuits becomes a problem especially with temperature. Even though my receiver was in an air conditioned Van, there was a temperature rise in the cabinet due to associated equipment that generated heat.

One solution is Automatic Gain Control (AGC). I also found by using negative coefficient thermal resistors placed in the emitters of each transistor some gain control with temperatures was provided.

These resistors introduced another problem. As the resistance value increased with temperature it caused the internal stray capacity of the transistor to vary, which are, essentially, part of the tuned circuit of the amplifier. I chose this approach as a compromise to save precious time in the design period and it turned out to be satisfactory. I did not use AGC but the operator had a volume control on a remote panel to set the audio level. If the gain went too high or low on his "S" meter he could simply control it manually.

6d. Local Oscillator

The Local Oscillator required five frequency selections over a 30 mHz range. The local oscillator frequencies were each Crystal Controlled. The design of the oscillator was standard but selecting the different frequencies made it a bit difficult.

In today's world there are at least two things an engineer should not design himself, Power Supplies and Crystal Oscillators. These are abundant on todays market and can be made to suit anyone's needs.

Switching the frequencies to one local oscillator line can be done with RF switching or by simply switching the dc power to the desired oscillator. I chose switching the power approach because it was easier to switch dc voltages than the RF at a remote control station.

Emitter followers were used in each oscillator output and were tied together to two separate emitter followers drivers. Remember, there are two channels in this receiver so each driver and detector has to be isolated.

Switching was then accomplished by powering up the desired oscillator. This left the circuit intact so that the emitter followers would not see a varying load. Also, I didn't have to worry about pick up from the other oscillators because their power was turned off. The Local Oscillator circuits were placed in a separate brass can mounted on a plug-in board and located between the IF pre-amplifiers and post-amplifiers.

6e. IF Amplifier

The main advantage I had in this design was that the required 3 mHz

IF band pass was determined by a specially designed and purchased passive Butterworth Filter that contained very steep side slopes. My approach was to use a wide band IF pre-amplifier (10 mHz) in the front end to drive the Butterworth filter and a second wide band IF post-amplifier after the Butterworth filter to make up the loses in gain through the filter. This provided me with the advantage of not having to be too precise on tuning the IF strip band pass and I could stand some degradation due to temperatures.

In the IF pre-amplifier, I used the Chebishev staggered tuned approach. The problems that I encountered here was ripple due to the Chebishev and Butterworth filters, the insertion loss from the Butterworth filter, and the overall loss in gain.

Creating such wide band of 10 mHz with a center frequency of 30 mHz was not an easy task. To keep the total ripple down to less than 3db and maintain the gain, I had to add five stages of amplification and tuned circuits in each IF strip. The Passive Butterworth filter already had 1 db of ripple, so in the IF amplifier I had only 2 db to play with.

The Butterworth filter had an input impedance of 50 ohms and an insertion loss of 10 db. This meant that the pre-amplifier also had to drive the 50 ohm filter with an emitter follower and had to be less than 0.1 volt rms signal level or distortion would be introduced.

The pre-amp gain had to be approximately 30 db to overcome the losses due to the superheterodyne bridge rectifiers converting the 60 mHz RF frequency to 30 mHz and the insertion loss of the filter. The post-amplifier also had to be 10 mHz wide, staggered tuned and had to drive the AM detector with at least 2.0 volts rms.

6f. Detector

The detector I used was a Crystal Diode half wave type. Crystal diodes were made for RF detectors and to my knowledge they are still being used. Since I did not use AGC this design became fairly simple to demodulate the 30 mHz signal.

6g. Audio

The audio output of the demodulator was composed of analog dc information and digital data. The dc was required to provide a signal response to an "S" meter to monitoring the phase for tracking. The digital data was composed of all the sensor information provided by the missile transmitter as it went through re-entry. The "S" curve was produced by comparing the levels of the two channels through a differential amplifier.

A differential amplifier in those days was a very complicated design. Because of the short schedule, I was very fortunate at that time to be able to use another engineer's design and not having to start from scratch. I still had to modify it to meet my needs such as balancing the front end to meet the temperature requirements. I accomplished this by using a dual transistor in one package so that both transistors would have the same temperature coefficient.

I'm not going to elaborate on this design because one can purchase a differential integrated circuit amplifier today that would easily do the job. The original design of this Diff-Amp took more than two months to accomplish.

This concludes the design of the receiver and I hope some of this information will help the reader in some way. The main things I want to stress is; packaging, decoupling and again keeping it simple. Take advantage of what is available and what is important systems wise. For example, I didn't need AGC so why put it in the design. In packaging today, RF circuits are now mounted on multi-layer printed wire boards. This is fine as long as one knows the procedures in laying out RF circuits and high speed ICs on PWBs. These techniques were discussed in Chapter II. I had a month and a half to complete this receiver design and in the end the receiver performed magnificently. I was proud of it.

7. Outer Space Communication

When our department was transferred to a Spacecraft Division I received an IRAD contract to investigate the latest techniques for transmitting and receiving information to and from a spacecraft, a space station and satellites. This division was just newly formed in 1963 and it was a great opportunity for me by getting into it on the ground floor and to learn all there is to know about outer-space communications.

I will not get into all of the details of this program because it took two years of extensive research to investigate the capabilities at the time and for me to recommend some new approaches. The final report for this project was 113 pages long. I did not work on this program full time because I was still involved in other black box designs. I will, however, touch on a few subjects that may be of interest.

It must be remembered that in the early 1960s satellite communications were in their preliminary stages, but much of the theory still applies.

The investigation can be broken down into 3 stages; Transmission, Receiving, and Data Processing. The main performance characteristic under consideration was the amount of information that can be transmitted over the communication system. Major efforts have been made to design circuitry that would increase the information capacity of these systems. Improvements in antennas, high efficiency RF power amplifiers, and low noise receivers have brought the performance of these components near theoretical optimum. Little increase in systems capacity was expected through further component improvements.

Where does the improvement then come from? Special modulation, data processing, and signal detection techniques had to be employed. Techniques such as Coherent Phase Lock Loops, Correlation Detection, matched filters, Synchronization, and

Data Processing were some of the areas where improvements can be realized. In addition, Orthogonal Bi-Phase modulation would permit reception of signals below the noise level which allowed minimum error probability per bit. The Data Processing technique would transmit only that information that passes a priori or posteriori condition.

The amount of information that may be transmitted through space is limited by the system bandwidth and signal power. It is assumed that the system had to operate at threshold or the lowest signal-to-noise ratio possible for a given probability of error. As a result the important parameters become transmission power, antenna gains, losses in the transmission medium (range), system losses, frequency, noise figure of the receiver front end, modulation, coding methods, and data processing.

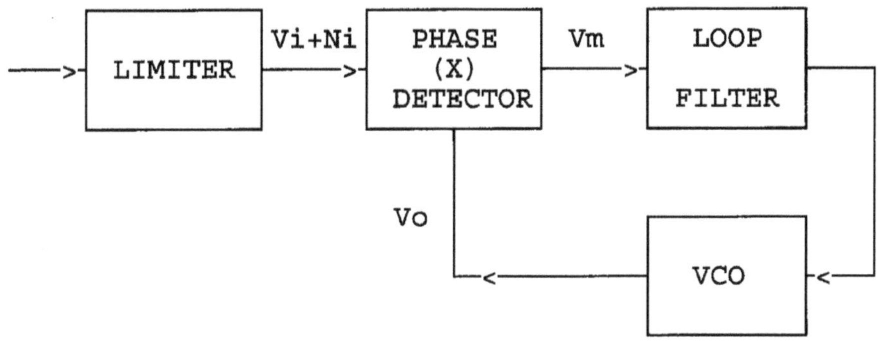

FIGURE 14
PHASE LOCK LOOP

*5. C.E. Shannon, A Mathematical Theory of Communication," BSTS, Vol. 27, July & October 1948.

*6. M.Schwartz, "Information Transmission Modulation and Noise," McGraw-Hill, 1959.

*7. A.J. Viterbi, "On Coded Phase-Coherent Communication," IRE Trans. on Space Electronics and Telemetry, March 1961.

*8. C.S. Weaver, "Thresholds and Tracking Ranges in Phase-Locked Loop," IRE Trans. Sept., 1961, pp. 60-70.

*9. R. Jaffee and E. Rechtin, "Design and Performance of Phase-Lock Circuits Capable of Near-Optimum Performance Over a Wide Range of Input Signal and Noise levels," IRE Trans., March 1955, pp. 66-76.

*10. J. Kalith, "Correlation Detection of Signals Perturbed by a Random Channel," IRE Trans., Vol. IT-6, June 1960, p. 361.

*11. G.L. Turin, "An Introduction to Matched Filters", IRE Trans., Vol. IT-6, June 1960, p. 310

*12. D. Middleton, "On New Classes of Matched Filters and Generalization of the Matched Filter Concept", IRE Trans., Vol. IT-6, June 1960, p. 349

*13. E.A. Marcatile, "Errors In Detection of RF Pulses Embedded in Time Crosstalk, Frequency Crosstalk, and Noise", B.S.T.J., May 1961, p. 921.

7a. Channel Capacity

The expression that relates the available channel capacity to systems parameters of an ideal communication system is given by Shannon's (*5) equation:

C = B Log (1 + S/N)

where
C = Binary bits/second
B = Bandwidth, Hz
S = Average signal power
N = Average noise power

The channel capacity (C) is the theoretical limit and is defined as the maximum rate of information transmission. The term information is defined as a message which has been minimized according to its probability of occurrence and contains only new information (*6). This means that all redundancy has been removed from the signal to be transmitted.

There is one other parameter that is involved and that is information rate H (*6) and is given by:

H = 1/t bits per second

where
t = transmission time per bit

The information rate is determined by the transmission-probability, that is a transmitted word was received correctly. The channel efficiency is then:

E = H/C.

The way to increase the amount of information is with time. Since the Channel Capacity equation is in bits/sec one can see if the time were increased between bits, then more information can be received over a narrower bandwidth. By transmitting from outer space at a rate of say 2 bits per second, for example, it would require a lot of waiting around at the command center to receive a message. But if time is no object then this is one way to get more information using a smaller Channel Capacity. Also by using digital bits to represent analog data allows the system to use Pulse Modulation (PM) communications.

7b. Phase Lock Loop (*8) & (*9)

A Phase Lock Loop is an ideal demodulator for PM systems. The loop consists of a Limiter, a multiplier, a filter, and a VCO (voltage controlled oscillator) as shown on Figure 14 on page 255.

An input signal (Vi) plus noise (Ni) is impressed on the multiplier, (Phase Detector).

Vi = /2 A sin [w(t) + 0i (t)]

Ni = N (t) sin wt
where
A = RMS Signal Amplitude
w = Angular Frequency
0i (t) = Represents the phase angle

/2 = square root of 2

The output (Vo) of the VCO has an amplitude C and a phase 0o (t).

Vo = /2 C cos [wt + 0o (t)]

The multiplier combines the input signal with the VCO output given a low frequency output proportional to 0i - 0o, or:

Vm = Km AC sin [0i(t)- 0o(t)] +

/2 KmCN(t) sin 0o(t)

The Loop Filter passes only the difference frequency. A voltage that is proportional to this difference is applied as a correction voltage to the VCO. This error voltage forces the VCO output phase, 0o, to be equal to the input phase, 0i. Under this condition, the error voltage (0i - 0o) is minimized and the Feed Back Loop is considered locked. Because the Loop Filter has a narrow band width, the amount of noise passed to the output of the Phase-Locked Loop is proportionally smaller and the Signal to Noise Ratio (SNR) is greatly improved.

The SNR improvement using a Phase-Lock Loop is computed by:

$$SNRo = \frac{SNRif \times Bif}{Bf}$$

where

SNRif = signal-to noise ratio out of I.F. amplifier.

Bif = I.F. Bandwidth

Bf = Loop (filter) bandwidth

A SNR improvement of 20 db can be realizable using a Phase-Lock Loop system.

In addition, the Phase-Lock Loop also provides a lower input threshold than conventional FM or PM systems. With Phase-Lock Loop detection the threshold limitation is improved by 6 db over straight FM or PM.

7c. Correlation Detection (*7) and Matched Filters.

In the optimum threshold theory there are three principal conditions of operation:

1. Coherent Reception
2. Incoherent Observation
3. Partially Coherent Observation.

Coherent Reception involves an averaged cross-correlation of the received data, with the known signal waveform. Incoherent Reception is the auto-correlation of the received wave with itself. The detection of signals whose form is exactly known at the receiver in white Gaussian noise is essentially cross-correlation. The receiver cross-correlates the received signal with each possible transmitted signal. This operation can be performed by a multiplier-integrator combination, or by a "matched filter".

In the general concept of a matched filter the "matching" is a form of optimization, which attempts to enhance the reception of the desired signal in the presence of noise. Reception itself may consist of a detection process, or an estimation operation, where the intent is to measure a parameter of the signal. The recognition of reception as a decision process (a Yes or No), and the use of statistical decision theory to provide criteria of optimality, is called Bayes matched filters. In a decision system numerical value judgements are assigned to the various possible correct and incorrect outcomes. Bayes reception is defined as that operation on a receiver which minimizes the average cost of decisions.

7d. Synchronization

A PCM telemetering system requires time synchronization of the receiver equipment with the transmitting equipment. This amounts to keeping two clocks in sync by means of a separate radio link or by using Manchester code, where the clock is transmitted with the data.

In a two way system the master clock at the command station, has to be extremely stable. This is because there are additional phase errors due to doppler shift, etc., on the return link. A spacecraft clock would be the "slave" of the master clock. The required master clock stability can only be accomplished by using an atomic clock.

7e. Communication Link

Example of a Communication Link for a two-way system is shown in the block diagrams of Figures 15 and 16. Figure 15 on page 282, shows a process of converting analog to digital information as well as the modulation method used in a ground station.

Figure 16 on page 284, shows only the spacecraft receiver section of the transponder. It demonstrates the demodulation process and the technique used for converting the digital information into analog.

The detailed operation of this system may best be described by following the transmission path through the system: Assume that the input to the transducer of Figure 15 consists of a voice command. The signal is processed and converted to digital in the encoder. The encoder output is applied to a bi-phase modulator together with the sub-carrier frequency (fs1). Here, the sub-carrier is bi-phase modulated by the input signal and the resultant signal is applied to the summing amplifier. Another sub-carrier frequency (fs2) is bi-phase modulated by the synchronization signal and this signal is also applied to the summing amplifier. The carrier frequency is phase modulated by these two sub-carriers. The composite carrier is then amplified in a power amplifier and radiated via the transmitting antenna. The sub-carrier frequencies and the synchronization pulse are all derived from an atomic clock source.

As shown on Figure 16, the input signal plus noise are both coupled to the Spacecraft receiver by the antenna. These R.F. signals are then mixed with the local oscillator signals in the mixer stage. The mixer output is the difference frequency of the multiplied first VCO frequency and the input signal (fc1 + noise).

The signal frequency is further reduced by another mixer, which mixes a lower multiplied frequency with the output of the first IF amplifier. The AGC signal is taken from the last IF amplifier to control the gain. The IF output is limited and applied to a phase-lock loop discriminator. Since all of the LOs are derived from the first VCO the receiver is completely locked under normal operation. This VCO signal is also used to provide the carrier source for the transmitter section of the transponder as shown by (A) on Figure 16.

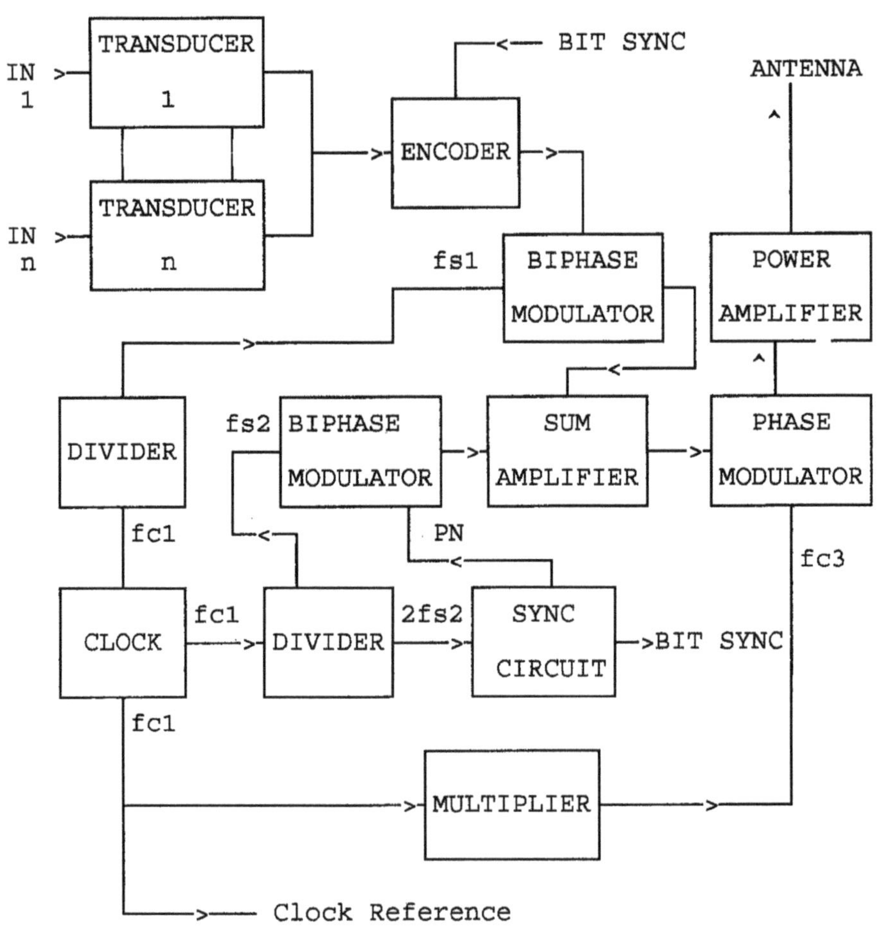

FIGURE 15
OUTER-SPACE GROUND TRANSMITTER

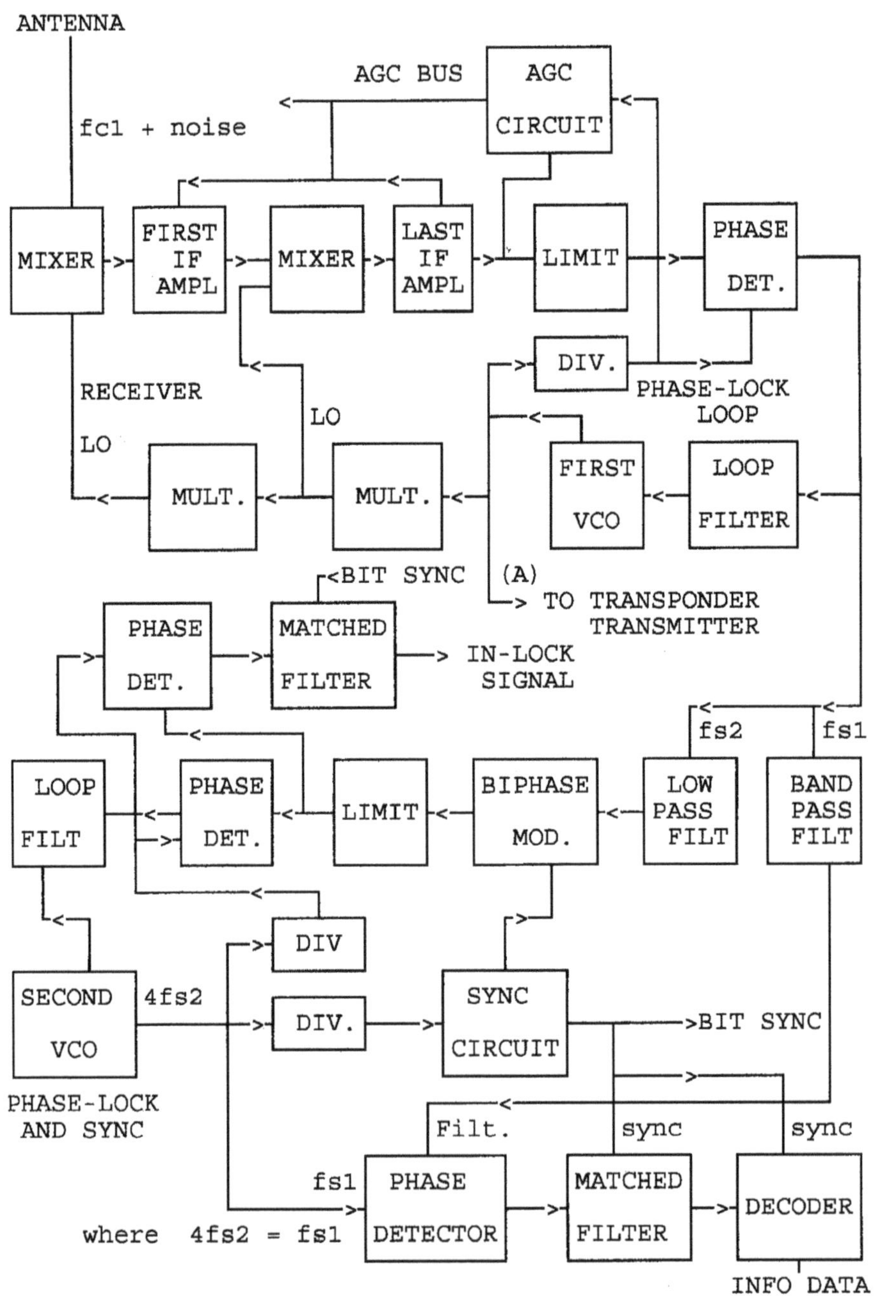

FIGURE 16
SPACECRAFT RECEIVER-TRANSPONDER

The sub-carrier frequencies (fs1 + fs2) are phase detected and separated by a high and a low pass filter. The low pass filter passes only the sync modulated sub-carrier fs2. This sync signal is applied to a bi-phase modulator, another phase-lock loop, and a sync circuit. The Synchronization sub-system provides bit sync and word sync for the matched filter and the data processing section.

The band pass filter passes only the information modulated sub-carrier, fs1. This sub-carrier is mixed with the second VCO frequency, in the phase detector and the information pulse is reproduced. This pulse is correlated in the matched filter to produce a high SNR. The decoder then converts this pulse signal to analog information or Data.

8. Sonobouy Digital Display

The next major assignment I had was a Sonobouy Digital Display System (SDD) in 1964. Actually, this system was designed by someone else in my company who couldn't get the equipment to pass environmental testing. The SDD was a system that was part of an airborne ASW program. There were three major boxes associated with this system that contained the control logic which was used to select the Sonobouys to be dropped at sea and the communication channels to be used. It was actually a very complicated system. After two years of trying to pass environmental testing the original designer left the company.

The Navy, at this point, refused to accept any new airplanes from my company unless this system was fully functional. The president of my company panicked and became heavily involved. As a last resort the Program Manager finally asked my boss if we could salvage this design and save the program. I was given the task, as lead engineer, to first make a study and determine if it could be fixed. My investigation showed that, yes it could be fixed, but it would require major modifications. At this point in time company funding was no problem. It had to be fixed.

When I looked into the design I couldn't believe what I saw. This designer was using Silicone Control Rectifiers (SCR) to provide the digital logic. There were thousands of them being used. Other chips such as RTL, DTL or TTL were not invented yet, but to use SCRs for logic instead of transistors was unbelievable.

In those days logic circuits were accomplished using discrete transistors and an array of "OR" diodes. SCRs were new on the market and an engineer in my company got the bright idea that SCRs would make ideal switching circuits for logic design. Wrong! They are absolutely the worst solid state component to be selected for this application.

This was another example of an engineer, with little experience, tackling a design with new and untried components. Any experienced designer would never have selected SCRs to perform digital combinational logic equations.

The reason I am using this example is to further show how an inexperienced engineer would try to design hardware not knowing all of the consequences. Today, of course, SCRs wouldn't even be considered in logic designs with all the TTL logic chips available.

SCRs do, however, have a very good application. An SCR is actually a latching diode with a triggered gate. When a pulse or dc voltage is applied to this gate the SCR fires or saturates "ON" where the anode goes from a high impedance to a low impedance. It would remain in the fired state until the anode voltage is removed or there is a minimum current applied. The application for SCRs were intended for AC power controls or as power switching latch circuits, never for logic.

I suppose this engineer was going to revolutionize the whole world of digital electronics. But can anyone imagine what a chance he took. Here he had a 2 million dollar program in trouble and to jeopardize the sale of several hundred aircraft just to prove a point that SCRs can be used as logic circuits is unforgivable. He and his manager that approved that approach, should have been fired. But of cause they weren't. The designer left the company but the manager remained holding the bag. I had to bail out that department. I didn't even get an "atta boy" for my contributions. He violated the basic rules of design; in this case the 2nd, 3rd, 4th, 7th, and 8th rule as shown below:

Basic Rules for Military Design:

1. Keep the electronics simple
2. Don't use new or exotic components
3. Use components that have been on the market for at least 2 years.
4. Don't do a marginal design. Always over-design to allow for environmental conditions.
5. Provide for testability
6. Consider oneself as the repair person for maintainability
7. Don't be a loner in a complicated design. Ask for help. (no one man decision)
8. The engineer must have experience
9. Package for access & maintainability
10. Use EMI filtering whether it needs it or not
11. Use SEM or standard Printed Circuit Boards
12. Always purchase the power supply
13. Design for ease of production

The reason I took this assignment was because our department was on the verge of being disbanded. I knew if we were successful we would be reborn again. In a way I was also taken a chance, but as I have said many times, electronics design is really simple and as long as one knows what he is doing he can almost fix anything. I have never come across a bad electronic design that couldn't be salvaged or corrected.

It took 5 EEs and 6 MEs to modify the SDD design and in 6 months we passed environmental testing. To my knowledge these units are still in operation after 40 years.

It was quite a feat considering how sensitive SCRs are and for us to continue try to make them perform logic functions was a major challenge. I was not allowed to eliminate the SCR approach because someone had to save face.

As I expected, when we passed environmental testing we did become a major design department again in 1967 which lasted to 1990. The Project Engineer of that program never forgot me after that and he did give me new work when I became a manager.

After that design fix our department come to be known as fire fighters because we had to salvage so many poor designs that were done by either in-house Project Engineers or from outside suppliers. It seems we were only called in when there was a design problem. If we had been given the design contract in the first place, many of these problems would never have happened. But that's another political story. A good design organization is essential in a major company and should be supported by upper-management. We were obviously not.

9. Frequency Translator Design

The Frequency Translator was an ASW system that was used to analyze Submarine activities at much lower frequencies than the existing sound detector system that was presently on board our aircraft. The key to this design was crosstalk and minimum Phase differential between two channels. The low frequencies that were involved was from 1 Hz to 100 Hz. Pretty low.

To translate the higher frequencies down to these low frequencies, RF superheterodyne detection technology was applied. The front end amplifier required 40 db of AGC and very low distortion. The standard radio AGC circuits could not be used because they would not meet this distortion requirement.

To accomplish this 40 db AGC action I used Raysistors. Raysistors are variable resistor devices whose resistance would vary 1000 to 1 over a controlled range of dc voltage applied to a separate heater line. The only weakness with these units were that the response time is very slow, but in this application speed was no problem.

AGC was accomplished by detecting the output of the last stage of an amplifier and feeding a DC component voltage back to the Raysistor heater line. With the resistance portion connected as a resistor pad in the front end, 40 db of AGC was easily attainable with little or no distortion. Minimum crosstalk was accomplished using isolation and Mu-metal shielding between channels.

The next major problem was phase shift between channels. In my first design approach I was out of Spec on phase shift. I then began using 1% resistors and selected transistors and other components to minimize phase shift. I was still a little out of Spec and I didn't like the approach of selecting components to accomplish this task because it becomes a production and maintenance problem.

In another part of the design another engineer used an adjustable capacitors to tune a notch filter that was used to reduce (notch out) the third harmonic of the fundamental frequency. This was another major requirement in the Spec that could distort the signal.

I noticed that as I adjusted this capacitor there was a phase shift in the channel. I quickly went to work on this circuit and made some modifications and, low and behold, I had a method of adjusting the phase between channels to almost 0 degrees. This was a great feat. This phase differential error has always plagued many manufactures in the ASW business.

When I demonstrated the phase differential to the Navy they couldn't believe it. They thought I was cheating. I never told them how I accomplished it, so I gloated on being a hero. It was all by accident. It turned out that we didn't even need the notch filter anyway because the third harmonic was already low enough.

Sometimes I felt somebody up there or down there was looking after me. This has happened to me many times where I would accidentally discover something when I was trying to solve another problem. The secret is recognizing these peculiar phenomenons and to use it to one's advantage.

10. Frequency Divider

My first invention at BTL was another one of those cases of discovering a phenomenon by accident. The patent was issued Nov. 12, 1963, five years after I submitted the disclosure. I had already left BTL in June, 1961.

At the time I was trying to produce an analog frequency divide-by five circuit that was originally invented by another BTL engineer. He had previously used vacuum tubes but I had to use Transistors.

As I was making adjustments in the bias voltage on the transistor I noticed different frequency divisions occurring. I stayed with this peculiar occurrence and perfected it, so that with one transistor I was able to have an adjustable frequency divider up to 15. Stability wise though, over the military temperature range, the highest that was practical was divide by 9. BTL patented my design and, as far as I know, it is still being used in that system today.

In today's technology, of course, divide-by circuits are a dime a dozen and can be done easily with one or two chips. All one has to do to divide the analog frequency down, is to first use a Schmitt trigger such as an SN54LS132 to square the sine wave off, use a SN54LS92 or equivalent to divide the frequency down, and then tune to the desired frequency with an LC network to recover the sine wave. Simple. If C-MOS is required the CD4018 can be used as a divide-by "N" frequency synthesizer. It sure is much easier today.

In another situation, I had a design requirement where I had to provide 20 trigger points that had to be delayed and produce sub-frequencies that were all derived from a 1 mhz clock, the lowest being 100 hz.

I started out by using many CD4018 divide by chips but the amount of circuitry was getting out of hand. Also the trigger pulses were not in sync with the clock. I thought about the design for a moment and knew there had to be better way.

I finally came up with the idea of using an E-Prom as my divide by circuit. I programmed the E-Prom to provide pre-determined trigger pulses at specified clock periods and, again, using a Schmitt trigger I was able to produce a 100 Hz frequency that was locked in and in sync with the clock. The easiest divide by circuit I had ever worked with.

11. Auto-Correlation (*14).

In Teletype communication, noise is one of the biggest problems, especially in the old days of 1955. In Teletype communication the operator would tune his radio receiver to a selected channel and as long as there was a carrier frequency being transmitted from somewhere there was no major noise problem. Today, Teletype transmissions are going on all the time from many different sources so there always seems to be a carrier present, but up to 1965 there were only a few stations transmitting over military channels.

Because of not having a carrier on at all times, the Teletype printer would type reams of useless paper from receiver noise. The receiver had to be left on throughout a flight in case someone tried to communicate with the aircraft using the teletype-writer.

Because of this noise problem, I was asked to perfect an Auto-Correlation technique that was invented by someone else, to stop the paper from being wasted on board the aircraft and to improve the signal-to-noise ratio (SNR). To enhance the SNR, I had to use three noise reduction techniques; filtering, threshold detection, and auto-correlation.

In order to understand the application of auto-correlation with teletype communication, a brief description of teletype theory is now discussed.

The Teletype system (TTY) operates on 2 audio frequencies which are detected by a radio receiver. The 2 frequencies are 1757 Hz "space" and 2425 Hz "mark". These 2 frequencies (depending on the polarity of a Baudot code) are converted to wide pulses, where a plus pulse is a "space" and minus or zero pulse is a "mark". The width of each pulse consists of 16 bursts of the space or mark frequencies.

In the Baudot code 5 bits are used to determine what letter is being transmitted and the first and last bits are used as the start and stop pulses for a total of a 7 bits of Baudot code.

The object of auto-correlation is to examine the burst of the mark or space signals, send the signal to a delay line and also to one end of a multiplier as shown in Figure 17 on page 266. The output of the delay line is then sent to the other end of the multiplier which combines the two signals.

The basic theory of this circuit is, if the delayed signal pulses line up with original signal there would be an additive function, but if there is noise present the pulses would be randomly added and subtracted and would be rejected because they would not reach a preset threshold level.

To implement this theory into the TTY system four of the first 16 periods of the sine waves (each of the 5 Baudot codes contain 16 bursts of "space" or "mark" signals) would be wasted because the delay line is set to delay 4 sine wave periods. This means that the fifth period of the original signal would be compared with the first period after it is delayed, and so on.

With little or no noise the two signals would peak to twice the height, but when noise is present, say on the first pulse, the peak may never get high enough to set the threshold. This applies to all of the remaining 12 periods of the sinusoidal signal. The threshold circuit must remain set for the remainder of the "space" or "mark" period or it is rejected. If only noise is present the preset threshold level would never be achieved so no printing action on the typewriter would occur.

With this system I was able to improve the signal-to-noise ratio by 3 db. It stopped the paper from running on the deck due to noise and it made less printing errors. Today, there are even better ways of detecting Teletype Communication information which I will not get into. I just want the reader to understand what Auto-Correlation is by illustrating one application.

Auto-Correlation is a technique that is still being applied today in communications, especially in RADAR to improve the signal-to-noise ratio of the RADAR pulse. In RADAR, however, there are no bursts of sine waves as there are in teletype communication and the application is somewhat different.

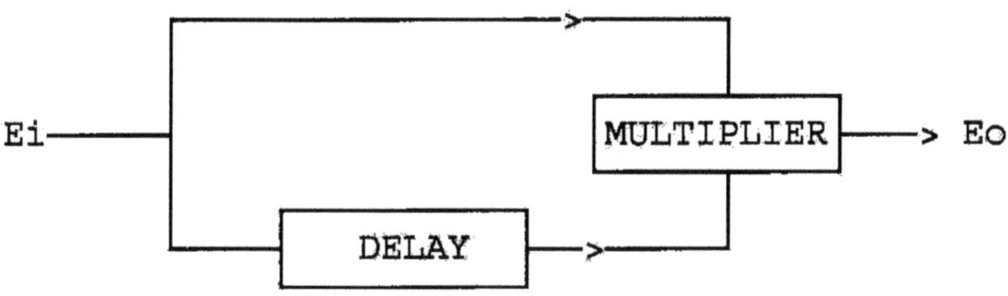

FIGURE 17
AUTO-CORRELATION

*14. Auto-Correlation, "Information Transmission, Modulation, and Noise", By M. Schwartz, page 429 McGraw-Hill 1959.

12. Helicopter Lighting Power Control

The lighting for the overhead and panel displays for our Helicopter for the Army (1968) had to be adjustable for day and night flying. The only power source on board was 28v dc. Chopper power supplies were not acceptable because of EMI. The required current requirements varied from 1 amp to 100 amps.

The power supply design I chose was the standard emitter follower type with feedback for regulation. The biggest problem was heat dissipation which my ME packaging engineer solved. The second most difficult requirement was that the output had to be short circuit proof. Today, there are many regulators to choose from that provide short protection, but in those days they were not available.

To accomplish this protection I decided to use the crow-bar SCR approach to turn off the power supply in case there is a short circuit. A special circuit would sense when there is short circuit in the load which triggered the SCR gate to fire its anode. The SCR was connected from the base of the power transistors to ground, and when it was triggered it would actually short the base to ground turning the power transistors off. The SCR is a latching device and would stay that way as long as there is a positive voltage present.

Instead of manually switching the SCR line open to see if the short went away, I chose to use a 10 second timer that would periodically open the line. If the short was still there the SCR would fire again. When the short was finally removed the power supply would return to normal after 10 seconds.

In those days this solution worked fine, but, as I had said, today one would use a regulator chip for this application. The reason I included this design now is to illustrate that one has to be able to use other alternatives to accomplish a task.

This particular supply originally went out for bids and had no responses because of the difficulty of meeting the EMI requirements. There is always a way to overcome a problem. Actually, the power supply requirements were much more complicated than what I had just expressed and I didn't want to get into all of the details.

The packaging design, however, is something that should be considered in this design. The worst case conditions for this power supply was when the unit was in its most dimmest mode. It then had a power dissipation of 150 watts. Since no cooling fans were aloud (fans produced EMI), all of the heat had to be dissipated through radiation and convection.

One rule of "thumb" I used as a method of measuring the temperature of a black box is the hand test. As long as I can put my hand on an enclosure or a finger on any component at room temperature without being burned the chances are I would pass the high temperature environmental conditions of +71 degrees C. One has to be careful with this method because there may be some hot spots not available to the finger. This is when one might want to use thermal couples, or temperature sensitive discoloring patches and finally Thermal Imaging equipment.

In this design the ME first made some calculations on how much surface area was required to maintain the temperature. Since we were limited in box size (15" x 6" x 6), in order to meet temperature radiation requirements the ME said the complete enclosure had to be made of heat sink extrusions with fins on all sides. We finally settled on heat fins on all sides except the bottom, which would be mounted on an airframe plate to provide heat transfer to the ship by convection.

The extrusions were standard purchased heat sinks and we Dip-brazed welded them together. The power transistors and diodes were mounted on the outside heat sinks in areas that were machined out. The key to this package was it went together like a book for easy access to the wiring and components. There were three screws on the top and three on the bottom and when they were removed the whole unit opened up like a bible.

The main point I want to stress is simplicity. In the beginning of this design I had numerous suggestions, all were very complicated. Each suggestion would have worked, but the final approach that I took was to make it a simple design, easy to assemble in production and would last forever. That was in 1968 and it's still flying today. I have never received any field failure reports on these units in over twenty years. No news is good news.

13. Fault Warning System

The Fault Warning System (FWS) was another one of those programs that was originally given to an outside vendor that couldn't do the job. (I discussed this box before in Chapter II). That vendor worked on the development of this system for two years and they still didn't have a working prototype. I was asked to investigate them and find out what their problems were and if they could ever produce a system.

I spent a month traveling to their facilities listening to their presentations, going through schematics and examining their equipment. Then I wrote my report on how bad it was but I wasn't getting any response. The last straw came when their Annunciator Panels melted during a test. Upper-management finally became serious with my report and asked me to attend a meeting with the big boys to discuss their problems.

There were many problems with their design, but some were due to my company's specification. What this vendor did wrong first was to hire many foreigners who didn't know anything about military packaging or Mil Specs. They were trying to package too much electronics of Flat Packs on the back of each annunciator panel. Heat has to go somewhere. They should have put the electronics in a separate enclosure to drive the annunciator panel. The panel by itself generated a lot of heat.

On the Voice Warning System, my company imposed an impossible Specification on them which required them to have 80 channels of voice on a one inch tape recorder.

When the vendor bid the job they said they could do it because they had a new invention that would revolutionize the recording industry (beware of new and untried inventions). What their approach was; by using an FM recording system and lowering the voice level on each channel down into the micro-volts range they discovered they could increase the number of channels on a one inch tape.

The major problem with this theory was there was crosstalk on all channels. After my upper-management read my report and listen to my explanation they immediately canceled the contract and gave the program to our department.

The first thing we did was to have the Equipment Specification rewritten to be more practical and remove the requirement of 80 channels on one tape. We replaced the old recorder requirement with two purchased units that were tried and proven but with only 20 channel each for a total of 40. We designed the electronics and package and in eight months we had a prototype system in operation for that Helicopter.

I will not go through the details of this design because it was not my direct responsibility and was designed by another engineer. I didn't really care for his approach either, but it did work. I was later given the responsibility for the second phase of this program but I was not allowed to steer away from the original design. The design was kind of unique, in one respect, because it used an old fashion pulsing technique commonly used in RADAR systems with vacuum tube electronics.

Each annunciator display was assigned a capacitor which was charged with 28 volts. When a fault occurred the capacitor assigned to that fault would be momentarily shorted to ground. This provided a negative going pulse to the sensors of two Master Caution Lights which would alert the Pilot and Co-Pilot that there is a problem.

The Pilot or Co-pilot would then look at his annunciator panels to determine what and where the fault is. If it was a major fault a Voice Warning message would have immediately told the Pilot and Co-pilot that the Helicopter has a serious problem and what it was. Other less serious faults would be displayed on the annunciator panels and the pilot could make his decision to return to the base or go on with the flight.

Seems simple enough, but because there are 28 volt spikes floating around the aircraft from other sauces, there is bound to be an EMI problem which could triggered the Master Caution Light on. Pilots don't like false alarms. I finally licked that problem or, at least, reduced it by EMI shielding and filtering but, as I said, I didn't particularly care for this design approach. We did have an occasional incorrect fault displayed and sometimes multiple faults. Today, a micro-processor would be ideal for this application.

The next major problem I encountered with this system was the connectors. Our company Project Management decreed that all connectors shall be purchased from one supplier. It was a center locking type. I and my ME discovered a major problem when we put it on a vibration table. When I hooked up the annunciator panel with this system and began the shake and bake test the panel lights looked like a series of Christmas lights blinking all over the place, The connector manufacturer was brought in and eventually this problem went away for another reason which I will discuss in the next paragraph

14. Digital Voice Warning System

The Voice Warning units that were finally used in the above case were too large and heavy for the Helicopter. As a result, I invented a Solid State Digital Voice Warning System of which my company filed and received a patent in April 30, 1974.

To this day I maintain that all of those Solid State Synthesizer Voice Warning Systems used in today's automobiles and aircraft are infringing on my patent rights, but my company didn't want to pursue the legal hassle of fighting it, so I lost some income. I think if my patent were for an aircraft part such as a screw my aerospace company would have fought the patent infringements all the way up to the Supreme Court. That's life, I guess.

With this patent the above Voice Warning system could be installed in a much smaller package with the capability of having more than 100 channels of voice warning messages. I had a small suitcase unit built under a research program, to demonstrate its capability to the Army and to the Patent office. I was trying to win a contract from the Army to replace those recorders in the Helicopter. I was unsuccessful because the Helicopter contract was canceled. Win some and lose some. This is how the connector problem we had above went away.

15. Control Logic Assembly (CLA)

In Chapter II, I discussed the mechanical enclosure of the CLA design of Figure 5. This particular program also came to my department because of a poor Equipment Specification. See also Figure 22 on page 291.

This job went out for bids and twelve companies responded. An Equipment Specification was released to all companies and when it became time for them to make an oral presentation I was asked to participate in the reviews. I was told not to evaluated their cost but merely to review their technical presentation.

It took a month to hear all of those presentations and when it was over I wrote my report. I said in my report, basically, that none of these companies could design this equipment as specified in the ES. They were all lying just to win the contract. Even I couldn't design the electronic in this box because of the impossible specification requirements. When I asked these suppliers how they would accomplish certain tasks they would weasel word me to death.

The main problem, as I mention in Chapter II, was that the specification had called for solid state relays to drive large solenoids and power relays. These solenoids and relays were required to provide switching of up to 60 amps of AC and DC currents. There were many of these 60 amp circuits and, in addition, several 30 amp circuits, thirty 10 amp circuits and hundreds of circuits requiring 5 amps or less.

Now, there is no solid state switch in industry that provides a lower voltage drop across its switching points than a relay. I would estimate that a 60 amp solid state switch would have a drop of at least 2 volts. This means that this one solid state switch would dissipate 120 watts of power. A 30 amp circuit would have a 1.5 voltage drop and dissipate 45 watts. Circuits of 5 amps or less may get by with a 1 volt drop.

If I added up all of the power dissipations in the box just for relay loads alone I would say the net results would be on the order of 5,000 watts. This does not include all of the other cumulated power dissipations that would be generated by other circuits such as power supplies and lamp drivers. Imagine trying to cool this amount of heat in a small box. For comparison, a home television set generates only 80 watts or less. The Equipment Specification was unbelievably bad.

I haven't even discussed how this little box was going to house all of these huge solid state switches. The box size requirement was on the order of 6 x 14 x 12 inches. Just to house all of those solid state switches it would take a box ten time as large.

When upper-management read my report, especially my old Division Engineer, they canceled the bidding and gave the job to me. My old Division Engineer knew me and my reputation and said to the Vice President, "If Panicello says it won't work, that is good enough for me." They, of course, investigated further and agreed with my analysis. The companies that bid on the original box claimed "foul." They figured that my company was just stealing their design by doing it in-house. They almost went to court until they saw that my design was nowhere near their versions.

My design approach was to first eliminate all of those solid state switches from the Equipment Specification and to use good old fashion relays to do the job. There is a finite voltage drop even in relays through the contacts of 0.01 volts. This is insignificant when one compares it with solid state switching. The second thing I insisted on was to place all of the relays outside of the box. They were finally installed on a rely panel like they did in the good old days.

The object of this box was to control all of the power functions of an S3-A aircraft (see Figure 34, page

315). This was the first time this approach was used on an airplane. Up until now the controls of flaps, bombay doors, an other power function on an aircraft were done directly with electricity, hydraulics and controlled by hand switches.

By using combinational logic circuits in this new approach, the pilot, for instance, can select any operation he desires and doesn't have to worry about the proper sequence of flipping many switches in a given order. The electronics ensures him that many of the initial functions have automatically been accomplished first before an operation can be executed. The pilots loved it.

Before they would have to use a tabulated chart which tells them the sequential order of switching that has to be done in order to select a particular function. If the sequence was not done in the proper order there could be a problem.

The way this system worked was there were many sensors all over the airplane. The majority of sensors were Proximity switches. In other words, a pilot couldn't drop bombs from his bomb bay until his bomb bay doors were completely open. Proximity switches were located at certain positions that would sense that the doors were completely open or closed. The Prox switches were magnetic sensors so there was no metal contact. The output of the Prox switches were bi-level of 28 volts or zero which would be fed to my box.

The combinational logic in my box would compare all of these inputs and if the right combination was true, an output signal would be generated either to turn on a relay, lamp or provide an output signal to another system. The output signal could be 5 volts dc, 28 volts, or an analog sign wave. There were many timing devices in the box such as 60 minutes, 30 minute and several 3 and 1 minute timers. Timing was necessary to ensure a function had plenty of time to happen before the next sequence could proceed.

The final design required two boxes mainly to provide redundancy and to have secure signal controls because many of these functions were life supporting for the crew. We also didn't want to inadvertently drop a torpedo or atomic bomb on a populated city.

When I got the go-ahead for this job there was no new Equipment Specification prepared. The old one didn't apply anymore. The problem I was faced with was time. I had to produce a prototype of both boxes in six months but I didn't have any of the Boolean Algebra equations submitted to me as part of the initial specification. They began coming to me in dribs and drabs.

Originally they wanted me to put each function on the same board. This way if that functioned failed all they had to do is replace that board. Sounds great but I didn't have any of the equations yet. How am I going to start the design?

I decided to break up the functions using separate receiver cards and driver cards. This way I can at least start the design.

There were three types of receiver cards; 28 volt inputs, 5 volt inputs and analog inputs. There were also three types of output cards; relay drivers, lamp drivers and analog output cards. I left room in the center of the enclosure for Logic cards. These would be designed as the equations arrived.

The systems engineer knew approximately how many sensors and relays were required. That didn't change from the original specification. So, from that information I knew about how many receiver circuits were required and how many relay and lamp circuits were required. I always allowed extra room. Now I can estimated how many boards would be required leaving room for %20 expansion. I could start my PC designers in motion and also the ME could begin designing the enclosure.

One of the problems I had was that the equations came in as manual switch functions in hand written form. Example; relay 24 shall come on if Prox switch A and B are also on but cannot be turned on if proximity switches 26 and 33 have not been energized.

This is, of course, a simple example, but now I had to convert their wordy sentences into boolean equations. I then had to reduce all the combinational logic equations down to its simplest form using the Veitch, Karnaugh or Mahoney Map technique. Many of the functions that came in were sometimes written over several pages like they were writing a letter to me.

Mahoney Map is just a variation of the Veitch diagram. It is a technique that is used in Boolean algebra to reduce all the variables on a truth table to its simplest form, thereby reducing the number of NAND or NOR gates in the design. It did cut down quite a number of chips.

Anyway, by using this three circuit approaches I could assign any number of receivers to a logic card as required, perform the algebraic reduction on the logic board and provide an output on a relay or lamp driver card. Since I was using a wire wrap mother board, shown on Figure 31 on page 309, it was easy to make changes which came in by the day over a few months.

I was asked, many years later, why I didn't use a micro-processor to do this job. The answer was simple. First, Micro-processors weren't invented yet, and second, each function had to stand on its own. In other words, if a bullet destroyed one function in the box the others should still work. That was the reason there so much redundancy in the two boxes. There were even two separate Power Supplies in each box in case one was damaged. A single micro-processor would not provide this redundancy. If both boxes were blown up the pilot could still fly home under emergency conditions.

The main point I'm trying to make with this design is, as a manager one may not always know what the design is going to look like and he may never have all the inputs he needs. One has to have that gut feeling on what the box would contain from experience and go on from there. Also, even though an Equipment Specification is supposed to be the bible, it may not be correct and may have to be updated periodically. In the CLA case, the Equipment Specification was being produced as the equipment was being designed. A little backwards, but sometimes that's the way it has to be done.

16. Control Logic Assembly Test Set.

A picture of the CLA Test Set was illustrated in Chapter I and is shown again on Figure 23 on page 316. It was partially discussed before and I will now review it further to point out some sound design decisions that had to be made.

The first thing that I was asked years later was, "Why didn't I use a Micro-processor." In this case a Micro-processor would have been ideal. But, alas, they were just invented at that time and only a new four bit commercial version was on the market. Well, the reader by now knows my position on using new and untried products, plus the fact I didn't know too much about the chip, so I decided to design the tester using discrete components.

It turned out that this tester became a special purpose computer that automatically tested the Power Control Logic Assembly. In Chapter I, I elaborated on how the tester grew to twice its size because my upper-manager, without consulting me, told a Navy Captain, who was permanently assigned to our company, that the box would also be used to test airplane wiring. Not only did he cause the box to grow to twice its original size but he made me overrun my budget. Upper-management can sure help one to death.

The design of this tester was made up solely of discrete components; meaning NAND and NOR gates, transistors, shift registers, and Programmable Read Only Memories. This design was probably the most complicated box that my department has ever produced. It was actually an invention that shocked everyone over its capabilities and its performance. Even the Navy were astonished. My company didn't want to pursue its inventive possibilities, however. They claimed it would come under the heading of "good engineering."

In Chapter II, I discussed the competition between an Automatic Test Station versus Ground Support Equipment (GSE). I said that in today's world of micro-processors, a GSE would be no bigger than the Interface Device needed for the GSE test station. With a Test Station the Government is paying twice the cost to test a piece of equipment; the cost of the test station and the cost of an Interface Device (ID).

I still agree with my statement because my CLA tester, for example, replaced an ID that would have been used on a Test Station. For the same size as an ID that would have been required, the Test Set completely tested the Power Control Logic Assembly. It was so good, for that matter, that it even replaced an HP Computer bench tester that was used for production testing the CLA.

The CLA Tester computer design used an eight bit system that would exercise every variable input of each Boolean equation that the CLA received. In other words, it not only checked all the go functions of the equation but it also checked the no-go functions. This was the problem the production HP computer programmer didn't attempt to do. He checked only one true condition of the combinational logic equations and only one of the false conditions. He also never checked the redundancy circuits nor the timing circuits in the CLA. If a function worked it was deemed Ok. But which redundant circuit was working and which one wasn't?

He also failed to check the tolerances of the different parameters; such as input signal levels of 15 volts + or - 2 volts, timing limits (1 to 60 seconds), and frequency limitations (30Hz). He didn't even check that both dual Power Supplies were working. These supplies were connected in parallel in the CLA and if one failed the other would maintain the power to the box. This individual programmer finally left the company after this disaster. He bailed out and found a new job before he was discovered.

This is the kind of thinking that I want to impress upon the reader as a prospective designer and manager. In this case the production Functional Test Engineer ignored engineering and produced his own test program. If he had even allowed us to review his program, as he is supposed to, many countless manhours would have been saved. Functional Test spent two years trying to perfect the HP computer program to test the CLA only to have it scraped. Check and balance once again was disregarded.

Its not that my tester could be any better than a large HP computer. Let's face the facts, HP had one hundred times the capability that my little tester could ever muster up. It was just because a stubborn programmer thought he knew it all and wouldn't work with engineering.

The way this tester works is, a technician simply depresses a start button after he hooks it up to the CLA. As one can see from Figure 23, there are hundreds of inputs and outputs on three large connectors that are hooked up to the CLA. It takes about two minutes to run through all of the tests once, mainly because it has to wait for the 60 and 30 second timers to go through their cycles. There were also many 5 to 1 second timers that had to be tested and be within their tolerances.

To check the redundancy circuits the technician had to repeat the automatic testing by turning one power supply on at a time. So, the whole test took 4 minutes because it had to be done twice.

If there was a problem in the CLA the test sequence would stop indicating a fault and a test number would be displayed. This number would be looked up in the manual which tells the technician which circuit had failed. It would go even further by telling him if it was a receiver card, a driver card or a logic card. The faulty card could then be replaced with a known good one and the test sequencing could continue or he could start over again. The bad card may now be tested separately off-line or with a CLA using an extender board.

I tried later to add a sub-routine software which could be used to locate a faulty card automatically without using the maintenance manual, but I was turned down. It would have been no major design problem but it woe it was my only prototype. We allowed him to use it but we kept borrowing it back due to equation changes in the CLA from Project.

They finally had the Navy bail-back a tester for production. Remember; Black Box production for the aircraft, in many companies, are a different organization than Product Support production for test equipment. So, we now caused a conflict between these two production houses who do not support each other and are under different management. The only way Black Box production could solved this problem was to go directly to the Navy and have them bail-back a field test unit. It's a funny business.

17. Communication Switching Matrix

A CSM was originally designed in my department for a large Naval aircraft, the P3-C that was made by my company. Twenty years later, Canada decided to purchase several of these aircraft but with modern electronics and to include some of their own GFE. I partially discussed this design in Chapter II.

The problem was, the old plane used an RCA radio for communications and it had a older Navigation system. The new Canadian plane required Collins radios and a modern version of the navigation unit. Also the Canadians wanted to use their own equipment on board that directly interfaced with the CSM. A new CSM had to be designed.

Incidently, because of these changes a Navigation Junction box that we originally designed also had to be modified. This caused a major modification in this box as well because the junction box now became a catch-all to save cost. A junction box is supposed to contain no electronics but the rules were bent here. This box became so big and heavy it took two strong men to lift it. In the end my organization did received a contract to modify the Nav. J Box for the Canadians.

The new CSM was originally scheduled to go to outside suppliers for bids. We were told we were out of the bidding. Project didn't even want to hear our sob stories on why we should get the job. Somehow, my boss managed to wangle Project to invite us to a preliminary meeting and to discuss our version. Project didn't initiate this meeting but our contact in upper-management forced the issue. They didn't want anything to do with us. Fortunately, I had spent at least two weeks going over the old design and knew most of the requirements. I still had no idea what would be discussed at this meeting.

My boss began his presentation using a black board and doing the best he could, but I could see that half the audience was unimpressed. My boss finally asked me to discuss the technical aspect of the design. I wasn't getting anywhere either until the Project Manager asked how much would our design cost. My mind was turning, fast. I could tell from his reaction that the lowest bidder was going to win this job.

I had the suspicion that he had already received prices from other vendors but he never mentioned that to us. He knew that this factor would eliminate us completely because we are always too high. It is true. We do bid high. We had to. I didn't have the luxury that some suppliers have who can purposely bid low and later ask for more budget or pray for ECPs. I had to use company policies and employees who don't come that cheap. In addition our overhead costs were much higher than most small vendors.

The old CSM consisted of two very large boxes; a panel box full with electronics and a second communications box having even more electronics in it. These two boxes provided all of the switching required for selecting Sonobuoys frequencies, voice communication, inter communications, Secure voice, TEMPEST, and it had to interface with the Navigation unit as well as the main computer. It required Manchester communications between boxes and to the transmitter/receiver. It was definitely a difficult design.

As I surveyed the situation I knew I had to come up with something revolutionary. I had an idea that would impress them so much that they had to give us the program. Without previously informing my boss I said, "Our proposal is based on the cost of using only one electronic box. The Panel box would have no active elements in it. It would be a very inexpensive design to produce because it would only contain switches, readouts, and possibly some relays. All of the complicated electronic would be contained in the main box which would be no larger than the original enclosure. So, the main cost would be only for one major box. In other words, we would cut our cost in half."

My boss was stumped and had a worried look on his face, "Wait a minute, Joe," He said. "We never discussed this before. Are you sure you can do this?"

I said, "Absolutely. I spent two weeks studying this design and I know it can be done this way."

The Project Manager immediately stood up and said, "If you can guarantee that you can put all of the electronic in one box you got the job!"

My boss and I walked out of there with a 2 million dollar design job. He was still not convinced and when we got back to his office he read me the riot act for not talking to him first. To be honest, I wasn't a hundred percent sure I could do it, but I had that gut feeling that comes only with experience. I knew it could be done. It turned out to be a piece of cake. See Figure 24 on page 295 for the final design.

As a result of our winning this contract, Project decided to give us another box, the Teletype Interface Unit for another million, which is shown on Figure 25 on page 297. Then they gave us the task of modifying the Nav. J box for another 500 grand. With these boxes we had enough work to maintain our organization for the next two years.

Nobody gave me an atta-boy for my effort either. As a matter of fact I was reprimanded by my upper-manager for supposedly designing a hot thermal box. It was this design that I had so much trouble fighting the Thermal Department over power dissipation and with my upper-management. Because I was so frustrated over this treatment, the first chance I had I transferred out of this department to the Skunk Works.

My immediate boss tried to block the transfer because he knew of my capabilities. I had worked for him for over twenty years but, fortunately, his new boss thought I was a nobody and approved the transfer right under his nose. I was gone for six years on different programs and later returned to take my old bosses place when he retired and I was promoted to Design Manager.

18. Display Control Panel (DCP)

The DCP was first initiated as a research project. In a laboratory demonstration it proved to be an ideal display unit. It would replace a large Vacuum Tube Display system which was being proposed for an update version of our P3 aircraft. I must admit it was impressive at first.

I was given the assignment as temporary Department Manager of this research project and, in addition, I had to monitor my existing aircraft programs. This was quite a load for me. I also had to manage twenty other IRAD programs. Even though the DCP research program was company funded to demonstrate its capabilities there was an excellent chance it could be used on the update aircraft. That's when all hell broke loose.

If one would notice on Figure 28 on page 303, an operator is working with our prototype DCP. This part was alright but when one examines the electronic board in Figure 29 on page 321, it becomes obvious that a very large board was used. In the commercial world large boards are accepted for all personnel computers, but not for the military.

When I began looking into this design I found too many faults with it. First of all, for military applications, this design is unacceptable because every Vendor doing a military design MUST use the standard Sem-E size boards. Secondly, the chips and the display being used were not Mil-approved, and were recently invented. Some of the major chips being used were made in Japan with no guarantee of production. This alone was taboo. Thirdly, the portion that was to be used on the proposed aircraft update required only a forth of the electronics that appears on that large board. Fourth, the power supplies were very large because it had to provide power for the extra electronics. This PS had to be invented and produced in-house.

The approach my boss took was to use one design for many applications on the aircraft, but these other applications were still not accepted by Project and the Navy. In addition the whole board would have to be stuffed and tested in production even if only a small portion is used. (I mention in Chapter II, that I was against using one board for different applications).

Fifth, the package, as it stood, would never pass environmental qualification of shake and bake, and EMI.

Sixth, the contract for the update plane required us to produce 130 units over a six month period or one every working day. My best estimation was that we couldn't produce one unit a month.

We eventually lost the contract to another aircraft company because of this design and some other poor management decisions. The Navy, for example, discovered that a special processor chip was being proposed would be used on many applications throughout the aircraft, but the chip was not going to be produced by the supplier. The manufacture had too much trouble with the yield in production. They said another more reliable version would eventually replace it, but no date was given.

My main purpose of re-discussing this design was to show how the people in charge sometimes ignored military requirements and ignore those who were in the business for years. When I addressed this problem to my new boss he simply said, "I don't care what the Navy wants. This is my program and I will do with it as I please." I couldn't believe what I was hearing. The Navy is the customer and they tell us what they want, not what he wants.

It was obvious to me that this man had never produced a box for the military. Even in his bidding he neglected to include our support organizations. Non-Standard Parts would have killed the program because he had so many new untried chips in the design.

I will admit, again, the design was impressive as a research project and that's where it should have remained. The military version should have been turned over to experienced and qualified military designers who would package the unit properly and meet all military requirements and specifications with the proper data submittal.

19. Component Selection

In the Military world one doesn't always have the freedom of selecting his own components. Most of the parts have to be a Mil Approved Standard part, they prefer that the part has a dual source, it cannot be an obsolete part, it has to be USA made, and it should to be selected from the Program Parts Selection List (PPSL) that is specified in the SOW. If any of these requirements cannot be met a waiver must be applied for to the appropriate government agency.

It also must be understood that a waiver may be denied, so one had better know in advance that the part he is dealing with has a good chance of being accepted as a Non-Standard part. It may even take six months before the waiver is accepted. Many a program went into the red because an engineer selected a part that was not approved. The Government may take the stand that the he violated the contract and the company has to pay for the re-design, if needed, or they may just cancel the contract.

When one is involved in the design there would always be many Non-Standard Parts, so be prepared to submit these parts for approval as soon as possible. On an ATE program that I was no longer involved with, they made the fatal mistake of not submitting the Non-Standard Parts List to the government until they were into the program by a year and a half. Talk about re-work.

Remember again, the Non-Standard Parts person in the Military may not accept the list from the design department. He would want the list submitted from the company's Non-Standard Parts Department or equivalent. This subject was also discussed in Chapter II, and is repeated here to further emphasize how important the whole picture of military electronic design and data submittal is. Many managers would fall on their faces whenever they try to by-pass the system.

20. Search Water Radar (SWR)

Search Water Radar was a Radar system that was developed in England. It is a unique Radar system that could detect any small craft on the ocean. It was so sensitive that it could even detect a submarine's periscope. England was trying to sell there SWR to the United States and we were commissioned to develop an interface unit so that the Radar would be able to communicate with our present Navigation unit, with our main computer and with the aircraft's existing Radar displays.

I briefly discussed this design in Chapter III and will now cover it further. Our prototype unit had its own 4-bit micro-processor in it. Our 4-bit version was denounced by upper-management and they were afraid that the follow on design might be too complicated for us if we used an eight bit version.

I must make a statement at this time. I wasn't manager of this design group when they were given this contract but I came on board as the new manager and after the experimental unit was tested in the ocean. The test results were passable but not that great. There were errors in detection and location of a target. It turned out that these errors were explainable and could be corrected.

First, the 4-bit processor was not adequate to do the job and second the Navigation unit introduced a phase error on one of its vector readouts. We had ample proof of these limitations and knew how to overcome them.

Project, on the other hand, disregarded our findings and the word went around throughout the company that we screwed up. Once we got this bad label on our department it stuck forever. I was also accused of being too expensive and when I challenged the Project Manager on this he told me that's the rumor that's going around. I have to admit we were a bit high on other programs but not on this one.

On the SWR Interface Unit the opposite was true. We were the lowest bidder even for production units. The Project Manager, however, had predetermine that this design was going to go to one particular outside supplier whom, I think, was giving him a kick-back. He wasn't even going to let other suppliers compete which is against all government contracts. He was shrewd, however.

In the next development phase he asked me to bid the job and practically guaranteed me that we would get the job if our quote was reasonable. What he was doing was playing games with me simply to get an idea on what it would cost so he could use it as leverage against this other company.

I was invited to a meeting to listen to this other company's presentation. I took my key engineer with me and after we listened to their presentation we discovered they didn't know much about the technical aspects of the program. It was pitiful as a matter of fact to listen to them wangle through their presentation. I figured we were a shoe-in to win this contract. But then a strange and unethical thing happened. This Project Manager said to the representative of this Vendor that their quote of 4 million dollars was too high. "Our internal quote was for 2 million dollars. Can you match that?"

This representative simply made a call to their eastern plant and after conferring with his manager for a few minutes he said yes, they could match my proposal. Right there in front of me this Project Manager committed an unethical crime and awarded them the contract. I couldn't believe my ears. Using my in-house quotation to bargain down an outside supplier is completely against company policy. And to award the contract without going out for bids is against all Government laws.

The worst part of this whole episode was we were told by this Project Manager that we are to give them all of our expertise and design data and help them throughout the program. I would have to send my key designer to their plant back east to help them produce a unit. What gall.

I didn't sit down for this, however. I made a lot of waves all over the place. I wrote Inter-department letters to everyone I could think of including the Vice President. It was now a fight to the finish. I would get phone calls from different upper-managers chewing me out for my letters. One manager said that we fell flat on out face on the experimental unit and we should shut up about it. I thought, for sure, I was going to be fired. If I did, I certainly would have gone to the Government with my case.

Fortunately, or maybe unfortunately, as the case may be, the Navy decided not to use the British

Radar version. They had an American version that could do the job, they said. Actually, the American version was not as good, but there was enough lobbing going on in Congress and the British Radar was canceled. After that there were no more threats or phone calls directed at me and that Project Engineer immediately bailed out and left the company. Being a Design Manager is sure a crazy position

21. Navigation Interface Unit (NAIU)

I briefly discussed this unit in Chapter III and will now expand on it further. Our S3-A aircraft was designed as a submarine chaser that is capable of flying from or landing on an aircraft carrier. The Navy brass wanted to use this plane to move people around and carry cargo. By removing all of the mission electronics there would be ample space to transport people or wounded military personnel and carry parts.

This was a great idea, because this plane had a range of over three thousand miles from the carrier, about three times as far as their present people carrier could. But an electronic problem was discovered later on with Navigation. Because the main mission computer was removed no navigation information could be displayed to the Pilot. So, we received a job to design a Navigation Interface Unit (NAIU) as shown in Figure 26 on page 319.

This was another one of those jobs I inherited after I was made manager and the first two prototypes were already installed. When I took over the program I discovered that the design wasn't bad but the program was very late and was way over budget. Also, because the Navy was going to covert only two planes into what they called CODs, only a total of three interface units would be assembled. These were done in our own Lab.

Now the fun began. When it was initially quoted the previous manager didn't include complete documentation submittal, such as non-standard parts, drawings to micro-film quality, maintenance manuals, etc. His argument was, why does the Navy need a full blown drawing submittal when only three units were ordered. I can see his reasoning, but that's not what was in the contract. It was for a full Mil program.

Getting more budget from Project was a bear. That's when I especially got the reputation of being too costly. But the problems didn't go away. Two years later the Navy decided to order three more units. They decided to convert two more planes but the aircraft conversion would be done by Navy civilians instead, at their base.

Now what does one do with a design that is over three years old. The first thing we had to do is order parts, but some of the parts were now obsolete. There was an important module that was no longer in production and the man that designed the original box was transferred to another department. Before we got the new order I had a hell of a time finding him a position within the company because he was scheduled to be laid off.

To make a long story short, we had to design our own module and piggy-back it on its original board using hay-wires to hook it up. Other obsolete parts had to be purchased from suppliers that specialized in producing obsolete chips but for five to ten times the price. We had to pay the price or redesign the cards using the new standard chips on the market. It was a nightmare.

To top it off, two years after we completed this contract, the Navy ordered three more units. The cycle almost repeated itself. By now our delivery quote had doubled. The Navy was kind of mad at us, but we had no choice. Tooling up in dribs and drabs was costly. All together the Navy ordered sixteen Navigation Interface units over a ten year period.

I mention in Chapter III, how the Navy steals work from industry because no one can outbid them. The COD was one of those programs. They rewired the aircraft at a Navy Base with civil service personnel. How could my company out-bid them when they have no overhead, no taxes, no building rent, etc. Fortunately for us, they didn't want to take over our NAIU because it was too complicated and important. The pilot could not fly the S-3 COD without our NAIU.

22. Packaging

I discussed packaging in Chapters I & II. I will now touch on it further. Figure 30 on page 307, illustrates the modular assembly of the CLA-Test Set. As one can see all the sub-assemblies are separated from the main chassis which provides easy off-line assembly and testing. Production loved this approach because the whole package goes together like a glove and it's easy to maintain and repair. This particular unit also has self test capabilities allowing it to be serviced in the field.

The CLA unit of Figure 31, is an example of a wire-wrapping technique. This is the bottom cage of one of our CLA units. The way it's done is, the bottom cage is sent separately to a wire-wrap house who uses a Gardener-Denver machine which automatically wire wraps the cage. They also completely check the connections with a computer before they deliver it. As I mentioned in Chapter III, wire wrapping should only be used when the clock frequencies are less than 5 Mhz, otherwise radiation pick up and cross-talk would drive one crazy.

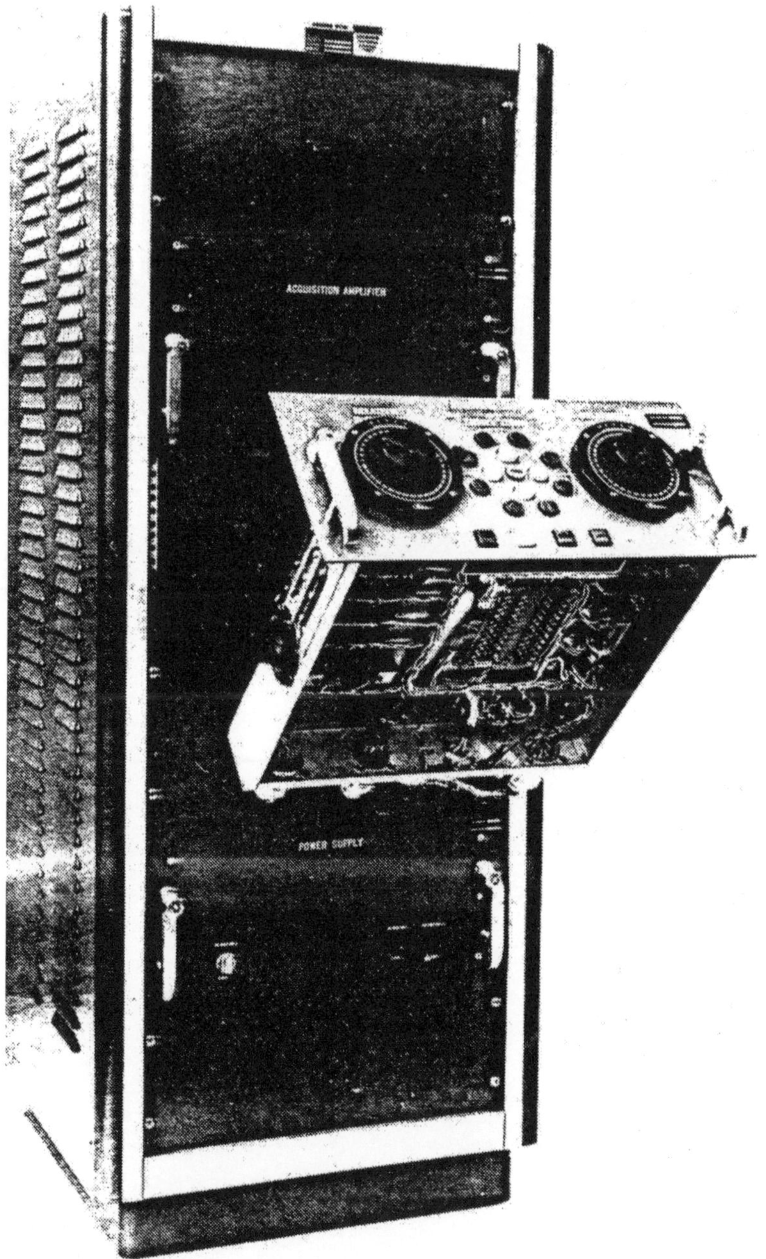

Figure 18. IR Acquisition System

Figure 19. Polaris Tracking System, DRIFT SYSTEM

Figure 20. Shipboard DRIFT SYSTEM

Figure 21. Polaris Tracking Receiver

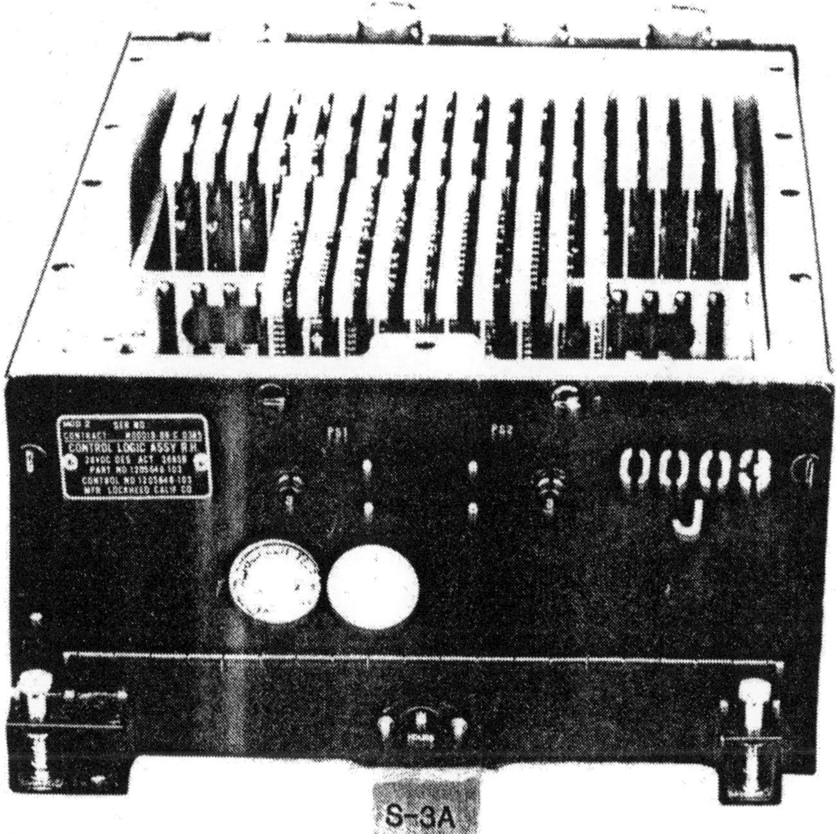

Figure 22. Power Control Assembly, CLA

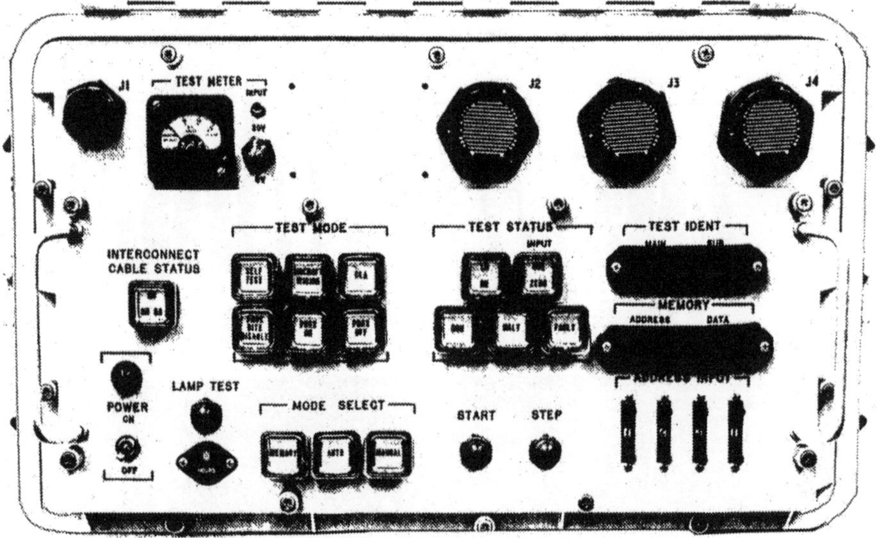

Figure 23. Control Logic Assembly Test Set

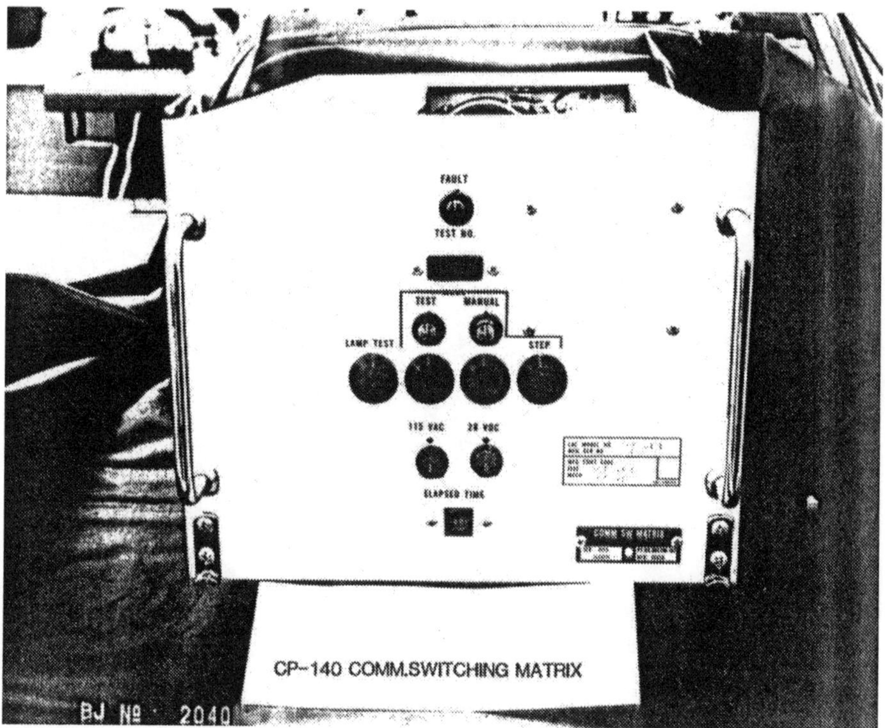

Figure 24. CP-140 Comm. Switching Matrix

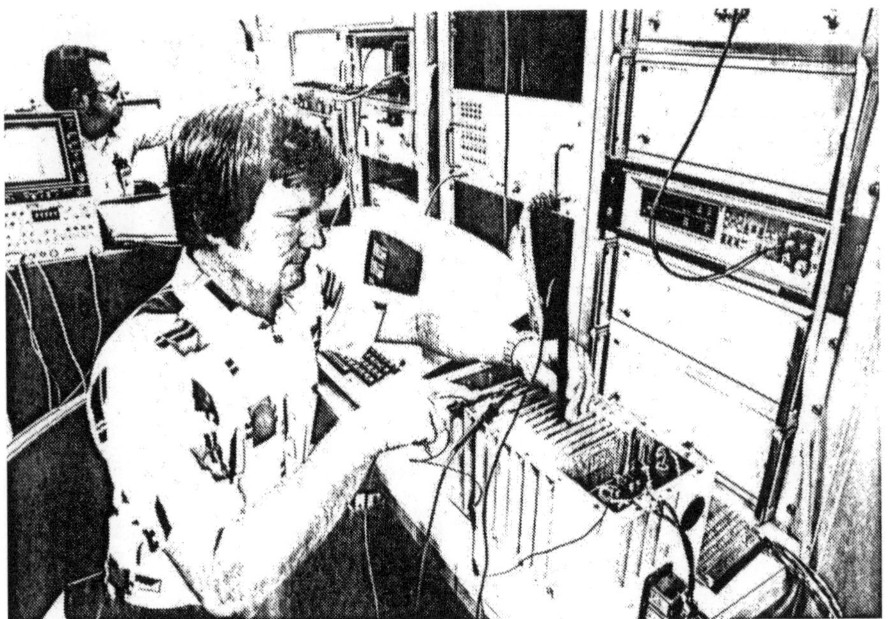

Figure 25. CP-140 Teletype Interface Unit

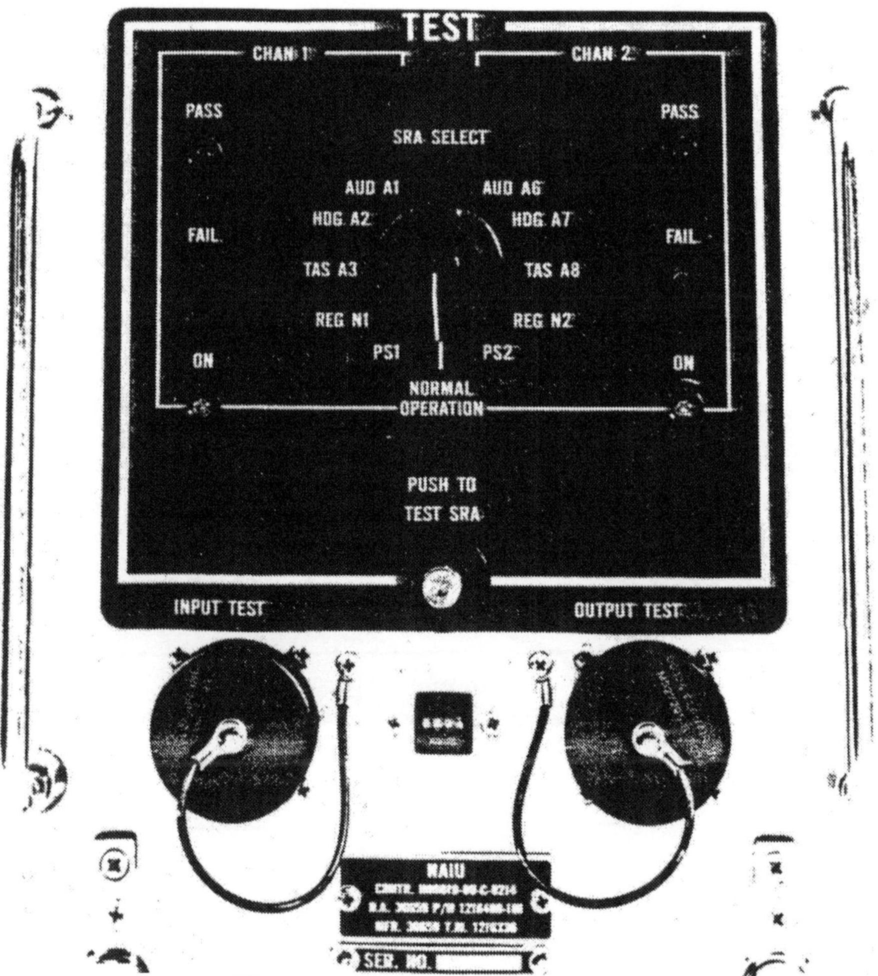

Figure 26. Navigation Interface Unit

Figure 27. Search Water RADAR Interface Unit

Figure 28. Display Control Panel

Figure 29. DCP Large Printed Wire Board

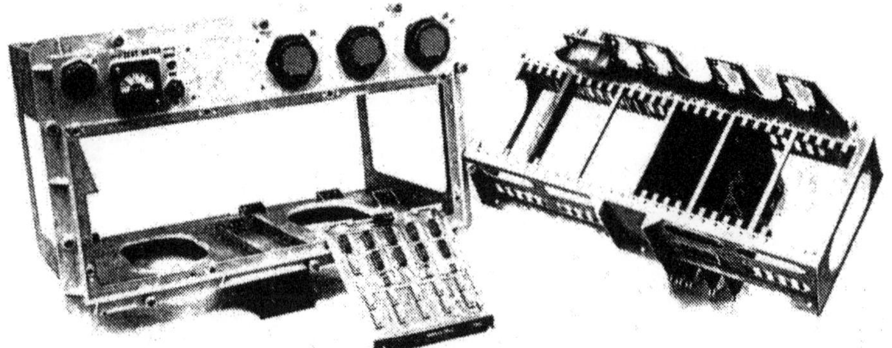

Figure 30. CLA Test Set Modular Assembly

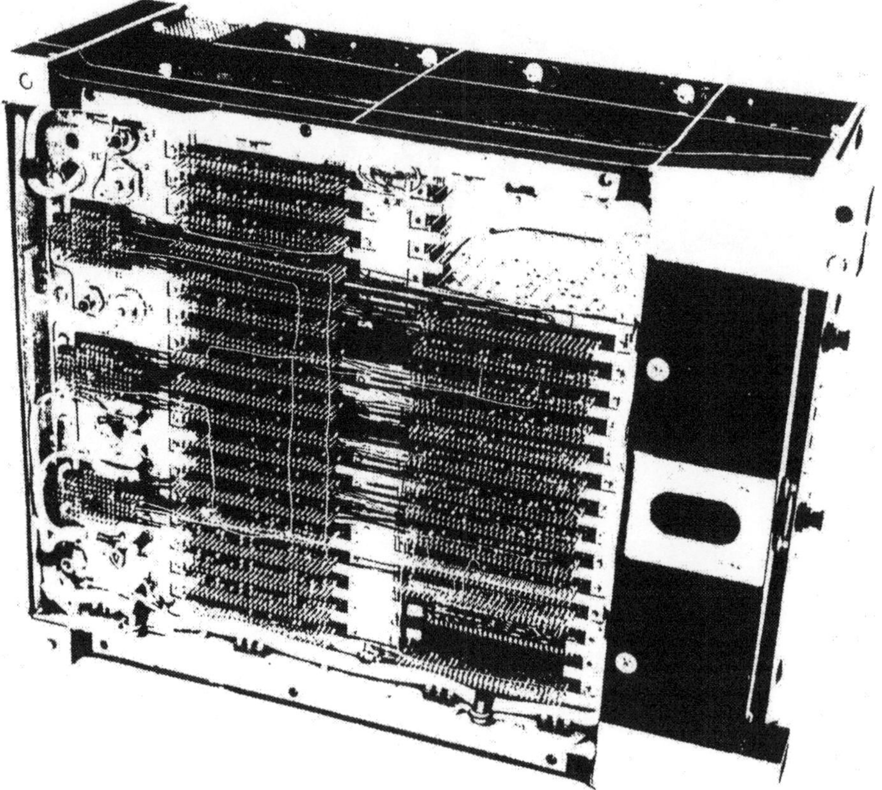

Figure 31. CLA Wire Wrap

SSAB0064 ENGINEERING DRAWING LIST

INDENTURE															TITLE	CHG	SIZE	REMARKS
1	2	3	4	5	6	7	8	9	10	11	12	13	14	15				
B	U	-	9	1											INTERCONNECTING BOX, J-4754/USM-429(V)			COMMON TO
	1	3	1	1	3	5	4								INTERFACE DEVICE ASSY - BU91		J	---
		1	3	1	1	3	4	3							MULTIPLEXER/RELAY BOARD ASSY		E	---
			1	3	1	1	3	4	4						MULTIPLEXER/RELAY FAB		E	---
			1	3	1	1	3	4	5						MULTIPLEXER/RELAY MASTER PATTERN			---
			1	3	1	1	3	4	6						SCHEMATIC MULTIPLEXER/RELAY BOARD		J	---
			1	3	1	1	3	3	4						HANDLE-PWB		D	ALL ELEX BDS
			1	3	1	1	3	5	9						PWB ASSY - MICROPROCESSOR NO. 1, BU91		E	---
			1	3	1	1	3	6	0						PWB FAB - MICROPROCESSOR NO. 1, BU91		E	---
			1	3	1	1	3	6	1						PWB MASTER PATTERN-MICROPROCESSOR NO.1			---
			1	3	1	1	3	6	2						SCHEMATIC- MICROPROCESSOR NO. 1, BU91		J	---
			1	3	1	1	3	3	4						HANDLE-PWB		D	ALL ELEX BDS
			1	3	1	1	3	6	3						PWB ASSY - MICROPROCESSOR NO. 2, BU91		E	---
			1	3	1	1	3	6	4						PWB FAB - MICROPROCESSOR NO. 2, BU91		E	---
			1	3	1	1	3	6	5						PWB MASTER PATTERN-MICROPROCESSOR NO.2			---
			1	3	1	1	3	6	6						SCHEMATIC- MICROPROCESSOR NO. 2, BU91		J	---
			1	3	1	1	3	3	4						HANDLE-PWB		D	ALL ELEX BDS
			1	3	1	1	3	6	7						PWB ASSY - MICROPROCESSOR NO. 3, BU91		E	---
			1	3	1	1	3	6	8						PWB FAB - MICROPROCESSOR NO. 3, BU91		E	---
			1	3	1	1	3	6	9						PWB MASTER PATTERN-MICROPROCESSOR NO.3			---
			1	3	1	1	3	7	0						SCHEMATIC- MICROPROCESSOR NO. 3, BU91		J	---
			1	3	1	1	3	3	4						HANDLE-PWB		D	ALL ELEX BDS
		1	3	1	1	3	7	1							MOTHER BOARD ASSY -BU91		J	---
			1	3	1	1	6	3	8						MOTHERBOARD,FAB-BU91		J	---
			1	3	1	1	6	3	7						MOTHERBOARD,MASTER PATTERN -BU91			---
			1	3	1	1	3	7	4						WIRE LIST,MOTHERBOARD-BU91		A	---
			1	3	1	1	2	5	0						CONNECTOR SHROUD		E	BU89,BU90,BU91&BU92
		1	3	1	1	3	3	3							SYSTEM INTERCONNECT DIAGRAM BU91		J	---
		1	3	1	1	3	5	7							WIRE HARNESS ASSY-BU91		E	---
			1	3	1	1	2	2	8						INTERFACE ASSY - AMP CONNECTOR		E	BU89,BU90&BU91
				1	3	1	1	2	2	9					MOUNTING BRACKET		E	BU89,BU90&BU91
		1	3	1	1	2	5	1							CABLE ASSY-(J1)		J	ALL EXCEPT BU94
		1	3	1	1	5	1	2							ALIGNMENT FIXTURE,PDP UUT		E	BU89,BU91&BU92
		1	3	1	1	5	1	4							ALIGNMENT FIXTURE,ELEX BD		E	ALL
		1	3	1	1	3	1	7							POWER SUPPLY ASSY		J	BU90,BU91&BU92
			1	3	1	1	3	2	1						POWER SUPPLY BOX ASSY		E	BU90,BU91&BU92
	1	3	1	1	3	5	8								OUTLINE AND MOUNTING-BU91		E	---
	1	3	1	1	5	1	3								HARNESS CONFIGURATION		E	BU89,BU90&BU91

BU91

Figure 32 Engineering Drawing List

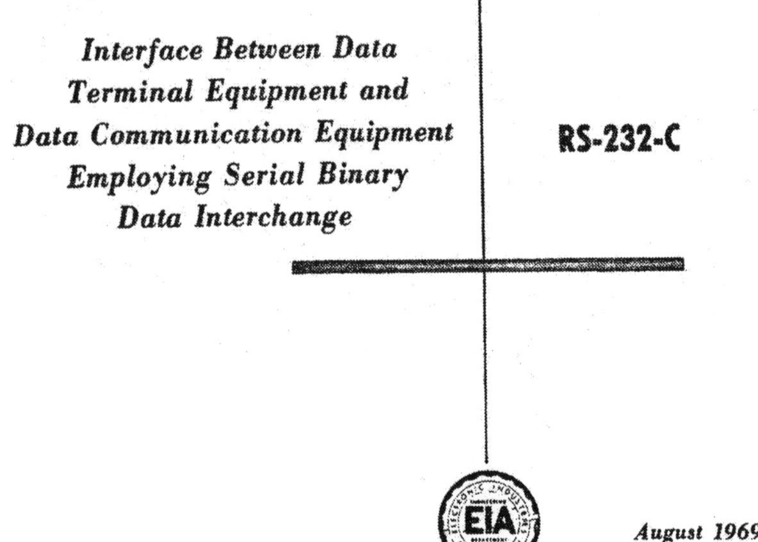

EIA STANDARD

EIA RS-232-C

Interface Between Data Terminal Equipment and Data Communication Equipment Employing Serial Binary Data Interchange

RS-232-C

EIA

August 1969

Engineering Department

ELECTRONIC INDUSTRIES ASSOCIATION

Figure 33. RS-232-C Interface Bus Line

Figure 34 S3-A Aircraft

Figure 35 Integration Laboratory

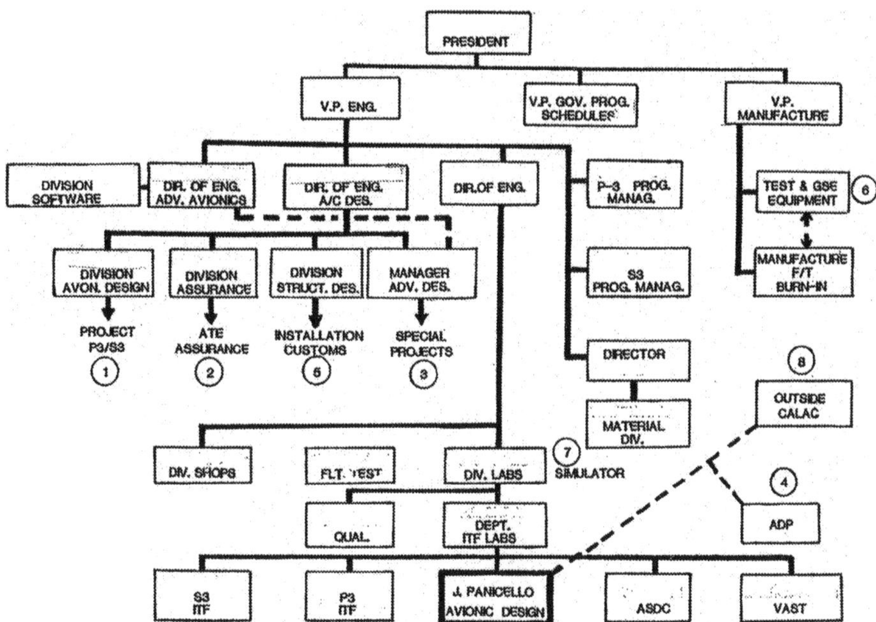

Figure 36 Organizational Structure

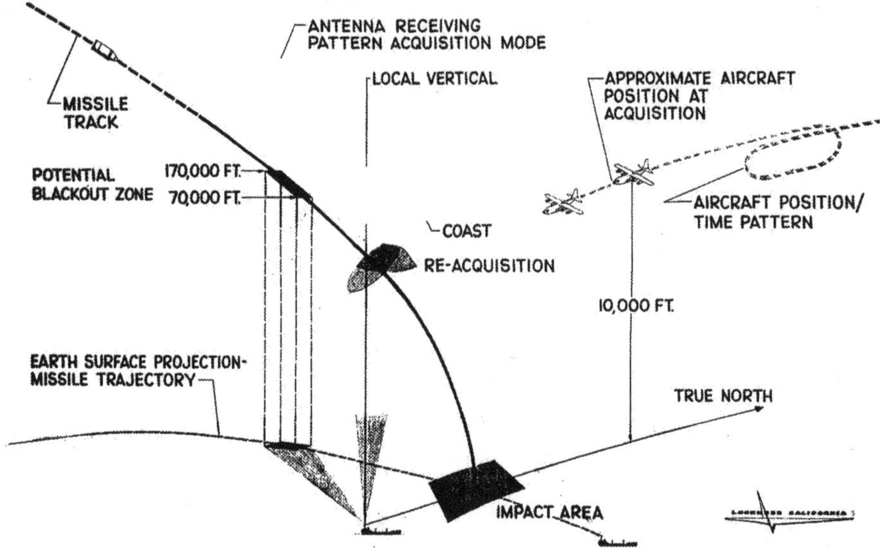

Figure 37 Polaris Missile Tracking Method

MANAGEMENT COURSE OUTLINE

Note: The following course outline covers only Chapters I & II

Week 1. - 3 hours
Course Introduction
Electronic Design Manager
Design Team
Quotations & Procedure
Fact Finding
Quote Format Procedure
Prototypes
Manhour And Tracking Charts
Manloading Charts
Week 2. - 3 hours
Electronic Engineer
The Design Engineer
Design Frauds
Learning The Design
Lead Design Engineer
Support Organizations
Steps In Electronic Design
Week 3. - 3 hours
Systems Engineer
Definition
Equipment Specifications
Software Engineer
Week 4. - 3 hours
Mechanical Engineer
Enclosure Design
ARINC/ATR Enclosure
Example Construction
Drawings
Non-Standard Parts
Week 5. - 3 hours
Printed Wire Engineer
Design Procedure
Design Changes
CADD Operators
Technicians
Development Shops
Liaison Engineer
Week 6. - 3 hours
Review/Questions
Midterm Exam - 1 hour
Week 7. - 3 hours
Support Organizations
Testability/Reliability
Thermal
Maintainability
Non-Standard Parts
Week 8.- 3 hours- Support Organizations
Quality Assurance
Inspection
Electromagnetic Interference (EMI)
Weights
Stress
Producibility
Qualification
Functional Test
Production
Safety
Human Factors
Week 9. - 3 hours
Drawing Release
Check
Integration Laboratory
Flight Test
Material
Purchasing
Master Scheduling
Administrations
Finance
Contracts
Computer Services
Week 10.- 3 hours Product Support
Automatic Test Equipment (ATE)
Ground Support Equipment (GSE)
Logistics
Publications
Field Services
Training
Week 11. - 3 hours
Summary/Questions
Final Examination

Total Course Hours: 33

ACRONYM GLOSSARY

ACQ Acquisition (processor)
AFC Automatic Frequency Control
AM Amplitude Modulation
ATE Automatic Test Equipment
ATP Acceptance Test Procedure
ATR ARINC 404 case sizes
BIT Built In Test
BITE Built In Test Equipment
CADD Computer Aided Drafting
CDR Contract Data Requirements
CHIP Integrated Ckt or Transistor
CMOS Comp Metal Oxide Silicon
CM Configuration Management
COMINTComm. Intelligence
CW Carrier Wave (no mod)
DAC Digital To Analog Converter
DDM Diagnostic Diagram
DEPOT Military Test Station
DGAR Des. Gov. Acceptance Rep
DID Data Item Description
DF Direction Finding (Processor)
DM Design Manager
DTL Diode Transistor Logic
DVD Design Verification Demo.
ECL Emitter Coupler Logic
ECO Engineering Change Order
ECP Eng. Change Proposal
ECM Electronic Countermeasures
EDL Engineering Drawing List
EDM Electronic Design Manager
EE Electronic Engineer
ELINT Electronic Intelligence
EMI Electromagnetic Interference
EPM Engineering Project Manager
EPP Engineering Project Plan
ER Equipment Reliability
ES Equipment Specification
EW Electronic Warfare
FAT First Article Test
FSD Full Scale Development
FM Frequency Modulation
FTP Functional Test Procedure
FTR Functional Test Requirements
GFE Gov. Furnished Equipment
GSE Ground Support Equipment
I Level Intermediate Test Level
IC Integrated Circuit
ICS Intercommunication System
ID Interface Device
LE Liaison Engineer
IF Intermediate Frequency
ILS Integrated Logistics Support
LS Low Powered Schottky Logic
ME Mechanical Engineer
MH Man Hours
MP Maintenance Plan
Mod Modulation
MS Military Standard
MTBF Mean Time Between Failure
MTTR Mean Time To Repair
O Level Organizational Level
OTPS Operational Test Program Set
PCE Printed Circuit Engineer
PDD Program Design Data
POD Preprod OTPS Demonstration
PM Program Manager
PP Program Plan
PS Power Supply
RF Radio Frequency
PS Project Slip (same as ECO)
RFP Request For Proposal
RFQ Request For Quotation
ROM Read Only Memory
ROM Rough Order Of Magnitude
SE Systems Engineer

SID Systems Interconnect Diagram
SOW Statement Of Work
SRA Shop Replaceable Assembly
SWE Soft Ware Engineer
TAAF Test Analyze And Fix
TBD To Be Determined
TTL Transistor Transistor Logic
UUT Unit Under Test
WAG Wild Ass Guess
WBS Work Breakdown Structure
WRA Weapons Replaceable Assem

Author's Major Experiences

Bell Laboratories
Experience in Anti Submarine Warfare (ASW), 1952 to 1961.

1. 1953. Designed a vacuum tube high impedance Sono-Buoy 100 db gain amplifier
2. 1955. Designed same as 1. above but using Transistors.
3. 1956. Designed Variable High Frequency Precision Oscillator (1 to 5 MHz)
   a. Using a high impedance boot strap Emitter Follower
   b. Power Amplifier design to drive 50 ohm coax cable line at RF frequencies
4. 1957. Invented an transistorized frequency divider (Patent No. 3,110,820).
5. 1958. Designed a wide band multi-feedback video amplifier
6. 1960. Designed a transistorized oscilloscope test equipment used on a submarine
7. 1961. Left BTL. Hired by Lockheed Aircraft Company in Burbank, California

Lockheed Experience

* 1. 1961. Air Force. Designed the electronics for an Infra-Red Acquisition System
* 2. 1962. LMSC. Designed a transistorized Polaris Missile Tracking Receiver

3. 1963. F-104 aircraft. Designed a Infra-Red Gun Sight Pre-Amplifier
4. 1964. IRAD. Researched and Published - Information Transmission in Space
5. 1965. NASA. Responsible for the Outer Space Station Display

* 6. 1966. P3B Designed the electronics for a Jordan Navigation Interface Box.
* 7. 1967. P3B. Designed the electronics for the Sono-Buoy Data Display system.

8. 1968. P3B. Designed the electronics for the Freelator ASW system.
9. 1968. P3B. Responsible for the Broad Band Analyzer ASW system design
10. 1969. P3C. Designed the Noise Reduction Unit for the Teletype Writer

* 11. 1969. P3C. Designed the electronics for the Signal Data Converter (TTY)
* 12. 1970. P3B & C. Submitted a $1 million cost savings for a Son-Buoy Receiver

* 13. 1971. AH-56 Helicopter. Designed a special Power Supply for Displays
* 14. 1972. AH-56. Designed the Fault Warning System for the Pilot/Co-Pilot
15. 1972. IRAD. Invented a Solid State Voice Warning, Pat No. 3,808,591.
* 16. 1973. S3-A aircraft. Designed the Control Logic Assembly (CLA), Fig. 5
17. 1973. S3-A. Designed the CLA computerized Test Set, Fig. 6.
* 18. 1974. CP-140 Canadian aircraft. Designed the Comm Switching Matrix
* 19. 1974. CP-140 Department designed the TTY Unit having Manchester Interface.
20. 1975 to 1979. ADP Skunk Works. EPM on EW & electronic designs
21. 1980 & 1981. EP-3. EPM for EW Elint programs in the ADD Depart.
22. 1981 to 1987. Engineering Design Manager (EDM) - Design/Develop Dept.
    * a. US3 aircraft. Navigation Interface Unit (NAIU)
    b. Search Water (Anti-Submarine) RADAR Interface Device for England
    c. Drone. Special Platform for Electronics (Secret)
    * d. 444. (Secret) aircraft. Steering Decoupler
    * e. 444. (Secret) Pump Control
    f. IRAD. Digital Control Panel (DCP)
    * g. S3-A. Bomb Bay Decoder
    * h. S3-A. Wing Decoder
    i. S3-A. Automatic Test Equipment (ATE)
23. 1988. S3-A. EPM for the Communication Interface for the Teledyne Company
24. 1989. S3-A. Staff Engineer for the ATE program for Lockheed
25. 1990. Retired. 30 years with Lockheed. From 1980 to 1990 in management

* Represents programs that were rescued or redesigned by the Author or his department. There were several programs that had to be redesigned by me or my department because a contractor went "Belly Up," or couldn't do the job. A contractor was cancelled for poor workmanship. There were no outside bids for a system. An original designer left our company because of poor design. Contractors hired Job Shoppers and couldn't follow up the design. Other departments couldn't meet environmental qualifications. Contractors couldn't meet the Equipment Specifications, or they designed to a poor Equipment Specification.

About The Author

Joseph Francis Panicello was born in Queens, New York, on October 31, 1927. He was married for 30 years to his late wife, Rose, and is now married to his lovely wife Barbara who has eight children. Joe has three daughters, Jo Ann, Marie, Teresa, eight grandchildren and one great grandchild.

Joe Panicello is a World War II veteran who served in the Navy on LST 533 for two of his three years of his enlistment. He has a brother, Carl, who was in the Army during the war is now living in Long Island. Another brother, Thomas, who was also in the Army, was wounded in the Battle of the Bulge. Both Tommy and Joe's step sister, Marie, passed away in the nineties.

Prior to pursuing his writing career, the author maintained a successful 40-year career as an Electronic Engineer for Lockheed Aerospace in Burbank, California and Bell Telephone Laboratories in Whippany, New Jersey. He received his college education under the GI bill at Newark College of Engineering in New Jersey in 1959 and at RCA Institute in Advance Technology in New York in 1952.

Mr. Panicello is a member of The American Legion, The American Fiction Society, the National Writers Association, the United States L.S.T. Association, and the Navy League of the United States.

He has published six books thus far, three are fiction; *Vindicated, Brian's Comet*, and *The Wheeler Dealer,* two historical novels, *The Great Sicilian Norseman,* and *A Slow Moving Target, The LST of World War II*, and a non-fiction *Engineering Manager.*

Both *Brian's Comet*, *The Wheeler Dealer, A Slow Moving Target*, and *Engineering Manager* may be purchased through the Internet via, www.1stBooks.com, through Amazon.com, and at Barnes & Noble or other book stores. His other books *Vindicated* and *The Great Sicilian Norseman* may only be purchased from North Hills Publishers by calling (818) 894-6729.

www.ingramcontent.com/pod-product-compliance
Ingram Content Group UK Ltd.
Pitfield, Milton Keynes, MK11 3LW, UK
UKHW041949190726
13854UKWH00004B/1868

9 781410 723789